# Photovoltaic Design & Installation



by Ryan Mayfield

**Photovoltaic Design & Installation For Dummies®**

Published by
**John Wiley & Sons, Inc.,**
111 River St.
Hoboken, NJ 07030-5774
www.wiley.com



For general information on our other products and services, please contact our Customer Care Department within the U.S. at 877-762-2974, outside the U.S. at 317-572-3993, or fax 317-572-4002. For technical support, please visit https://hub.wiley.com/community/support/dummies.

Wiley publishes in a variety of print and electronic formats and by print-on-demand. Some material included with standard print versions of this book may not be included in e-books or in print-on-demand. If this book refers to media such as a CD or DVD that is not included in the version you purchased, you may download this material at http://booksupport.wiley.com. For more information about Wiley products, visit www.wiley.com.

Library of Congress Control Number: 2019932064

ISBN 978-1-119-54435-7 (pbk); ISBN 978-1-119-54433-3 (ebk); ISBN 978-1-119-54437-1 (ebk)

Manufactured in the United States of America

C10008330_021919

# Table of Contents

# Introduction

Solar electricity is one of the most accessible forms of renewable energy available, and it can be adapted to fit anyone's specific needs. The source of it, sunlight, allows photovoltaic (PV) modules to produce electricity and perform useful work, such as running electrical loads or putting energy back into the utility grid.

Globally, there has been an incredible increase in demand for solar modules and PV systems since the turn of the 21st century. No longer is PV technology limited to NASA spacecraft and backwoods cabins. Nowadays, PV systems are within reach for many people. This global demand and increased accessibility and affordability have helped transform the PV industry from a small cottage industry into a worldwide megabusiness.

As a renewable energy educator, advocate, and user, my primary goal is to see the solar industry grow successfully and sustainably — two things that require consistent high-level performance from the people "in the trenches" or, more appropriately, "on the roofs." PV system designers and installers are tasked with providing high-quality systems that can benefit clients for more than 20 years. If you follow the information, advice, and tips I present in this book, you'll be well on your way to completing this task with the best of them. (Of course, no perfect power source is out there. To be a good PV system designer and installer, you have to realize the limitations of PV systems, which is why I help you consider those limitations in this book, too.)

## About This Book

As the PV industry expands and captures the attention of more and more individuals, so does the amount of information about these systems. *Photovoltaic Design & Installation For Dummies* adds to this ever-expanding mass of information, but it stands out in its ability to provide you with practical, hands-on specifics. My intention with this book is twofold: to give you the ability to jump headfirst into the PV industry and to arm you with the knowledge you need to competently install a system (more or less) on your own. I use my real-world experience and desire to see every roof holding a well-designed and properly installed PV system to help you.

My approach is to give you the information you need in easy-to-understand sections that are relevant to the work you'll be expected to perform as a PV designer and installer. (I don't expect you to read this book from cover to cover, but if you do, I'll take it.)

*Note:* This book is focused on the applications used to power homes and small businesses — with or without the electrical utility present — but you can take the principles and guidelines I present and adapt them for just about any application you need.

## Conventions Used in This Book

Even though the title of this book has the word *photovoltaic* in it, most folks in the industry don't throw that word around day in and day out. Instead, we prefer to say *PV* — it's simpler, and it creates far fewer tied-up tongues.

Also, throughout this book you'll see references to *PV array* and *PV system.* The former term refers to the collection of PV modules that makes up the PV array, and the latter term refers to the PV array and all the associated components (disconnects, inverters, batteries, and so on) that make it operational and useful.

In addition, all time references throughout this book are based off solar time (which I cover in Chapter 4). *Solar noon* is the time of day exactly between sunrise and sunset, when the sun is at its highest point in the sky that day. Your longitude, the time of year, and whether you're subjected to daylight saving time influence the relationship between your clock and solar time.

Last but not least, the following conventions help you more easily sort through all the new ideas I throw at you:

- The key words in bulleted lists and the action steps I ask you to follow in a specific order appear in **boldface.**
- New terms that are important for your understanding of PV systems are in *italics,* as are words I choose to emphasize.
- Web addresses appear in `mono font`. (Note that when this book was printed, some Web addresses may have needed to break across two lines of text. There's no extra punctuation, though, so just type in exactly what you see, pretending as though the line break doesn't exist.)

## What You're Not to Read

So many interesting side notes about PV technology, politics, and installation techniques exist that I could easily use all the pages of this book for nothing but various tangents. I think I've done a good job of sticking to the main points to give you the most important information you need, but some stories and bits of knowledge are just too good to keep to myself. These stories and tidbits are included in the gray-shaded boxes (sidebars) and paragraphs marked with the Technical Stuff icon. Don't feel bad for skipping over this text if that's what you choose to do; you aren't going to miss out on the essential information a PV system designer and installer needs to know.

## Foolish Assumptions

Simply by picking this book up, you've already proven my first assumption, that you have an interest in finding out more about PV systems, to be true. I also make some other assumptions about you; here they are, for your viewing pleasure:

- You're one of the many individuals ready to make a career move that includes PV design and installation; you're either already a building-trade professional or you're enrolled in classes and want to secure an entry-level position in the PV industry.
- Perhaps you're a homeowner looking for a better understanding of PV systems; you're thinking about having one installed and want to be as knowledgeable as possible when working with a PV system designer and installer.
- You're interested in technically challenging projects, and you can accept that good designs and installations don't happen overnight. You want to prepare yourself and gain some knowledge first.

## How This Book Is Organized

*Photovoltaic Design & Installation For Dummies* is arranged in five different parts. Each part can stand alone or be read in conjunction with the others. If one of the following parts (or a particular chapter within that part) excites you more than the others, by all means, head there first.

## Part 1: Here Comes the Sun: Shedding Some Light on PV Systems

Get ready to discover the fundamentals of PV design and installation in this part. Here, I introduce you to the major components used in the systems you'll most likely be working with. And in case you need a refresher course on (or an introduction to) the fundamental electrical principles, this part has you covered. It also explains the specific relationship between the sun and the earth and how this relationship affects where you should position a PV installation.

## Part 2: Digging into Complete System Details

Everything you ever wanted to know about the major components of PV systems can be found in Part 2. If you've been wondering how a PV module works and how the acid inside a battery can make electricity, here's your chance to find out those things. This part also covers the basic functions of the other major components of PV systems: charge controllers, inverters, and wiring. And because safety should always be your top priority, I also get you acquainted with essential safety devices (think fuses and circuit breakers).

## Part 3: Sizing a PV System

In this part, I walk you through the different methods for *sizing* your PV systems (figuring out how many PV modules you need), whether they're grid-direct or battery-based, and the necessary wiring and safety devices. Sizing is the area that can make or break your system. If you properly size and *specify* (select) all the system components, your system will likely pass inspection much faster, and your client will be left with a safe, efficient way of capturing and using solar energy. If, however, you make major errors in your preparation and design, fixing those errors after the system is already installed can be a difficult (and sometimes dangerous) task.

## Part 4: Installing a PV System

Part 4 is where I cover the physical installation process, as well as the activities that precede and follow it. I kick things off by walking you through the permitting process so you have the permission you need to install the system. Next, I cover the safety considerations you need to have in mind when you're out installing systems. (After all, PV construction sites offer a number of hazards for everyone on-site; by preparing for these hazards, you can avoid them altogether.) After that,

I break the installation process into the mechanical and electrical portions. Then I give you a process to follow when you're ready to *commission* (turn on) the system and get the modules working prior to the mandatory building department inspection. Finally, I address proper PV system maintenance.

### Part 5: The Part of Tens

The Part of Tens is designed to give you small chunks of information for a quick and easy read. Prepare to find out some of the common mistakes made in PV installations so you can do your best to avoid them, as well as some quick tips on how to keep the PV systems you design and install working as well as they should.

## Icons Used in This Book

Throughout this book, I use the following icons to bring your attention to interesting, important, or even downright critical information.

REMEMBER

Although I'd love it if you remembered everything in this book, I know that's practically impossible. That's why I mark ideas I want you to keep in mind no matter what with this icon.

TECHNICAL STUFF

Every now and then I tend to "nerd out" on you with interesting but nonessential information. I promise to keep these asides (relatively) short and to warn you that they're coming with this icon.

TIP

This little bull's-eye brings your attention to information that may make your job a little easier.

WARNING

Don't let anyone fool you. PV systems pose real dangers. This icon alerts you to a situation that requires special attention and caution.

## Where to Go from Here

Wondering where to start reading? If it were me, I'd jump to Chapter 6; this is where I get into the PV modules themselves — far and away the most interesting part of PV systems (in my humble opinion). If you'd rather choose the approach my children take — which is to decide what you think is best despite my advice — flip to the table of contents. There you can breeze through all the chapters and see what jumps out at you.

After you get what you want out of this book, I strongly suggest you find a way to put that knowledge into action. Numerous organizations that provide hands-on training and experience are available, and these classes are an excellent way to get your hands on some equipment and have access to individuals who are knowledgeable and eager to help you master PV systems. Then, if you have the means, install a PV system on your own home. Doing so lets you run the show and gives you some personal insight into what it takes to install PV systems (plus it provides you with a slightly less stressful situation for your first solo installation because you don't have a client or supervisor looking over your shoulder).

Wherever you end up going first in this book, realize that you're jumping into an exciting and rapidly evolving technology and industry. No matter where you go within the pages of *Photovoltaic Design & Installation For Dummies*, you'll be well on your way to becoming involved with one of the most dynamic and exciting industries you can imagine.

# 1 Here Comes the Sun: Shedding Some Light on PV Systems

## IN THIS PART . . .

Understanding the fundamentals of photovoltaic (PV) systems is the key to designing and installing high-quality systems that'll perform beyond your (and your clients') expectations. That's why this part provides the foundation you need to move forward with designing and installing PV systems.

Chapter 1 gives you an understanding of where PV systems started, where they are now, and where they're headed; Chapter 2 gives you a feel for the major components used in all types of PV systems and how they relate to each other. Chapter 3 provides an overview of the electrical fundamentals that drive system designs and installations, and Chapter 4 outlines the relationship between the sun and the earth and guides you on assessing the solar resource. Rounding out the bunch, Chapter 5 walks you through the process of selecting where on a client's property to install a PV system; it also helps you quantify the solar resource at a particular location.

IN THIS CHAPTER

- » Taking a look at the PV timeline
- » Refreshing yourself on the basics of electricity and the solar resource
- » Selecting a site for a PV system and outlining the different system components
- » Making the sun work on your client's behalf with proper system sizing and installation
- » Getting to know Code

# Chapter 1 The Photovoltaic Revolution

In recent years, photovoltaic (PV) systems have popped up on people's homes and businesses all over the United States. Believe it or not, this desire to use the sun to power people's electricity needs is anything but new. Ever since scientists discovered materials that can produce electrical current by simply being exposed to light, people have been excited about this energy source. Get ready to dive into the exciting and ever-expanding world of solar power with the overview I provide in this chapter.

## Peeking into the Past, Present, and Future of PV Installations

To truly appreciate PV, it helps to have an understanding of where the technology came from, where it's at now, and where it's going.

The operating principles for modern PV cells were first discovered in 1839 by a French physicist named A.E. Becquerel. After that, a number of scientists played with and improved on Becquerel's original discovery. In the 1950s, Bell Labs created the first piece of PV technology designed for use in space. This technology soon found its way back down to earth for use in telecommunications applications in remote areas. In the 1970s and 1980s, people began using PV modules to charge batteries and then used those batteries to run various lights and appliances in their remote homes. These early PV pioneers helped set the stage for today's PV industry.

The first PV cells weren't very efficient or widely used outside of space programs. They were also quite costly. Yet over the years, researchers and manufacturing companies increased efficiencies and reliability and managed to drive down costs drastically. All of these contributions have led to the widespread use of solar modules and their availability to you and me. In the following sections, I describe some common PV applications, a few brief pros and cons of PV systems, and the future of the PV industry.

## Acquainting yourself with typical PV applications

Modern PV systems can be found in a wide variety of applications. They power calculators, pump water, help offset the energy used by floodlights along highways, and, of course, power homes and businesses.

For you and me, electricity is available nearly everywhere we go, and PV systems are able to integrate with the existing utility grid. In remote, developing areas, PV systems provide valuable energy for powering lighting systems, running refrigerators, and helping deliver clean drinking water.

## Checking out PV pros and cons

REMEMBER

PV systems have some serious advantages on their side. Producing electricity from the sun has environmental benefits because the power source is an abundant renewable resource that's available every day (even though PV systems aren't as effective during cloudy weather, they still produce a small amount of power on those days). PV is also a highly adaptable power source. You can use individual cells to power small electronics and individual panels to power specific loads. You can build small arrays to power homes, or you can build utility-scale projects to send massive amounts of power into the utility grid. And after PV systems are installed, they can provide many years of clean, reliable power at virtually any location on earth.

On homes and businesses connected to the utility, PV systems are considered *distributed generation*, a power source that produces electricity close to the location where the power is used. They're able to offset the requirements on the central power plants sending out the electricity most people use.

WARNING

PV systems aren't the right answer for all applications. They have some disadvantages too. For example:

» **The sun isn't a continuous power source.** At night, the PV modules can't produce power, so in some scenarios, you have to use a method to store the energy for later use (adding cost and complexity to the system design and installation).

» **The amount of area required to produce power is large in comparison to other sources of power.** For large-scale projects, significant portions of land or roof space are necessary. Not every homeowner or business owner has access to such space.

## Looking into the future of PV

Most people accept that "alternatives" to the "conventional" power sources acquired by burning fossil fuels must be developed, and so, to use a really bad pun, the solar industry has a bright future ahead of it. The worldwide demand for solar electricity continues to grow on a global level, and the amount of time, effort, and money being put into the industry is amazing. Many research and development projects are underway that will help drive down costs further, increase efficiency, and deliver better PV systems.

I'm reluctant to make any bold predictions about the future of PV technology; ideas I had just a few years ago that I wrote off as too far-fetched and nearly impossible are proving to be real solutions in the PV industry. However, I *am* comfortable saying that the overall acceptance of PV systems and their integration into the built environment will do nothing but increase in the years to come. Solar power is no silver bullet, though. It'll take a movement on many fronts to make the global shift away from over-reliance on fossil fuels.

# Introducing PV Components and Systems

PV systems can be ridiculously simple (connect a module to a load and use the load as you wish), but they can also seem overwhelming when you first look at them. Every time I go to design and install a PV system, I look at the whole process as a real-life puzzle that must be arranged and put together just so. Each project is unique in its own way, and that's part of the fun.

REMEMBER

Yet despite the differences in the details, the PV systems you connect to homes and small businesses have some specific and very necessary components (all of which have an important role to play in the system).

- **PV modules:** The individual units that you place in the sun to produce electricity from the sun are called *PV modules.* A number of modules connected together in different configurations form a *PV array.*
- **Battery bank:** Batteries provide a way of storing the energy produced by the PV array. Individual batteries connected together make up a PV system's *battery bank.*
- **Inverters:** Devices that take power from the PV array or the battery bank and turn it into AC power used to operate loads are *inverters.*
- **Disconnects and overcurrent protection:** These components are necessary for ensuring the safety of the system and the people who come into contact with it.

Grid-direct PV systems, which send power back to the utility grid, have become the most popular type of PV system at locations where the utility grid is present. They offer increased efficiency and reduced maintenance as well as decreased costs compared to their battery-based counterparts. What they give up, though, is the ability to use the PV array whenever the utility power goes out. Luckily for most people who have utility power present, utility outages are relatively rare; when they happen, they're generally short-lived.

Battery-based systems operate independently from the grid (stand-alone) or as a backup to the grid (utility-interactive).

In Chapter 2, I run through all the major components you need to install in a PV system — whether it's a grid-direct one or a battery-based one — and what their relationship is to the other pieces of the puzzle. There, I also explain how to determine which type of system is best for any given client.

# Knowing Your Electricity A-B-Cs

A good understanding of electrical concepts and fundamental equations is vital to designing and installing PV systems correctly. My goal is to have you feeling comfortable and confident in your understanding of the basic electrical terms and equations (don't worry — no calculus here). If you design and install PV systems long enough, I guarantee you'll find yourself using these concepts on a daily basis.

If you've never opened an electrical box for fear of the unknown monsters lurking behind the cover, or if you just need a refresher on all the terminology, check out the information I present in Chapter 3. The jargony, industry-specific terms that the PV world is filled with (many of which are born from the electrical trade) will be much clearer if you do.

## Solar Resource 101

The *solar resource* is defined as the amount of solar energy received at a particular site. Following are some of the terms used to describe the intensity of the sunlight striking the earth (these terms are used pretty frequently in PV system design, so I suggest getting familiar with them; for specifics on the relationship between the sun and the earth, see Chapter 4):

- **Azimuth** describes the position of the sun (and the modules) in terms of how many degrees the sun or the array is from north.
- **Irradiance** describes how intense the sunlight is at a particular moment in time.
- **Irradiation** refers to the quantity of solar energy received for a given amount of time (a day is a typical time frame).
- **Solar window** refers to the portion of the sky where the sun appears at a particular location on earth. The solar window varies based on your latitude. You want to do your best to keep any obstructions out of the PV array's solar window.
- **Tilt** describes the number of degrees that the PV modules are off of the horizontal surface.

But that's not all you need to know about the solar resource. A PV system's location on the earth has a definite effect on the overall system installation. Also in Chapter 4, I help you look out at the sky and see the big window that must be kept open for your PV system to perform as well as it possibly can; I also describe the specific effects of the sun's path on the earth.

## Surveying a PV System Site

PV systems are wonderful, magical things, but they can't perform miracles all on their own. You have to give them a fighting chance for them to knock your client's socks off. In other words, you have to survey the area where the array will be located and make sure nothing will block sunlight's path to the array.

REMEMBER

Shade is your enemy, so take a critical look at the potential PV location when you conduct your site survey. A tiny bit of shade on the array may be unavoidable at certain parts of the year, but it's your job to predict and limit these scenarios to the best of your ability.

In Chapter 5, I show you what to keep an eye out for and how to properly assess your site's potential. I also outline the most common tools you should have on hand in order to perform a successful site survey.

# Delving into PV System Details

PV systems use a number of interesting components that can leave even the most seasoned electrician gawking in amazement. I walk you through the major components in the sections that follow. My goal is to boost your comfort level with the capabilities, limitations, and basic construction of all the major pieces of a PV system. I strongly encourage you to read these sections (rather than skip over them) so you can acquire some basic understanding; without it, any system you design is just a bunch of parts and boxes that likely won't get the job done.

## PV modules

PV modules are truly where all the magic of PV systems begins. They produce voltage and current, and, when wired correctly, they perform useful work. (To me, that's more magical than making entire bridges disappear.) In Chapter 6, I show you the module specifications you'll refer back to many, many times during the system design and installation processes. I also show you how the modules are at the mercy of their surrounding environment and will react to whatever sunlight and temperatures they're exposed to.

## Batteries

When you need to store the energy produced by a PV array and use it at a later time, you need batteries. I list your many options and go over the basic construction and operation of batteries in Chapter 7. (This information will prove invaluable when you start *specifying* [selecting] batteries for your battery-based PV systems.)

REMEMBER

Whenever you need to incorporate batteries in a PV system, you need to define the following parameters:

- The amount of energy the client needs daily
- The number of days the client wants to be able to go without having to recharge the batteries
- The amount of solar energy that's available for charging the batteries
- The temperature at which the batteries are stored

## Charge controllers

When you use batteries, you have to make sure they're properly charged by the PV array — that's where a charge controller comes in. In Chapter 8, I show you how to evaluate the different charge controller technologies that are used most often and explain how the different technologies interact with the array and batteries to deliver the maximum amount of energy to the batteries. I also outline the different feature sets commonly found on charge controllers and introduce the specifications to consider when selecting a charge controller for a battery-based PV system.

## Inverters

Because PV arrays and battery banks produce and store DC electricity, you almost always need to include an inverter in a PV system (or more than one, depending on the size of the array). An inverter takes DC electricity and turn it into the AC power used by most electrical loads in homes and businesses.

I go over the inverter categories used in PV systems in Chapter 9, but in general, you can classify inverters as either grid-direct or battery-based:

- All grid-direct inverters are also classified as utility-interactive (meaning they can take power from the PV array and send it into the grid).
- Battery-based inverters, on the other hand, can be classified as utility-interactive or stand-alone (meaning they can't send power into the grid and are meant for off-grid applications).

## Wiring and safety devices

Safety is one of the most important considerations you make during the design and installation process. To achieve proper equipment safety, you need to install a number of safety devices, not least of which is the right wiring. In Chapter 10, I introduce you to the different types of *conductors* (wires), the conduit that protects those conductors from damage, and the overcurrent protection devices you have to use to protect the conductors from having too much current flowing through them.

Another important piece of safety equipment is the disconnect switch that allows you to (no surprise here) disconnect the PV array from the inverter. When a disconnect switch is in the off position, you can safely access the components of a PV system for servicing.

Because other people (such as emergency personnel) may come into contact with the system, you need to make sure it's safe for them by installing labels that explain how to quickly and safely access the system.

As a final note on safety precautions, you need to keep in mind that at some point the wiring in the PV system may become damaged and pose a risk. Ground fault protection (GFP) devices help protect against fire hazards in the case of damaged conductors. In grid-direct inverters, this protection is preinstalled; for battery-based systems, you must install GFP separately.

# Sizing a PV System

REMEMBER

In order to have a PV system operate properly and meet your client's expectations, you need to spend some time in the design phase evaluating the individual components of the system and their interaction with all the other pieces. During this *sizing* process (which is when you determine the number of modules to use in a particular system), you have to consider the client's available budget for the project, the PV array location, and the specifications for the individual pieces of equipment.

In this section, I explain the basics of sizing and installing the two main system types and talk about sizing the safety equipment used in your systems.

## Grid-direct systems

Grid-direct systems offer more design flexibility than battery-based systems because you don't have to worry about storing any energy; the grid will usually be there to make sure all the loads can run when the user wants them to. Even with this flexibility, you need to carefully consider the PV array you design and the components you connect to it. In Chapter 11, I show you what you need to consider in the sizing and design process, from evaluating the energy consumption at the site to the utility requirements for connecting to the grid to the calculations used when matching a PV array to an inverter.

## Battery-based systems

Before you can size a battery-based system (either utility-interactive or stand-alone), you must evaluate the energy consumption used by the loads that will be powered via the battery bank. After you complete this step, you're ready to move on to sizing the different system components, specifically the battery bank, the PV array, the charge controller, and the inverter. Chapter 12 has the how-to on sizing these components. (If you need to incorporate a generator into any battery-based system design, Chapter 12 has you covered as well.)

### Conductors and safety devices

PV systems have some rather unique properties that need your attention when sizing wiring and safety components. The conductors are exposed to some extremely high temperatures, the current values passing through those conductors are at the mercy of the sun, and on top of all that, the DC current passing through the system requires you to seek out components that are specially listed for the application. Never fear. I walk you through the processes of properly choosing and sizing conductors, conduit, and overcurrent protection devices in Chapter 13.

## Bringing a PV System to Life

After you spend all the required time designing a PV system, you're ready for the real fun: going out in the sun and putting the modules to work. The next sections are dedicated to the different portions of the installation process to help get you up to speed. Of course, every situation you encounter will be slightly different from the last, but the fundamental ideas and processes behind PV system installation remain constant.

### Permitting

An important activity takes place before a PV system is ever installed: permitting. The local building department is responsible for providing permits for any PV installation. You must apply for this permit, just like you would if you were doing any other major construction project.

REMEMBER

Generally, you need to have two permits issued to you: an electrical permit and a building permit. The electrical permit is required so that the building department can ensure that the electrical portion of the system is safe, and the building permit helps keep the building safe from mechanical failure. (For full coverage of the permitting process, turn to Chapter 14.)

### Staying safe

The safety of all individuals on the job site should be the most important factor during any installation — period. When you're installing a PV system, you'll be on a construction site with numerous hazards. The Occupational Safety and Health Administration (OSHA) rules regarding construction-related trades will be in full effect, and you need to be sure to address these requirements. The OSHA Web site, www.osha.gov, is full of information and resources to help you make your work sites as safe as possible.

REMEMBER

Just because PV systems deal with solar power doesn't mean they don't possess the same electrical hazards associated with any other electrical system. Always keep this fact in mind. Also, because PV systems run on energy they obtain from the sun, you're going to be working on ladders, accessing roofs, and working in locations that are fully exposed to the elements — all of which add even more safety hazards you need to be aware of. (For a review of the major safety elements you need to consider, as well as methods you can use to keep yourself and others safe during a system installation, turn to Chapter 15.)

WARNING

It takes a single accident to not only seriously injure an individual but also cripple an entire business. So do whatever you have to do to keep safety top of mind during each and every installation job.

## Putting together the mechanical parts

For many PV system installers, the truly difficult part of any PV installation is the mechanical portion, which includes setting up the rack to hold the modules and evaluating the interaction between the PV array and the building. Depending on the location of the array (on a roof or on the ground), the installation of the mechanical components can represent a large portion of the time spent on the job site. Refer to Chapter 16 for a complete rundown of the issues you'll face as you install the mechanical components of a PV system, as well as effective solutions for them.

REMEMBER

A number of PV installations occur on rooftops, especially for residential systems. Whenever you're working with a rooftop system, you need to carefully consider the methods used to attach the racking system to the roof. For ground-mounted systems (including top-of-pole mounting), you have to evaluate how you plan to keep the array in place without overtaxing the racking system or constructing a structure that can support a small country.

## Adding the electrical parts

The electrical installation is always the portion of PV systems that receives the most attention — and rightly so given the numerous regulations you must comply with. The majority of these regulations are spelled out in the PV-specific portion of the *National Electrical Code®* (*NEC®*), specifically Article 690. I point out the highlights of Article 690 later in this chapter; consider this information a primer on the *NEC®*, not a complete review of the Code. (***Note:*** Some local building departments may have additional requirements, so be sure to check in with the local office to make sure you're on the same page.)

As I walk you through the different portions of the electrical installation in Chapter 17, I refer to the *NEC*® and point out specific requirements. I cover the required locations for different electrical components such as disconnects and overcurrent protection, and I guide you through the ever-fun topic of grounding the PV system (as well as grounding methods that are commonly used throughout the United States). I also note the requirements you need to follow when connecting the PV system to the local utility grid.

REMEMBER

You don't have to memorize specific passages of the *NEC*®, but you do have to be knowledgeable about the layout and format of the Code so you can effectively refer to it while designing and installing your systems.

## Commissioning, inspecting, and maintaining a system

After you install the system, you're ready to *commission* it (in other words, you're ready to flip the switch and release the magical electrons from the PV modules and put them to work). Before you get too excited, though, you need to take some time and make sure the system has been installed properly so you don't have to spend the next three weeks on-site fixing a problem that could've been discovered earlier. I walk you through this self-inspection, as well as the commissioning process, in Chapter 18.

When you know for sure that the system works, you need to turn it off and await the official inspection from the local building department. You must set up an appointment for an inspector to come out and look everything over. The inspector's job is to make sure you followed the basic requirements and installed the components you said you would. If he sees any major problems, he'll document them and require you to fix them before he'll pass the system and allow you to turn it on. Turn to Chapter 18 for the full scoop on inspection requirements and common problem areas in PV installations.

TIP

To avoid getting hung up for too long in the inspection process, speak with an inspector early in the process (even as early as when you apply for the permits). This way you can make sure you're clear on the local building department's guidelines and get initial approval on things like grounding and labeling while you're in the design phase (when it's easier to make changes).

Don't think you're done as soon as the system receives approval from the inspector. No matter how low maintenance it may be, every PV system still requires maintenance now and then. I outline the common maintenance issues you need to be aware of and share advice on how to approach them (as well as who's responsible for them) in Chapter 18.

# Introducing the Sections of Code You Need to Know

As you begin to install PV systems, including the electrical elements, you need to make sure you're familiar and comfortable with the *National Electrical Code®* (*NEC®*), also referred to as NFPA 70®. I reference the *NEC®* (also known simply as the Code) throughout this book, directing you to specific sections and applications.

REMEMBER

Here are some basics related to the *NEC®* that you may find helpful:

- If you've never referenced the *NEC®* or look at it only on rare occasions, I suggest you purchase a copy the *NEC® Handbook.* It contains the entire *NEC®*, along with some pretty pictures and helpful explanations that make the Code easier to understand. Perhaps the easiest way to obtain a copy of the handbook is to head to the National Fire Protection Association's Web site (`www.nfpa.org`) and click the Buy NFPA codes & standards link on the left-hand side of the page. From there, just search for the *NEC® Handbook.* The handbook is a little pricey, and you'll have to keep purchasing new editions periodically. Rest assured, though, that the cost is worth it.
- Always make sure you're using the correct version of the handbook for your area. The *NEC® Handbook* is released every three years, but the local *jurisdiction* (the office that issues building permits) may not be using the most current version. To find out which version the jurisdiction in your client's area is using, call the local building department.
- As you work more and more with the Code, you'll realize that people interpret the same section differently. The powers that be who author Code sections try for concrete language, but there's no way to achieve total clarity with a document like this. If you ever have a question, your best bet is to research your question and maybe even consult with your electrical inspector to make sure you're on the same page as he is (literally and figuratively). ***Note:*** The local *authority having jurisdiction* (AHJ) — in other words, the electrical inspector — has the final ruling on the Code interpretation; the Code is there as a guideline, but the enforcement is up to the AHJ.
- Although a specific section in the *NEC®* covers PV systems, you don't get to ignore the other sections. The entire *NEC®* applies to PV systems, so reference the other sections as appropriate. However, there may be situations where the requirements set in Article 690 are different than in other sections. In this situation, Article 690 supersedes the other Code sections because it relates to PV installations.

IN THIS CHAPTER

» Identifying the major parts of a PV system

» Distinguishing between battery-based and battery-less systems

» Helping your customers select the correct system for their needs

# Chapter 2

# Checking Out Common Components and Systems

This chapter may seem trivial and worth skipping when all you want to do is dive into a new-to-you technology and design systems that'll be the envy of your peers (not to mention major moneymakers). My advice to you? Remember that it's all about the details. Before you can design and install that envy-inducing, moneymaking PV system, you need to have a grasp on exactly what goes into a PV system. Each component has a unique and important role in the whole system. You, the designer and installer, have to understand these components and their purpose, as well as how they fit together as a whole, in order to help your customers select the best system for their needs. This chapter gives you the knowledge base you need to start working with and recommending PV systems.

## Introducing the Components That Make Up PV Systems

Even though you and I probably drive a different kind of vehicle, the major components inside each one are identical. Both vehicles have an engine, some wheels, and a steering wheel to help us navigate, although your vehicle may have a bigger engine than mine, and my vehicle may have bigger wheels and a smaller steering wheel than yours. You can look at PV systems in a similar manner. The basic components, such as the PV modules and the inverter, are the same even though the types of components differ based on the user's needs and what the local regulations require. In the following sections, I take a look at the major components of a typical PV system and explain how each component relates to the others.

*Note:* The following information isn't all-inclusive. Rather, it's a primer designed to help you understand PV systems as a whole. Each of these components receives much more attention in later chapters, so please don't feel overwhelmed with the descriptions I provide here.

## PV modules and racking

*PV modules* are the source of power in a PV system. They produce DC electricity, which is easily stored in battery banks but needs to become AC power via an inverter for use in homes. (I present the features and limitations of different types of PV modules in Chapter 6.)

*Racking components* hold the PV array to the location where you choose to place the modules. One typical way to mount the entire array is on the roof of the structure you're working with. Other options include *building-integrated mounting* (where the PV array becomes a substitute for the roofing or window materials) and mounting the array on a racking system down at ground level. Another common option is mounting the PV array on top of a pole rack. I give you the scoop on mounting methods and considerations in Chapter 16.

No matter where they're mounted, PV modules are always connected to these particular components: disconnects, inverters, and charge controllers.

TIP

The terms *PV module* and *PV panel* are often used interchangeably (I've been fighting myself for years on getting it straight when talking to people). You also see and hear *PV array* quite a bit. The fact is, different definitions describe these terms. If you're discussing modules, panels, and arrays with someone, you need to make sure you're both talking the same language.

- A *module* is a single unit consisting of PV cells contained in an environmentally secure package.
- A *panel* is a group of modules fastened and wired together.
- An *array* is all the modules (and panels) and racking components used to produce DC electricity.

## Battery bank

For systems that require energy storage (namely, any system that needs to operate without the utility grid; see the later "Battery-based systems" section for more information), a *battery bank*, multiple batteries wired together to achieve the specific voltage and energy capacity desired, is the best option.

The battery bank is typically housed in a container to keep the batteries safe. The PV array connects to it in order to provide charging, with a charge controller located somewhere in between. The battery bank is also connected to the inverter to provide power for the AC loads. If the system also uses DC loads, the battery bank is wired to a DC load center. (Don't know what loads and load centers are? I cover both later in this chapter.)

Just like your car's fuel gauge tells you when you need to fill up, a battery monitor tells you how full the battery is. In battery-based systems, a monitoring system is very important in keeping track of the battery level.

I describe the different battery technologies in detail in Chapter 7, but the most common batteries used for the typical PV system are as follows:

- **Sealed batteries:** These come in a sealed container that requires a reduced amount of maintenance by the end user.
- **Flooded batteries:** These come in an open (or flooded) container that requires a higher level of user interaction.

Sealed batteries are often referred to as "maintenance-free batteries." Don't believe the hype! Sealed batteries are indeed a reduced-maintenance option compared to flooded batteries, but they still need some attention. If you don't advise your clients to monitor and properly maintain a battery bank, the batteries will die prematurely. Relatively simple battery monitoring can not only keep the system owner updated on the batteries' status but it can also keep her from experiencing poor results when using the batteries. (Chapter 18 has details on the maintenance required for battery-based systems.)

All batteries give off gas when they're charging, releasing hydrogen. Sealed batteries release minuscule levels of hydrogen; flooded batteries can give off substantial levels of it. A wise choice (actually a requirement in most locations) is to keep all batteries inside a protective container that vents to the outside to avoid the possibility of hydrogen buildup and an explosion hazard.

## Charge controller

A *charge controller* is a piece of electronics that's placed between the PV array and the battery bank. As you can probably guess, its primary function in life is to control the charge coming into the battery bank from the PV array. Charge controllers can vary from a small unit intended to connect a single PV module to a single battery all the way to a controller designed to connect a multiple-kilowatt PV array to a large battery bank.

Flip to Chapter 8 for more information about charge controllers, including their features, technology, and sizing.

## Inverter

*Inverters* turn the DC power produced by PV arrays or stored by battery banks (in battery-based systems) into the AC power used in homes and businesses. They come in many different shapes and sizes. An inverter can be as small as the 100-watt unit you plug into the DC plug in your car or as large as a megawatt unit installed in conjunction with a utility-scale PV project. (I delve into the details of inverter specifications in Chapter 9.)

If the PV system is only going to supply power for DC loads such as lighting, water pumping, or small electronics, an inverter isn't a requirement. DC loads in standard residential applications don't show up much anymore, but they're common in recreational vehicles.

The inverters in PV applications fall into two major categories:

- **Utility-interactive:** These inverters can connect to a utility and either supply power to the connected loads or send electricity back into the grid, essentially running the meter backward.
- **Stand-alone:** These inverters aren't designed to interact with the utility and work by supplying power to loads.

Utility-interactive and stand-alone inverters have a similar appearance. If you were to look at two inverters from the same manufacturer, you wouldn't be able to tell the difference because manufacturers try to put multiple inverters in the same *chassis*, or outer shell, to reduce the number of parts they have to use and cut costs. The only way to know for sure which inverter is which is to look at the label on the side of each inverter.

Utility-interactive inverters come in two flavors:

- » **Grid-direct:** These are primarily used in systems for homes and businesses with utility power present. They don't provide the user with any energy storage, which means if the utility power isn't present, the inverter can't work. In grid-direct inverter systems, the PV array is connected to the inverter on the DC side, and the inverter is connected to the utility grid on the AC side.
- » **Battery-based:** These require a stable voltage source of power, typically a battery bank, to keep the inverter running. A PV system using this inverter requires more components and is generally more complicated, but it allows the user to have backup power for times when the utility goes out. Battery-based inverters use a PV array and charge controller on the DC side and connect to the utility on the AC side.

Stand-alone inverters are quite similar to utility-interactive, battery-based inverters. The main difference is the source of AC power. In a typical stand-alone system you have an AC engine generator to supply power to the loads and allow for battery charging when the PV array can't keep up with the user's energy consumption. On the DC side, a stand-alone inverter is connected to the same components as its utility-interactive, battery-based counterpart, but on the AC side, the inverter sends power only to AC loads.

# Loads

*Loads* are all the pieces of electrical equipment people want to use in their homes and offices. You can have DC loads or AC loads (and sometimes even both). You just have to make sure you supply the correct type of power to the load. For instance, you can't use a DC light bulb when AC power is provided.

Loads are served differently depending on the system used (you can find out more about different systems later in this chapter).

- » In grid-direct systems, loads are primarily served by the utility. The PV system can send power directly to the loads or back into the utility grid.
- » In battery-based systems, the inverters are designed to directly run the loads connected to the inverter in the load center. This requires sizing the inverters differently based on the inverter technology used. ***Note:*** Occasionally, DC loads are present in battery-based systems; when they are, they receive their power from the battery bank via a DC load center.

TIP

In a utility-interactive, battery-based system, the system owner defines the exact loads that get backed up. Although people would like to back up their entire suburban home, that's a very difficult task. Bring them back to reality with this list of loads that *can* be backed up with success: refrigeration, well pumps, lighting, and computer equipment.

## Load centers

A *load center* is the place where electrical loads (see the preceding section) receive their power. In a typical scenario where the utility is present, the utility sends power to a building at a single location. The utility's wires are connected to a meter; from that meter, a connection is made to the *main distribution panel* (MDP; also called a *main load center*). The individual wires running throughout the building are ultimately connected to the MDP via a circuit breaker.

When a number of electrical circuits are connected within a building, or if a detached structure is present, such as a garage, often an AC subpanel is used to help reduce the number of wires running all the way across the house. Figure 2-1 is an example of how a main panel and subpanel can be laid out in a residential setting. Usually, a circuit breaker in the MDP sends electricity to the subpanel. The subpanel then uses multiple circuit breakers to send electricity to loads and outlets that are physically close to the subpanel.

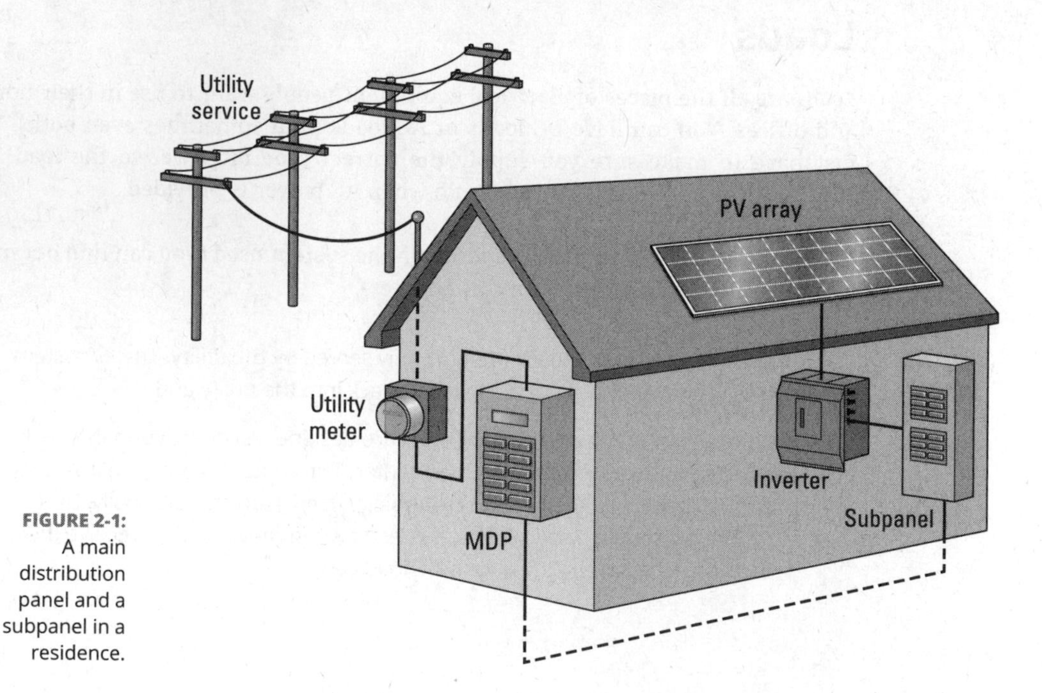


**FIGURE 2-1:** A main distribution panel and a subpanel in a residence.

So how do load centers come into play in a PV system? One possibility is to connect a grid-direct inverter to the AC subpanel, as long as the *NEC*® and utility requirements are met. In a utility-interactive, battery-based system, the AC subpanel can serve as the backup load center, allowing the loads connected to this panel to remain powered up during an outage. For stand-alone inverters, because no utility power is present, the inverter supplies the AC power and is connected to the MDP.

## Disconnects and overcurrent protection

A number of safety and servicing requirements exist for PV systems, including *disconnects* (the switches that can turn everything off, much like a big version of a light switch) and *overcurrent protection devices* (the pieces of equipment, typically a fuse or circuit breaker, that automatically turn a load off if it's drawing too much current). The requirements for the exact components and their locations vary depending on the specifics of your installation and the requirements of the local utility and electrical inspector, but here are the main considerations:

- Disconnects are placed in locations for system maintenance and to isolate the PV array and inverter.
- Overcurrent protection is placed in the system to protect the *conductors* (wires) from becoming too hot and starting a fire.

Disconnects and overcurrent protection are commonly combined in a single enclosure in the form of a fusible disconnect.

The *NEC*® sets the basic requirements that most inspectors follow, yet the *NEC*® is only a guide. Local inspectors are generally given the ability to enforce more requirements as they deem necessary. The *NEC*® has specific requirements for disconnects and overcurrent protection; I cover these requirements in detail in Chapters 10 and 17. However, many utilities have requirements that may exceed those found in the *NEC*®, so you need to understand those requirements and how they affect your installation.

## Utility interconnection

The point at which a PV system interconnects with a utility grid is often referred to as the *point of common connection* (POCC). This is an electrical connection that can take various forms depending on the size of the PV system and the electrical equipment already located on-site.

REMEMBER

When connecting an inverter to the utility, you need to let the utility know what you're doing in order to get its approval. Most utilities require that the system owner review the interconnection requirements and sign a document stating that she will adhere to the rules. As PV systems have become more popular, these documents are generally easy to obtain from the utilities and have standard language for all customers to follow. Thankfully, the requirements of most utilities aren't hard to incorporate into the system. Flip to Chapter 17 for more information on connecting to the utility.

The most straightforward way of making the POCC is to attach the inverter to a circuit breaker in the existing MDP on-site. Specific requirements for this type of connection need to be met, but that isn't a difficult task in most cases.

## MAKING THE UTILITY INTERCONNECTION WITH NET METERING AND THE FIT PROGRAM

*Net metering,* which is when the utility has to credit the system owner with retail rates for PV-generated energy (within some limits), makes utility-interconnected PV systems a reality. Under net metering, the utility is required to pay the system owner retail energy rates for every kilowatt-hour of energy put into the grid, meaning the system owner is credited at the same rate at which she is charged. The catch comes when the system generates more energy than its owner can consume. Each utility has its own way of handling this situation; it can credit the account or take the credit itself. ***Remember:*** The key is to know what the net-metering rules are with the utility and make sure the energy production and consumption at the site match well with those rules. A great place to find specific rules for your client's location is the Database of State Incentives for Renewables & Efficiency; you can access the database online at `www.dsireusa.org`.

Another popular method for connecting to the utility and maximizing the investment is known as a *feed-in tariff* (FIT). The idea is relatively simple: Governments and utilities recognize that energy produced from renewable sources such as PV has a higher monetary value than energy from fossil fuels. Therefore, the PV system is interconnected to the utility and metered separately from the building. The energy created from the PV system is bought by the utility at a higher rate than the standard retail rate, and the energy used in the building is charged at the standard retail rate. This system encourages customers to install high-quality systems that produce as much energy as possible to maximize their payments.

REMEMBER

If you need to connect to the utility in any method other than through an existing MDP, communicate this need to the utility early in the process. You may need to coordinate a time for it to purposely disconnect the building from its power lines in order to make the connection safely. Also, the utility may have additional requirements for safety equipment.

# Differentiating between PV System Types

You know the saying "the whole is greater than the sum of its parts"? It applies to PV systems too. After you have a picture of the major components of a PV system, you're ready to focus on the two main types of systems: grid-direct and battery-based. The sections that follow help you create a mental picture of both systems; they also get you ready to differentiate between the two for scenarios you're bound to encounter.

## Grid-direct systems

For those folks with access to the utility grid, a grid-direct system is the most common type of PV system being installed today. A *grid-direct* (sometimes referred to as *battery-less*) PV system requires the least number of components and is typically the most efficient in converting the sun's radiation into usable AC electricity for loads. These systems have the added benefit of requiring very little human interaction because the inverter is able to automatically control the flow of electricity from the PV array to the grid. In the following sections, I describe the configuration of a grid-direct system and explain its pros and cons.

### The configuration

REMEMBER

Grid-direct systems, like the one pictured in Figure 2-2, have a relatively straightforward configuration. The PV array is connected to the inverter, which in turn is connected to the utility. Grid-direct systems typically use

- A PV array with racking
- A junction box to get the wires in the building
- A DC disconnect
- An inverter (or more than one, depending on the size of the PV array and/or whether you have multiple arrays)
- An AC disconnect
- A meter to record the energy produced by the PV array
- A utility interconnection across a circuit breaker inside the MDP

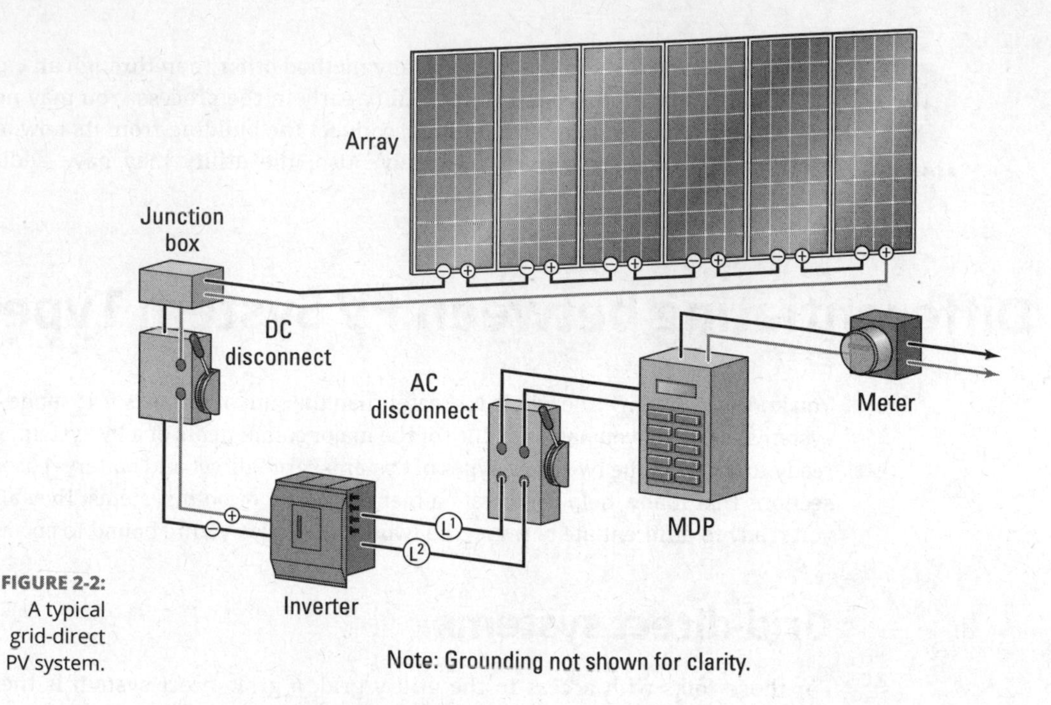


**FIGURE 2-2:** A typical grid-direct PV system.

As I explain in Chapter 17, the exact locations of all the equipment vary and are dependent on the specific system. However, Figure 2-2 is a good representation of the components most grid-direct systems use.

## The pros and cons

Since 2001, the grid-direct system has been the most popular type of PV system installed in the United States. Why? Because 2001 is when manufacturers started selling reliable grid-direct inverters. It's also when rebate programs came along.

REMEMBER

The pros of installing and operating a grid-direct system rather than a battery-based one are as follows:

- You have fewer components to work with.
- The design and installation are more simplistic (meaning less expensive).
- The system's energy output is greater.
- The maintenance requirements are fewer.
- The inverter can be located outside the building.

WARNING

The grid-direct configuration does have one major downside. The inverters used in these systems require the utility's presence in order to remain on. After the utility goes away (due to a power outage or a brownout situation), the inverter has to recognize that immediately and disconnect itself from the utility for safety reasons. Manufacturers call this an inverter's *anti-islanding feature*; it's a safety requirement of all inverters connected to the utility grid. Therefore, if you have a PV system using a grid-direct inverter and the power goes out, you'll be in the dark just as much as your neighbor who doesn't have a PV system. ***Note:*** If the inverters didn't have anti-islanding, they'd continue to send power back to the grid. If a utility worker was working on the line, she might get shocked (at best) or electrocuted (at worst).

## Battery-based systems

The second major type of PV system (in addition to the grid-direct system described earlier in this chapter) is battery-based. A battery-based system can be utility-interactive or stand-alone. The two types are almost identical, but their differences are important to know:

- Utility-interactive, battery-based systems require an inverter that's specifically listed as a utility-interactive inverter and has the same anti-islanding features as the grid-direct units. Although the stand-alone inverter requires safety listings, the anti-islanding portion doesn't apply. (I provide basics on different types of inverters earlier in this chapter.)
- Utility-interactive, battery-based systems allow the user to back up certain loads but not the entire house. Yes, it's possible to back up the entire house, but that's pretty tough. Major energy-consuming appliances (think electric water heaters and clothes dryers) can't be supported.
- Stand-alone, battery-based systems support all the loads in a home. The caveat, as I show you in Chapter 12, is that these loads must be very carefully considered and minimized as much as possible.
- In stand-alone, battery-based systems, the AC power source is an engine generator. Consequently, the inverter needs to be ready, willing, and able to accept power from the generator.

TECHNICAL STUFF

Battery-based systems can be used without an inverter to power DC-only loads, making them good for a PV array at a small remote cabin that needs only a few lights and a radio. Another possible location for such a system is a recreational vehicle that has a small PV array to recharge the house battery.

In the following sections, I fill you in on the basic configuration of a battery-based system and explain its pros and cons.

## The configuration

REMEMBER

Battery-based systems (see Figure 2-3) use many more components than grid-direct systems do. Both the stand-alone and utility-interactive varieties use the following components, unless otherwise specified:

- A PV array with racking
- A charge controller (or multiple charge controllers, depending on the size of the PV array in relation to the charge controller's size)
- DC disconnects (specifically, two for every charge controller and one for every inverter so you can properly isolate all the equipment)
- A battery bank in a vented battery enclosure
- Battery metering
- An inverter (you may need more than one if the number of loads run at the same time will exceed the ability of a single inverter)
- An inverter bypass switch
- A backup load center (for utility-interactive systems only)
- AC disconnects
- An MDP with overcurrent protection (this is where the inverter is interconnected to the utility in a utility-interactive system)

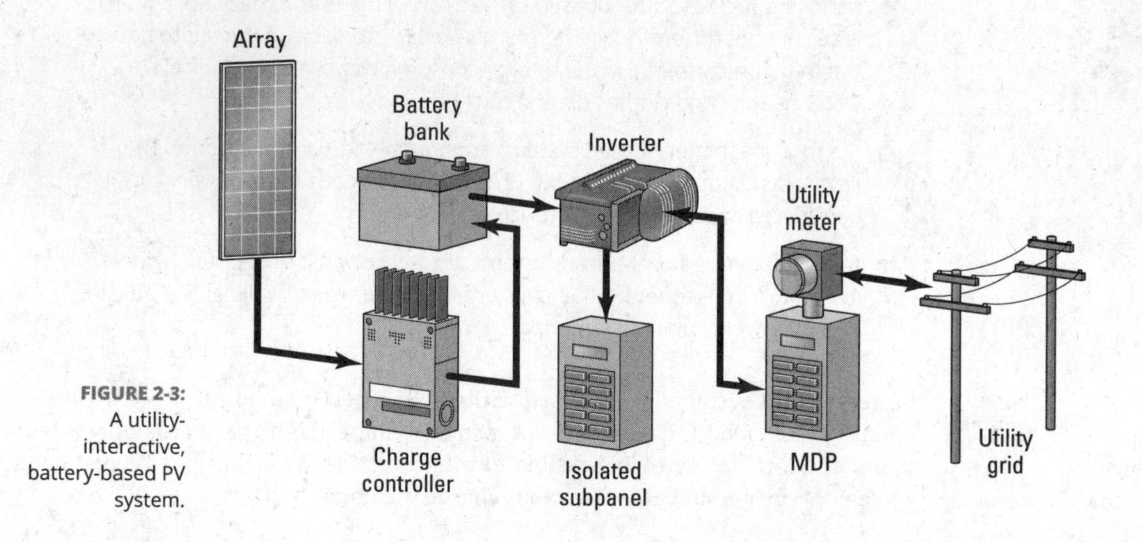


**FIGURE 2-3:** A utility-interactive, battery-based PV system.

The optional components include

- A DC load panel (located near the battery bank and DC disconnects)
- A backup engine generator (located away from the rest of the equipment to keep the noise level down but wired into the inverter[s])

## The pros and cons

Because battery-based systems can be used in utility-interactive and stand-alone scenarios, look at their pros and cons in light of where they're used:

- The major pro of utility-interactive, battery-based systems is that you have power when the utility is out. (Depending on the user, this may be more than just a pro — it may be an absolute requirement!)
- Stand-alone, battery-based PV systems are typically compared to an electrical system with power supplied by an engine generator. The pro to the stand-alone system in this case is that the PV array provides silent, clean power with far less maintenance than an engine generator.

The biggest cons to utility-interactive, battery-based systems are the added costs and increased maintenance requirements. Because a battery-based system contains more components than a grid-direct system, the battery-based option is always going to cost more, especially over time. Also, the more components a system has, the more often you have to check and make sure that everything is working properly, hence the increased maintenance.

The main downside to stand-alone, battery-based systems is the extra work because battery banks take a lot of monitoring and maintaining. And if the system also features a backup engine generator, there's even more to maintain. Fortunately, with a properly sized PV array and battery bank, the generator maintenance can be kept to a minimum.

REMEMBER

Although some users may be able to use a minimal battery bank to get them through a short power outage, others may require a battery bank to help them through multiple days of utility outages. This discrepancy makes estimating the cost of a battery-based system nearly impossible until you know more about the customer's electrical requirements.

# Figuring Out the Right System Type for Any Situation

After you know about PV systems and their components, you're better equipped to tackle the fun part — helping customers figure out which PV system is right for their situation. In the following sections, I separate the decision-making process into two main categories: when the customer is connected to the grid and when the customer is off the grid.

## The customer is connected to the grid

For customers who want a PV system, and where utility power is available, the options are two-fold. They can have either a grid-direct system or a utility-interactive, battery-based system. Sometimes customers automatically assume they want the utility-interactive, battery-based system because when the power goes out, they'll still have power for the television and cold food in the fridge. What they don't consider are cost and maintenance.

You can help your customer make an informed decision by taking the time to ask her several questions and encouraging her to answer honestly. Be sure to ask your customer the following:

- How often does the power go out?
- How long do the outages last?
- What activities (involving electricity) can't be interrupted due to a power outage?
- How much money is realistic for you to spend on this feature?
- How much time are you willing to spend maintaining a battery bank?
- Will you be willing replace that battery bank in five to seven years?

After thinking about these questions, the reality of the situation quickly sets in for most urban and suburban PV-system-owners-to-be, and they give up on the idea of a battery-based system for the easier to maintain and less expensive grid-direct system. Of course, some individuals (like those rural dwellers who are at the end of the electrical line and see outages that last for days) still opt for the utility-interactive, battery-based system.

REMEMBER

The most you can do is share your knowledge with customers to get them thinking about the pros and cons of different PV systems so they can make an informed decision. Unless the customer really has no preference, you can't make the decision for her.

## The customer isn't connected to the grid

If a customer is constructing a remote home or isn't connected to the grid in a current building but wants to make a change from her current system, the first type of PV system you should suggest is a stand-alone, battery-based one. This option makes the most sense for this type of customer because it offers an electrical supply that isn't dependent on an engine generator.

Provided the customer is willing to bring utility power in, you can also suggest a grid-direct or utility-interactive, battery-based PV system. The deciding factor here is how far the utility has to bring its lines and how much that's going to cost in addition to the cost of installing and operating a PV system. I've heard of utilities telling people it'd cost $40,000 to bring the utility a quarter of a mile to their residence — and then they get the pleasure of writing the utility a monthly check!

When working with a potential stand-alone client, you have to take some time with her early in the process to prepare her for the realities of living without the utility grid. Most of the time, making this commitment requires a major lifestyle change, at least as it relates to energy consumption. To help your client choose the right system for her, look at her life today as well as what her life will look like in the future and evaluate the different power sources that may apply for her.

TIP

During the design phase, you should establish a potential stand-alone client's commitment to living "off-grid" by asking the following questions (at a minimum) if you want the process to go smoothly:

- What's your budget for the electrical system? (It's nearly impossible to quote average costs with stand-alone PV systems due to the many variables involved.)
- Are you willing to maintain your battery bank one or two times a month?
- What are the electrical loads you want to run? (See Chapter 12 for more on this topic.)

» Where do you see your electrical demands in five to seven years?

» How much generator run time is acceptable for you?

» Do you have any other renewable energy resources (such as wind power or hydroelectric power) available to you?

REMEMBER

Keep in mind that relying solely on a stand-alone, battery-based PV system for 100 percent of a building's energy is a difficult task. I know of people who can pull it off, but they make major lifestyle changes that most folks wouldn't make. An engine generator is more or less a required add-on to a stand-alone, battery-based PV system because it allows the batteries a quick recharge when the PV system can't keep up, like in the winter during extended cloudy periods, which is when the solar resource is at its lowest and electrical energy consumption is high. If a PV system were able to maintain that level of energy consumption, the array would be so grossly oversized for the summertime that it wouldn't make a lot of sense.

IN THIS CHAPTER

» Understanding current, voltage, and resistance

» Getting to know Ohm's Law

» Differentiating between power and energy

» Calculating series, parallel, and series-parallel connections

# Chapter 3
# Powering through Electricity Basics

When you discuss electricity with someone, you tend to get one of two responses: the bobblehead response and the know-it-all response. The bobblehead is the person who just nods at everything you say because he has no knowledge about electricity so he can't challenge any of it. The know-it-all is the person who's just waiting for you to slip up so he can correct you.

You, as a PV system designer and installer, need to become the person somewhere between these two extremes. You absolutely must be educated in your language and understanding of the basic electrical terms and concepts. Some people will never catch you when you use the wrong term in conversation (it happens to the best of us), but if you continually mix up your terminology, you'll lose your credibility before you can even finish speaking.

In this chapter, I take you through the fundamental electrical terms and topics for PV systems. (Although I'm sure you probably have a concept of the majority of the terms, a quick review never hurts.) Then I show you how to apply those terms in a few important calculations. I wrap it up by covering the ways you'll configure PV arrays when you're out working with these systems.

# Going with the Flow: Current

Electrical *current* is typically viewed as the flow of electrons through a *conductor* (wire). Although this process is more properly described as the charge that's flowing through the conductor, most people can get onboard with the thought of electrons moving from one point to another and doing work along the way.

TECHNICAL STUFF

If you want to dive into the electrical theory aspects of current, by all means do. Plenty of resources exist, but I don't cover the topic any further here. I don't mean to belittle the importance of the concept of flowing charges versus electrons. It's certainly valuable to understand, but this topic is really beyond the scope of what most PV designers need to know.

In this section, I introduce you to the unit used to describe current, present the two different kinds of current, and explain how to measure current.

## Understanding amps

REMEMBER

The International System of Units (SI) describes current through the term *ampere* (A), or *amp* for short. Current is also represented by the symbol I, particularly in Ohm's Law (which I cover later in this chapter). *Ampere* represents a flow, or a rate of electron movement. The fact that this is a rate is very important because you have to measure and predict the amps flowing throughout your systems.

The base units for an ampere are coulombs per second. A coulomb contains approximately $6.24 \times 10^{18}$ electrons, which means an ampere represents a whole mess of electrons passing a point every second.

TIP

To better understand current, think of cars on a highway. Each car represents an electron, and the number of cars passing a single point over a period of time is the traffic flow. As long as all the drivers do their jobs, traffic flows smoothly, and the number of cars passing that point isn't restricted.

The other point to note here is that the electrons don't go anywhere after they do their work, meaning they aren't used up. They continue in the circuit and are pushed along by the electrons behind them. So what does this fact mean in terms of understanding amps? Well, after the electrons have done their work, they continue flowing and go back "home." This is an important fact to remember when you're sizing conductors based on the current flow.

## Distinguishing between direct current and alternating current

PV modules create a flow of current that's described as *direct current* (DC) because the electrons move in just one direction. As sunlight strikes the PV cells, the electrons move off the cells and through a *load* (any piece of electrical equipment) before heading back to the cells, all while staying in a straight line.

DC is also the form of electricity stored by and delivered from batteries. This characteristic is very convenient when charging batteries from PV modules, but it's rather inconvenient when you want to use a standard electrical load that runs on AC power. In Chapter 9, I detail the different inverters appropriate for battery-based systems.

WARNING

One of the main drawbacks of DC electricity is the limited distance it can be transmitted before significant losses make it too inefficient. In PV systems, these losses can be overcome by using larger conductors to carry the current, but this solution is costly and has some limitations. (I show you how to calculate losses associated with PV wiring in Chapter 13.)

The other form of electricity that's present in PV systems is *alternating current* (AC). In AC, which is the form of electricity delivered from utilities to homes and businesses, the electrons move in a back-and-forth pattern. So how does AC work in a PV system? In Chapter 2, I provide a basic description of the inverters that are connected in PV systems. These devices take the DC electricity in the PV array (in a grid-direct system) and batteries (in a battery-based system) and turn it into a form that the typical household loads can use, which is often AC.

REMEMBER

AC differs from DC mainly in the sense that the current alternates in flow at a specific interval of time. The process of alternating the direction of current flow can be viewed graphically, as shown in Figure 3-1. Notice the sine wave shape of the AC current flow curve? It indicates the back-and-forth movement of the electrons. In contrast, the DC current flow is represented by a line that remains straight over time.

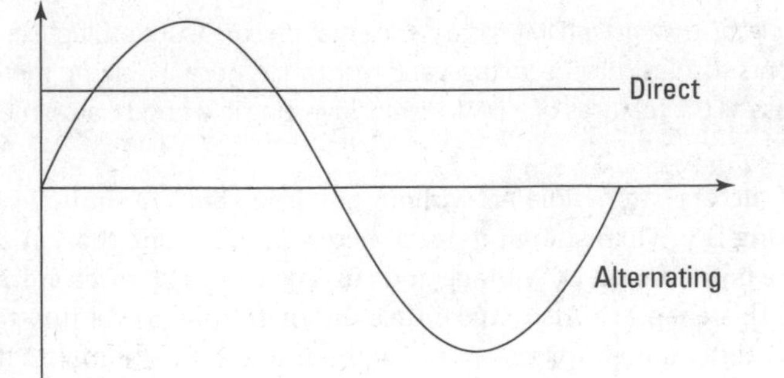


**FIGURE 3-1:** AC and DC current flow versus time.

## Measuring current with a meter

As you install a PV system, you must measure the DC output of an individual module (or sometimes even the entire array) in order to troubleshoot an issue or verify performance. You may also need to measure the AC output of the inverter in order to troubleshoot the inverter. In either case, to measure current correctly you must have a good, hand-held current meter (also known as an *ammeter*) that can read both AC and DC. In the sections that follow, I describe types of ammeters and explain how to use one safely and effectively.

TIP

Many inverters have a way to meter the AC side of the system. These meters generally report the power output (I delve into the details of power in the later "Pondering Power and Energy" section). Although this isn't the same as metering the current, it's generally helpful. As you discover in the later "Relating power to current, voltage, and resistance with the power equation" section, you can take the reported power value and calculate the current value.

### Checking out different kinds of ammeters

Two types of hand-held ammeters are available: inline and clamp.

- The inline meter (shown in Figure 3-2a) requires current to flow through the meter in order to be read. So to obtain a proper current reading with an inline meter, you have to place the meter in the circuit.

WARNING

  Only use an inline meter when you fully understand how the current is flowing through the system and the proper ways to disconnect and safely collect the current measurement. Without a firm grasp on this knowledge, you can really hurt yourself when using an inline meter.

- The other (and better) option is the clamp meter, shown in Figure 3-2b. Clamp meters have a "jaw" on the top that opens when you press a lever on the side of the meter. To measure current with one, place a conductor in the middle of the open jaw and release the lever, allowing the meter to determine the amount of current flowing through the conductor without exposing any portion of it.

TIP

  A *digital multimeter* (DMM) is a device that can measure multiple electrical components such as current, voltage, and resistance. Because clamp meters can encompass all the features of a DMM listed, investing in a good clamp meter is wise.

REMEMBER

A lot of meters can handle AC without any issue but are limited when it comes to measuring DC. When shopping for a meter, look for one that, at a minimum, can measure both AC and DC voltage and current values (I cover voltage in full detail later in this chapter). Also, make sure the meter has the ability to read DC levels that are high enough for the work you plan to do. I recommend a meter that can read at least 100 amps direct current (ADC). Many meters are limited to 10 ADC, which hinders your ability to measure many arrays.

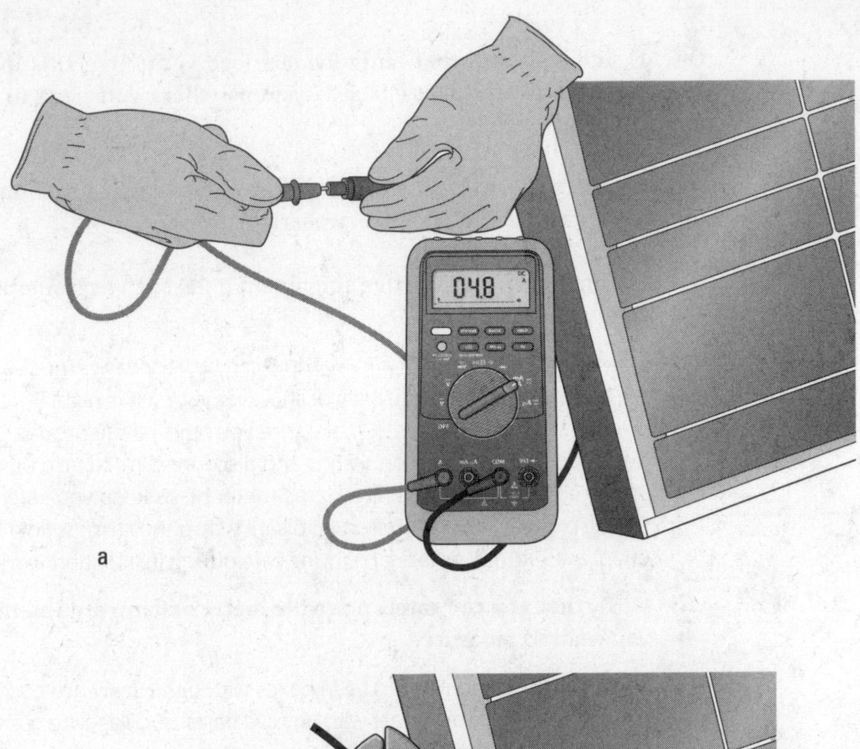


b

**FIGURE 3-2:** Inline and clamp meters.

## Walking through the measurement process

Typically, you measure DC current on a single module when you suspect that the module isn't working properly or on an array when you want to verify the proper power output. (For the full scoop on array commissioning and maintenance methods, flip to Chapter 18.)

On the AC side, if you suspect that the inverter is underproducing, then you need to check the AC current levels. Another common scenario for checking current on the AC side is to see what an individual load is pulling; this information can be especially useful in battery-based systems where you need to know how many amps each load needs.

As I explain in the preceding section, I recommend using a clamp meter when you measure current. Here's how to properly use one:

1. **Put on personal protective equipment (PPE) such as insulating gloves and safety glasses.**

WARNING

   Measuring the current of a PV module can be very dangerous work. Electrons are flowing, and the probability of shocking yourself is high. Before conducting any current measurements, make sure you have all the proper safety gear in place and are correctly connecting and disconnecting the circuit to reduce electrical hazards. Yes, using a clamp meter helps keep you safer during the testing process, but you still have to safely stop the current flow. (Refer to Chapter 15 for full details on staying safe during installation work.)

2. **Verify that you can safely place the meter's clamp around the conductor you want to measure.**

   These conductors will often be in boxes with tight clearances, so make sure you don't come into contact with any live parts, such as wire terminations, fuse holders, or exposed conductors.

REMEMBER

   Always expect that power is present and that a shock hazard exists.

3. **Set the meter's dial to read the appropriate range of DC current.**

   Some meters are *autoranging,* meaning they set themselves to the proper range. If yours doesn't, verify the amount of current you expect to see and set the dial to a value greater than that. For most PV modules, this amount is in the 10 ADC range.

4. **Zero the meter reading.**

   Many ammeters register a small value even if the meter isn't measuring anything. By zeroing out the meter, you give it a baseline that increases the accuracy of the reading. To zero out the meter, refer to the manufacturer's instructions that came with the meter.

5. **Open the meter's clamp and place it around a single conductor.**

WARNING

   Be careful not to place multiple conductors in the meter. If you do, the meter will read the sum of the currents, and you won't get an accurate reading.

6. **Read and record the current value from the meter.**
7. **Remove the meter from the conductor.**

WARNING

Never use an ammeter to connect the two terminals of a battery. At best, you'll ruin both objects by creating a short circuit between the two terminals. At worst, you'll place yourself in unnecessary danger.

# May the (Electromotive) Force Be with You: Voltage

The amount of push that the electrons have behind them is known as *voltage* (V); voltage is measured in volts (how about that?) and represented by the letter E in equations. Another way to describe voltage is as *electromotive force.* In the following sections, I explain the basic concept of voltage, define nominal and operating voltage, and tell you how to measure voltage.

## Grasping the concept of voltage

You can view voltage as the electrical pressure that encourages electrons to move; in other words, voltage makes current (the flow of electrons) happen. Very often you see voltage referred to as voltage potential because there must be a difference between the source of power's voltage and the load's voltage so that current can exist. For example, if you want to use a PV array to charge a battery, the PV array (the source of power) needs to have a greater voltage than the battery (the load). If the PV array has a voltage equal to or less than that of the battery, it can't push the current into the battery, which means the battery can't get charged.

REMEMBER

Both DC and AC voltages are present in PV systems. The concepts and properties for both types of voltage are the same. You just need to make sure you're thinking about and using the correct one depending on which side of the PV system you're working on. Common convention (and what I use in this book) designates the differences by giving a numerical value followed by VDC (for volts DC) or VAC (for, you guessed it, volts AC).

## Getting a grip on nominal voltage and operating voltage

A term you often hear as a PV system designer and installer is *nominal voltage.* This number represents a baseline for measuring voltage. Here's an example: Way back at the dawn of the modern PV industry (the turn of the 21st century), PV modules were available in one of two nominal voltages, 12 V and 24 V, because most PV modules at the time were being used to charge batteries that generally came in 6 V and 12 V nominal voltages. (***Remember:*** PV modules have to produce a voltage greater than batteries, or else there's no current.) So the actual voltage put out by the modules had to be higher than the battery nominal voltage to effectively charge the batteries.

That output voltage is called the *operating voltage*; it's the voltage value when the module is pushing current into a load. A typical 12 V nominal PV module has an operating voltage of 17 V to 18 V. In other words, it has 36 cells inside it that are wired to produce the 17 V to 18V. This allows the module to produce enough voltage to push the current into the battery under all conditions. So when you hear someone talking about a 12 V module, you know he's referring to the nominal voltage.

REMEMBER

The nominal voltages of PV modules have little meaning in grid-direct PV systems (although you still hear references to 12 V or 24 V modules). Where the distinction between nominal and operating voltages really comes into play is when you're installing a small PV array to charge a battery bank. In this case, you need to make sure the PV array's voltage is always high enough to charge the batteries. The easiest way to do this is to have an array nominal voltage that's equal to the batteries' nominal voltage.

TECHNICAL STUFF

The growth of the grid-direct, utility-interactive market has limited the demand for battery-charging PV modules. In fact, many of today's modules have no relation to the battery-charging systems of old. This move away from modules that are designed to charge batteries has given module manufacturers a greater amount of flexibility in terms of the number of cells they put in each module. The result? PV modules with a wide variety of voltage ranges.

## Measuring voltage

It's your job to be prepared to properly measure the DC voltage for individual PV modules and an entire array, which means you need to know how to use a digital multimeter (DMM). This is typically the same meter used to measure current (that is, if you're not using a dedicated ammeter; see the earlier "Checking out different kinds of ammeters" section for details on this tool).

REMEMBER

When you take the voltage measurement of a module, there shouldn't be any load on the PV system. Without a load to power, there's no current, and the modules are at open-circuit voltage (Chapter 6 details the different voltages associated with PV modules). This value represents the voltage potential that can be delivered from the module; knowing this number is important because it allows you to verify the proper voltage levels in your system before you start flipping switches.

Figure 3-3 shows a typical DMM that has been set up for reading DC voltages on a PV module. To make the measurement:

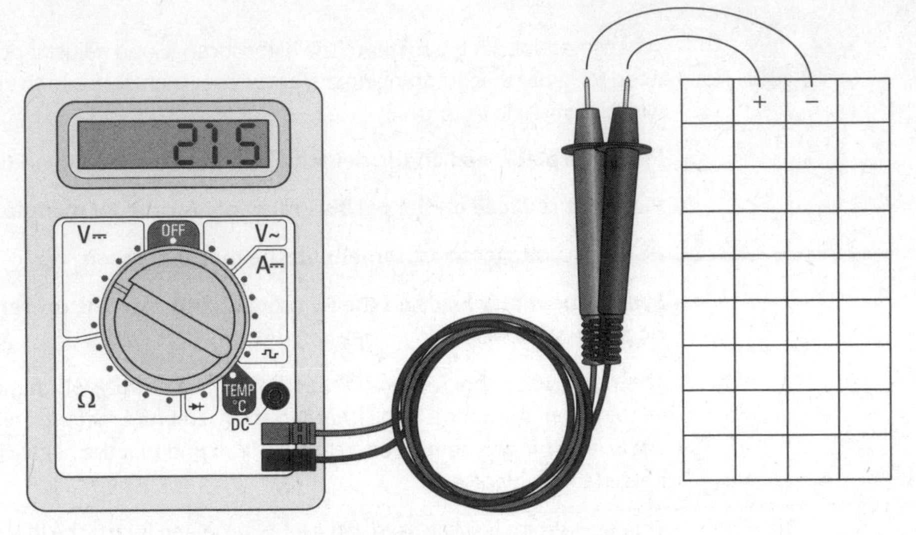


**FIGURE 3-3:** Measuring a PV module's voltage with a DMM.

1. **Put on personal protective equipment (PPE), including insulating gloves and safety glasses.**
2. **Make sure the disconnects are in the off position.**

   Performing this step stops any current from flowing, which keeps you from accidentally touching something that has current flowing through it.

   REMEMBER

   Don't give yourself a false sense of safety when making a voltage measurement. PV modules should be considered live as soon as they're in the sun; just because current isn't flowing doesn't mean you aren't at risk of getting shocked. See Chapter 15 for details on staying safe during installation work.
3. **Make the location where you want to take the voltage measurement accessible by removing the lids to the proper boxes, opening disconnect covers, or accessing the connectors on the backs of the modules.**

   You can take the reading off of an individual module or inside electrical equipment.
4. **Place the black meter lead in the connection point labeled Common on the meter.**
5. **Place the red meter lead in the connection point labeled V on the meter.**

   TIP

   Commonly, this point is also labeled Ω, indicating the ability to also measure resistance. (I cover resistance in detail later in this chapter.)
6. **Set the meter's dial to read the appropriate range of DC voltage.**

   REMEMBER

   The appropriate range of DC voltage is the estimated voltage you expect to see. You can arrive at this estimate by looking at the modules' ratings and where you're at in the system.

TIP

When in doubt, set the meter to the highest range and adjust it down if you need to. (Note that autoranging meters will set themselves to the proper range without any help from you.)

7. **Place the black lead on the negative terminal for the PV module.**
8. **Place the red lead on the positive terminal for the PV module.**
9. **Read the voltage measurement on the meter's screen.**
10. **Switch the meter leads on the PV module and note the presence of the negative symbol.**

    This negative symbol is your indication of *reverse polarity,* which means the voltage from the source is backward to the meter (the meter expects that the positive terminal is connected to the red lead and that the negative is connected to the black lead).

    This ten-second test can save you a lot of time and heartache in troubleshooting, which is why I strongly suggest you perform it every time you check DC voltages. Although some equipment claims to be reverse-polarity-protected, you can't rely on this "guarantee." If the equipment isn't able to protect itself, it'll let you know by releasing some smoke.

11. **Remove the leads from the module and pull them from the meter.**

TIP

I suggest removing the leads from the meter each time you're finished. Doing so forces you to think about the measurement you're about to take each time you go to use the meter. This is especially helpful if the meter can be used as an inline current meter. If you use an inline meter to check current and then go to check voltage without resetting the meter leads, you may be in for quite a surprise — namely, you'll allow current to flow through the meter when you don't expect it. Not a good thing.

# Making a Stop: Resistance

In the context of DC electricity, the term *resistance* (represented in equations as R) means resisting the flow of current. It's not much different than the resistance you encounter at home when you try and push that plate of broccoli in front of your child. You push the plate toward him, and he resists by pushing it back. In the next sections, I explain the unit of measurement for resistance and walk you step by step through the process of measuring resistance.

TECHNICAL STUFF

For AC circuits, the correct terminology for opposition to current flow is *impedance.* The calculations for impedance can get complex quickly and are beyond what I think you need to focus on. So for the purpose of the discussion here, I refer only to resistance.

## Introducing ohms

In electrical circuits, resistance is measured in *ohms* (Ω). The greater the number of ohms, the greater the resistance. You want the conductors in your PV systems to have little resistance so current can flow through them as efficiently as possible, but it's unreasonable to think there won't be any resistance. The best you can do is keep that resistance to a minimum.

TIP

The size of the conductor used in PV systems has the biggest effect on the overall resistance. The bigger the conductor, the more room the electrons have to move around, which equals less resistance. Think of the conductor as the highway that all the cars in your area are driving on. If the highway is just two lanes wide, only so many cars can move down it at a time, which leads to increased resistance. If the same number of cars were to travel on an eight-lane expressway, you'd have far less resistance. (Flip to Chapter 10 for the basics on conductors and other electrical and safety devices.)

## Measuring resistance

When you install a PV system, you may need to check the resistance in a given PV circuit (by this, I mean you may need to determine the resistance in the conductors you run from the PV array down to the inverter and/or battery bank). You look at resistance for two reasons: to verify that the conductors used allow the current to move with little opposition and to make sure that the conductors are properly connected to the electrical equipment.

REMEMBER

The easiest way to determine the resistance in a circuit is to find the total voltage drop across that circuit by using Ohm's Law (which I present later in this chapter) to calculate the total resistance. When resistance is present in the conductors, the voltage at the source will be greater than the voltage at the load (which is at the end of the line). I explain how to calculate this voltage drop value in Chapter 13.

Another common way of measuring resistance is to perform a continuity test. The goal of a continuity test is to find very little resistance in the circuit, indicating an easy path for the electrons to follow. A typical continuity test is to check the conductors connecting pieces of equipment. Think of two boxes mounted on a wall next to each other. You pull six conductors from Box 1 to Box 2, but after you're done, you can't tell which conductor in Box 1 is the same conductor in Box 2. Perform a continuity test by using a meter that's set on measuring resistance to positively identify the two ends of the same conductor. When you touch the ends of the same conductor with your meter, there'll be little to no resistance, and you'll know for sure that you have the right conductor. (Note that using a meter to verify the points where each conductor is connected is far more reliable than using your fingers and eyes to trace the path of each conductor.)

TIP

Most DMMs have a resistance-measurement feature that incorporates an audible alarm. This alarm indicates that a low-resistance connection exists between the two points you're touching. As soon as you remove one of the leads, the alarm shuts off, indicating a high amount of resistance between the two leads. This alarm helps you verify that the connections you made are correct before you turn on the power source(s).

Performing a continuity test doesn't take very long and can save you a lot of time during the commissioning process that I describe in Chapter 18. To use a DMM to measure resistance and check continuity between various components, just follow these steps (and take a look at Figure 3-4):

1. **Put on personal protective equipment (PPE) such as insulating gloves and safety glasses.**

   Flip to Chapter 15 for the how-to on staying safe during installation work.

2. **Switch the AC and DC disconnects to the off position to remove any power sources present.**

3. **Make the location where you want to take the continuity measurement accessible.**

   You generally take the reading inside electrical equipment, which means you have to open the covers to the disconnects and inverter(s).

4. **Place the black meter lead in the connection point labeled Common on the meter.**

5. **Place the red meter lead in the connection point labeled Ω.**

   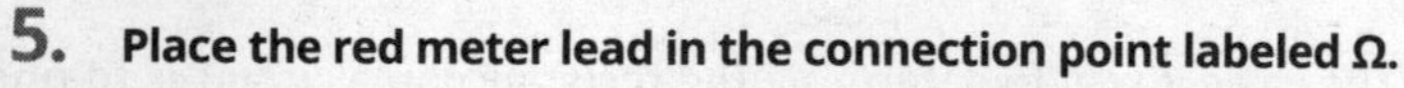

   TIP

   Commonly, this point is also labeled V, indicating the ability to also measure voltage. (I explain how to measure voltage earlier in this chapter.)

6. **Set the meter's dial to read the circuit's resistance.**

   Always start with the highest resistance values on your meter. If you need more precision on the readings, you can always turn the dial to the smaller numbers (unless of course you're using an autoranging meter that sets itself to display the proper scale automatically).

7. **Place the black lead on the first connection point.**

   This connection point is typically found inside a disconnect.

8. **Place the red lead on the second connection point.**

   This point is typically the other end of the conductor on which you placed the black lead.

9. **Read the resistance measurement on the meter's screen or listen for the alarm that indicates a low-resistance connection.**

If the alarm doesn't ring when you expect it to, you've either made the connection in the wrong location or placed the meter leads in the wrong spot. Investigate further by repositioning the leads from your meter to the points inside one of the boxes until you determine where the connections are being made.

10. **Remove the leads from the module and pull them from the meter.**

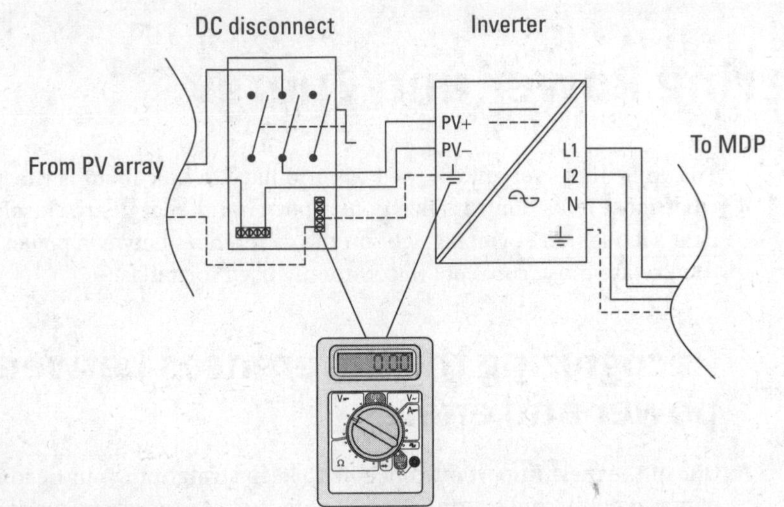


**FIGURE 3-4:** Measuring resistance with a DMM.

# Connecting Current, Voltage, and Resistance with Ohm's Law

As a PV designer and installer, you use three main electrical components in all of your calculations: current, voltage, and resistance. Knowing how these components relate to each other is critical to the success of your PV system designs. Consequently, you need to have a good understanding of Ohm's Law, which relates current, voltage, and resistance to one another. You'll use this formula in a variety of ways, from gauging conductor sizes and determining voltage drops to calculating power (I explain how to do that later in this chapter).

REMEMBER

Ohm's Law can be stated in many ways:

- Volts = Amps × Resistance ($E = I \times R$)
- Amps = Volts ÷ Resistance ($I = E \div R$)
- Resistance = Volts ÷ Amps ($R = E \div I$)

So, for example, to determine the resistance in a circuit for a PV array that's operating at 120 DC volts and 10 amps, perform the following calculation:

- Resistance = 120 DC volts ÷ 10 amps (or R = E ÷ I)
- Resistance = 12 ohms (R = 12 Ω)

# Pondering Power and Energy

The value that everyone in the PV world likes to talk about is the power value, but what most folks don't realize is that power and energy are closely related. In the next sections, I set you straight on the differences between power and energy and then relate those concepts to commonly used formulas.

## Recognizing the differences between power and energy

One of the most important concepts to keep straight in your head is the difference between power and energy. Even though these terms are misused everywhere, the differences between them are basic and even obvious, as you discover in the sections that follow. Using these terms correctly is crucial to your success as a PV system designer and installer.

### Power is a rate

Power is measured in watts (W), and 1,000 watts equals 1 kilowatt (kW). A *watt* is the measurement of the flow of energy, just like current (covered earlier in this chapter) measures the flow of charges. Simply stated, *power* is a rate; it's an instantaneous value. If you're talking to someone and he says his PV array produces 1,000 kilowatts per year, that makes as much sense as you telling him that you drive your car 65 miles per hour per year.

Part of the problem with the term *watts* is the lack of a time value, which people tend to associate with rates. They're used to seeing a rate given in terms of a quantity per rate of time: miles per hour, gallons per second, kilobytes per second, and so on. Truth be told, watts have a time value to them. You just don't see it.

REMEMBER

One watt is equal to one joule of energy per second (that's your time value).

### Energy is a quantity

*Energy* is the measurement of power multiplied by time; it's measured in kilowatt-hours (*not* kilowatts *per* hour). This is the number that your utility charges you for and, if you could hold it in your hand, the quantity of energy a load consumes.

REMEMBER

In PV systems, the power value (wattage) gets the spotlight even though the energy value (kilowatt-hours) is the value that makes the real difference. The energy value is the one you should concern yourself with the most because it's what people are consuming and you're selling. All PV projects are driven by a budget, and kilowatt-hours can be directly converted into dollars and cents. For example, if a PV array is installed in a location that doesn't allow it to produce enough energy (kilowatt-hours) on a consistent basis, the owner of that system will have to continue buying energy from someone, generally the utility. If that array were in a great location, it could produce a lot of energy, and the system owner would benefit from lower energy bills from the utility.

## Relating power to current, voltage, and resistance with the power equation

The power equation is relatively simple and much like Ohm's Law (which I present earlier in this chapter). In fact, it's really an extension of Ohm's Law because it relates voltage, current, and power.

REMEMBER

You use the power equation when determining the power output of your PV array when you're given the voltage and current specifications and when you want to figure out the amount of current your inverter will produce when the operating voltage and power value are known. You can also use the power equation to figure out voltage.

I like to present the power equation as you see it in Figure 3-5. Power is represented by P, current by I, and voltage by E. You need to know two of the values in order to calculate the third. See? Easy as pie.

**FIGURE 3-5:** A diagram of the power equation.

The easiest way to use Figure 3-5 is to cover up the value you want to find. What you're looking at tells you how to do the math. For example, if you want to find power, cover the P and you have I next to E. Mathematically, you can view this as follows:

Power = Current × Voltage ($P = I \times E$)

If you're after current, cover the I and you have P over E, or mathematically:

Current = Power ÷ Voltage ($I = P \div E$)

And, of course, if you want to figure out voltage, cover the E and you get P over I, which looks like this:

Voltage = Power ÷ Current ($E = P \div I$)

Figure 3-6 is a common graphic used to relate current, voltage, resistance, and power; it combines Ohm's Law and the power equation. To use this chart, choose the value you want to calculate from the four options at the center of the circle. Using your two known values, apply the calculation shown to obtain your answer.

For example, if you want to determine current flow (amps) through the wires in a 3,000 watt (3 kilowatt) PV system that operates at 200 volts, just check out Figure 3-6 to see that you need to divide power by voltage:

3,000 watts ÷ 200 volts = 15 amps ($P \div E = I$)

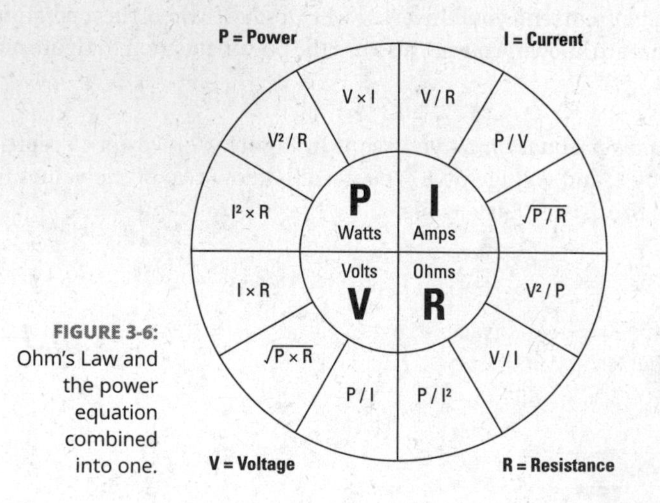


**FIGURE 3-6:** Ohm's Law and the power equation combined into one.

You can also use Figure 3-6 to determine the amount of power any load in a home will consume. For example, if a television requires 120 AC volts and draws 3 amps, consult Figure 3-6 to determine that you can calculate the power draw by multiplying voltage by current:

120 volts × 3 amps = 360 watts (V × I = P)

## Calculating energy in terms of watt-hours

After you know the number of watts a PV array can produce or the number of watts various loads will consume, keeping in mind that wattage is a rate, you can determine the energy production or consumption, which is measured in *watt-hours* (Wh). To do that, you need to know the number of hours the PV array will be operating or the load will be in use. Multiply this time value by the power draw to find the quantity of energy. Here's the equation:

Power in watts or kilowatts (W or kW) × Number of hours = Energy in watt-hours or kilowatt-hours (Wh or kWh)

So if I say that my PV array is rated at 3,000 watts (3,000 watts = 3 kilowatts = 3 kW) and the sun is out for 6 hours, then the energy produced would be:

3 kilowatts (kW) × 6 hours = 18 kilowatt-hours (kWh)

REMEMBER

There's a big difference between watt-hours (Wh) and kilowatt-hours (kWh), so be sure to always keep your units straight. If you're in too much of a hurry, you'll inevitably confuse the two and end up with some very funny numbers.

*Note:* This example is a big-picture view, but it's not entirely accurate because no system losses have been taken into account.

You can also calculate energy in terms of watt-hours in order to figure out what various loads consume each day. For instance, if a television draws 360 watts and the screen is on 2 hours a day (regardless of whether anyone is really watching it), then the television will consume

360 watts (W) × 2 hours = 720 watt-hours (Wh)

I show you how to take these energy calculations and apply them to sizing grid-direct systems in Chapter 11 and battery-based systems in Chapter 12.

## Introducing amp-hours, a companion to watt-hours

If you're using batteries, you should know that they're rated in terms of *amp-hours* (Ah) rather than watt-hours because you're focusing on the current flow.

As I explain in the earlier "Understanding amps" section, the term *amps* represents a flow of electrons at a certain rate. By multiplying that rate by a certain amount of time, you get the following formula:

Amps (A) × Hours = Amp-hours (Ah)

Amp-hours describe a quantity just like watt-hours do. Current is a rate, just like power, so all the term *amp-hour* means is the quantity of electrons that are available to do some work.

The conversion between watt-hours and amp-hours is a pretty simple one. Simply divide the watt-hours value by the voltage value to find the amp-hours value. The answer lies in the power formula that I describe earlier in this chapter.

TIP

If you're wondering why I want you to use the power equation for finding amp-hours, which is an energy value, allow me to explain. Amp-hours aren't truly an energy measurement; only watt-hours are because they include a voltage component. What you know from the power equation is the relationship of power, voltage, and current. So to switch between watt-hours and amp-hours, you need to know the voltage. By dividing watt-hours by a voltage value, you wind up with the amp-hours you're looking for. Here's the equation:

Watt-hours (Wh) ÷ Voltage (V) = Amp-hours (Ah)

Technically this equation isn't the same as the power equation, but you're applying the same principles, so it makes sense to think of using the power equation as the way of calculating amp-hours.

# Wrapping Together Current, Voltage, Resistance, Power, and Energy

Don't read this section until after you've digested all the terms in the previous sections. Why? Because this section is intended to serve as an all-encompassing review of electricity basics, not an introduction to the concepts. In this section, I use multiple analogies to help convey electrical concepts because thinking about things you can touch and feel is generally easier than thinking about something as intangible as electricity. Well, what are you waiting for? A list of helpful analogies awaits.

Imagine that you're taking a long road trip with your buddies. Like any good traveler, you decide to push the limits of both the car and everyone in it. Naturally, you run out of gas short of the next gas station. And just to add to the fun, you're out of cellphone range and have to push the car to the next station. You can apply the electrical terms from the previous sections to the process of pushing the car:

- The voltage is the pressure you exert on the back of the car to move it down the road.
- The amperage is the number of steps you take every minute.
- The resistance is the number of people who are sitting in the car refusing to help push — after all, it was your fault. (Yes, the resistance would also include friction, but just play along with me for now.)
- The wattage is the car's rate of speed creeping down the road.
- The watt-hours is the total distance traveled after pushing the car for some time.
- The amp-hours is the number of steps you take in an hour.

You can apply Ohm's Law and the power equation to this scenario to see how all of these values play a role in getting the car to the gas station. If five people are in the car and you're the only one who's pushing initially, Ohm's Law states that the amperage is equal to the voltage divided by the resistance. So with only one person pushing and four people sitting, the amperage will be small:

$I = E \div R$

I = The pressure of 1 person pushing ÷ The resistance of 4 people sitting

As your friends begin to realize that they're going nowhere fast, they get out of the car (reducing resistance) and start pushing (increasing voltage). This in turn increases the number of steps taken each minute (amperage).

Now that you have a voltage and amperage value, you can calculate how fast the car is moving (wattage) by multiplying the two together (thanks to the power formula). If you time yourself, you can figure out how far you moved the car (watt-hours). And if you really want to impress your buddies later, you can tell them all how many steps they took (amp-hours) because you know how many steps were taken each minute and how many minutes everyone pushed.

# Another Electricity Concept: Circuit Configurations

When you're wiring up a PV array (or battery bank), you must configure it in a way that the delivered voltage and current values are at the levels you need. I walk you through the process of determining these values in Chapters 11 and 12. In the following sections, I help you understand how voltages and current levels are affected based on the wiring configuration used, with a little assistance from the power formula I provide earlier in this chapter.

*Note:* To keep things simple, I focus on the DC side of the systems here and show you both PV modules and batteries. The values I use are for reference only; they aren't absolutes by any means.

## Series

*Series* connections are made by connecting the positive wire from one module to the negative wire of the next module. By taking a group of modules and placing them all in a series, you create a *series string* of modules.

TECHNICAL STUFF

Many people like to refer to creating series connections as *daisy chaining the modules.* A more common, industry-accepted way of describing this process is to say you're *stringing the modules together.*

These series strings, or simply *strings,* can be as few as 2 modules or greater than 20. The exact number depends on the PV modules, the components they're connected to, and the restrictions from the *National Electrical Code*® (*NEC*®). If I were to take five modules and place them all in series, I'd call the result one string of five.

When making series connections, the voltage values are additive, and the current values remain the same. If you look at Figure 3-7, you see five modules all wired in series. Each one is rated at 12 V and 4 A. If I were to put my DMM on the positive end of the first module and the negative end of the last one, I'd get a reading of 60 V (5 modules × 12 V per module). If current was flowing from the PV array and I placed my amp meter in the circuit, I'd see 4 A on my meter.

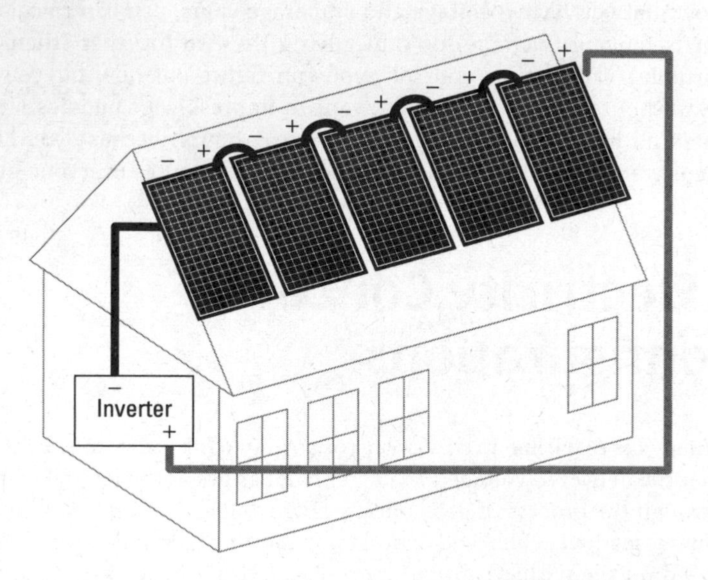


**FIGURE 3-7:** Five PV modules wired in series.

Now that you know the voltage and current vales for the string, you can calculate the power output with the help of the power formula I provide earlier in this chapter:

60 volts (V) × 4 amps (A) = 240 watts (W)

## Parallel

The number of modules that can be placed in series will be limited at some point, so if you need to create more power than one string can provide, you must place strings parallel to each other. *Parallel* connections are complementary to series connections: The positive wire from one module is connected to the positive wire of the next module; likewise, the negative wires are connected to each other.

The electrical characteristics for parallel connections also complement those of series connections. When modules are placed in parallel, the voltage remains constant, and the current values are additive. See Figure 3-8 for an example of five modules wired in parallel.

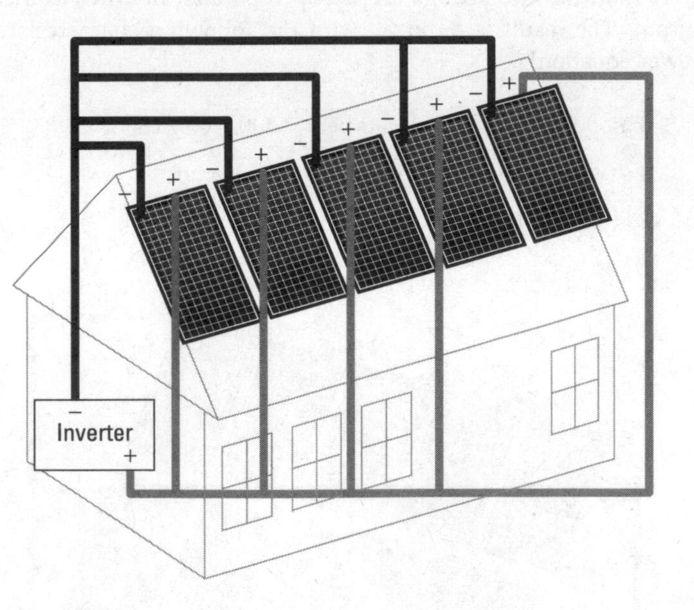


**FIGURE 3-8:** Five PV modules wired in parallel.

The modules in Figure 3-8 are the same as the modules used in the series example in the preceding section — each module is rated at 12 V and 4 A. If you were to take your DMM and measure the total voltage output, you'd see 12 V on your meter. If you then switched and checked the current value, you'd see 20 A

(5 modules × 4 A per module). The power output in parallel is the same as it was in series (as you find out with the help of the power formula):

12 volts (V) × 20 amps (A) = 240 watts (W)

TIP

So which type of connection is better: series or parallel? It really just depends. The voltage and current ratings of the components you connect the modules to dictate the configurations, and the performance issues round out the decision-making. In terms of which is better for power output, it doesn't matter. Five modules in series and five modules in parallel, for example, produce the exact same amount of power.

## Series-parallel

Most PV systems boast a combination of series- and parallel-connected arrays. Figure 3-9 shows how a sample PV array would be connected if you wanted two strings of five modules in parallel.

First, the strings of modules are wired together in series to increase the voltage. After that, the two strings are wired in parallel in order to increase the current output. The result is an array with the following characteristics (based on the power equation):

5 modules in series × 12 volts per module = 60 volts

2 strings in parallel × 4 amps per string = 8 amps

60 volts (V) × 8 amps (A) = 480 watts (W)

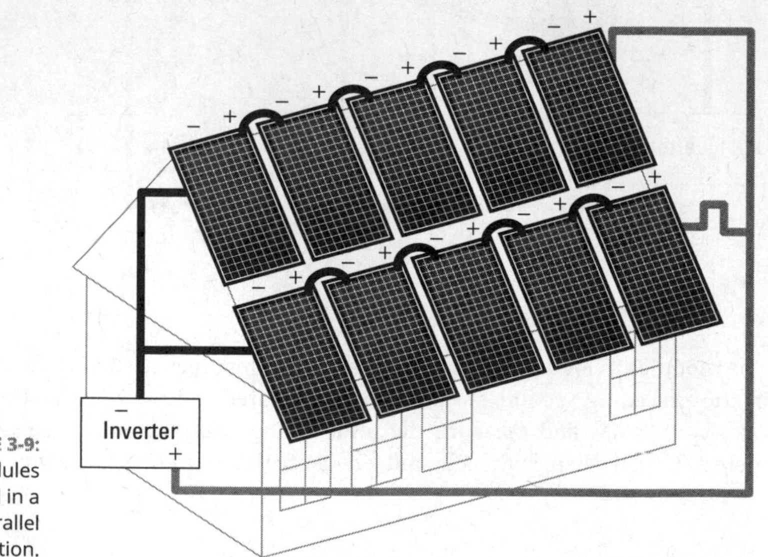


**FIGURE 3-9:** PV modules wired in a series-parallel configuration.

IN THIS CHAPTER

» Understanding solar radiation

» Seeing the effects of the sun's path

» Making the most of the solar resource with proper module placement

# Chapter 4
# Warming Up to the Solar Resource

A key consideration in the overall performance of any PV system is how the installed system can use the sun's energy most effectively. For the array to perform at its best, you need to install it so that it has access to as much energy as possible. Because sunlight is the "fuel" for any PV array, you want to make sure your arrays have full access to as much of the *solar resource* (the sunlight striking your site) as possible. Having a fundamental understanding of the relationship between the sun and earth allows you to properly site the PV array.

In this chapter, I introduce you to the different forms of solar radiation and how they affect the production of a PV array. I also present you with the keys to understanding how the sun's path affects your location. Finally, I give you language to use when describing the exact positioning of PV modules in relation to the solar resource.

*Note:* I don't expect you to become an astronomer or dissect solar radiation into its different energy values after reading this chapter. I just want you to focus on the crucial concepts that'll affect your everyday decisions when it comes to installing any PV array.

# High (Or Low) Energy: Solar Radiation

*Solar radiation* is the term I use to describe the energy that's sent to the earth from the sun. Having a good understanding of solar radiation is vital because solar radiation is the driving force for all PV systems. In the following sections, I get you familiar with the various components of the radiation that affects PV modules and help you differentiate between the power and energy values that come from it. I also explain how you can use readily available tools to evaluate the amount of energy received from the sun in the area where you're working.

## Distinguishing between direct radiation and diffuse radiation

The sun, that great nuclear reactor in the sky (and here you thought I wasn't going to acknowledge nuclear energy's role in the need to address the world's power sources), is constantly hurling radiation toward earth. Approximately eight minutes after that radiation leaves the sun, the surface of the earth is assaulted by it. The exact amount of radiation and how you describe it depends on the planet's atmospheric conditions. On a clear day, there's little to interfere with the radiation; on a rainy day, the cloud cover greatly reduces the radiation that can reach you.

REMEMBER

When you work with PV systems, two main components of the sun's radiation dominate your attention: direct and diffuse radiation. Both contribute to the overall radiation levels on earth, but it's important to remember that they're two separate pieces (see Figure 4-1 for a representation of these radiation components).

- **Direct radiation:** The *direct radiation* from the sun isn't intercepted as it travels from the sun to the earth's surface. It therefore makes the greatest contribution to a PV array and has the biggest effect on the array's ability to convert sunlight into electrical energy. On clear, sunny days, the vast majority of the solar radiation experienced is in the form of direct radiation.
- **Diffuse radiation:** The sun's *diffuse radiation* takes a more roundabout path to the earth. Typically clouds, water vapor, dust, and other small airborne particles block and scatter the diffuse radiation's path to the planet's surface. The diffuse component of solar radiation therefore plays a minor role in the power output of a PV module. However, on days when clouds cover the sky, all the radiation is in the form of diffuse light without any direct radiation component.

  Another form of diffuse radiation, *albedo radiation* (or *reflectance*), is part of the solar radiation at a site. It's light that's reflected from physical surroundings, such as a roof or the ground, and put back into the atmosphere as diffuse radiation.

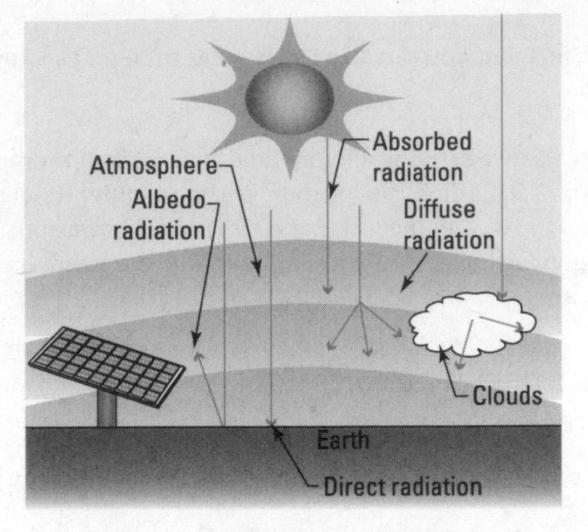


**FIGURE 4-1:** Direct radiation and diffuse radiation are the components of solar radiation.

The *total global radiation* received at a given site is the sum of the direct, diffuse, and albedo components.

TIP

To help you better understand solar radiation, think of it as money coming into your possession: Direct radiation is like the money you get from your paycheck, and diffuse radiation is like the money you get from an interest-bearing savings account. On days when you get a lot of work (the sunny days), your paycheck (direct radiation) is the major contributor to your bankroll; the savings account interest (diffuse radiation) contributes only a small amount. On the days when work is lean (those dark, gray days), your paycheck may contribute very little, if at all, to your wallet, whereas the savings account interest still contributes a small amount to your net worth.

TECHNICAL STUFF

One measurement that affects the solar radiation striking the earth is *air mass*, or the amount of atmosphere the radiation must pass through to reach the earth's surface. You don't need to spend too much time on the subject, but keep in mind that PV module manufacturers use a standard air mass value when rating their modules.

## Determining the intensity of solar radiation: Irradiance

After you know a little bit about the components of the radiation coming from the sun (which I share in the preceding section), you can throw some numbers into the mix. The term for the intensity of the solar radiation striking the earth is *irradiance*, and it's a measurement of power over an area (*power* is the rate of the flow of energy; jump to Chapter 3 if you need a refresher on power and other electricity basics). The standard units associated with irradiance are watts per square meter ($W/m^2$), but you can also easily refer to them as kilowatts per square meter ($kW/m^2$).

The amount of irradiance striking a PV module at any given moment is affected by a number of factors, including the location of the module, its position relative to the sun, the time of year, and the weather conditions. I cover most of these factors later in this chapter; in the next sections, I explain the irradiance-related basics that you need to know.

## Recognizing special conditions that can affect irradiance

Some common occurrences increase the irradiance levels striking PV modules:

- **Albedo radiation:** Also known as *reflectance,* albedo radiation comes from light-colored materials, such as snow or a white roof, in the array's vicinity. Increases due to albedo radiation generally aren't extreme, but they can have a measurable impact, with increases as much as 5 to 10 percent (see the earlier "Distinguishing between direct radiation and diffuse radiation" section for an introduction to albedo radiation).
- **The edge-of-cloud effect:** This is where a cloud passes over the PV array and prevents the sun from striking the array. As the cloud moves past the sun, the edges of the cloud actually concentrate the solar radiation and increase the irradiance values on the array. These occurrences are generally short-lived (the longest one I've ever seen recorded was 15 minutes), but they can increase the irradiance by more than 25 percent.

TECHNICAL STUFF

So should you try to increase irradiance through the use of mirrors or other reflective surfaces? The short answer is no. PV modules have been manufactured to operate without these enhancements. Adding them actually causes modules to deteriorate faster than normal, so the net effects are a shorter life span and a smaller energy output.

## Checking out charts to see irradiance differences

One way to really tell the difference between irradiance on sunny versus cloudy days is to examine charts that measure all the irradiance values, like the ones in Figures 4-2 and 4-3 (the values shown are in $W/m^2$).

- As you can see in Figure 4-2, the irradiance levels for a location on a perfectly cloudless day look a lot like a haystack curve. The irradiance values are at 0 $W/m^2$ at night and begin to climb as the sun rises in the sky. The irradiance values peak at *solar noon* and then decrease as the sun starts its descent to sunset. (Notice how the *y*-axis of the graph, which represents the irradiance values, doesn't have any numbers associated with it? That's because this graph is just meant to show you the typical look of the irradiance values over the course of one day.)

» In contrast, Figure 4-3 shows that the irradiance levels on a partially cloudy day are much more erratic. The amount of irradiance measured moves up and down, indicating the movement of clouds between the sun and the array.

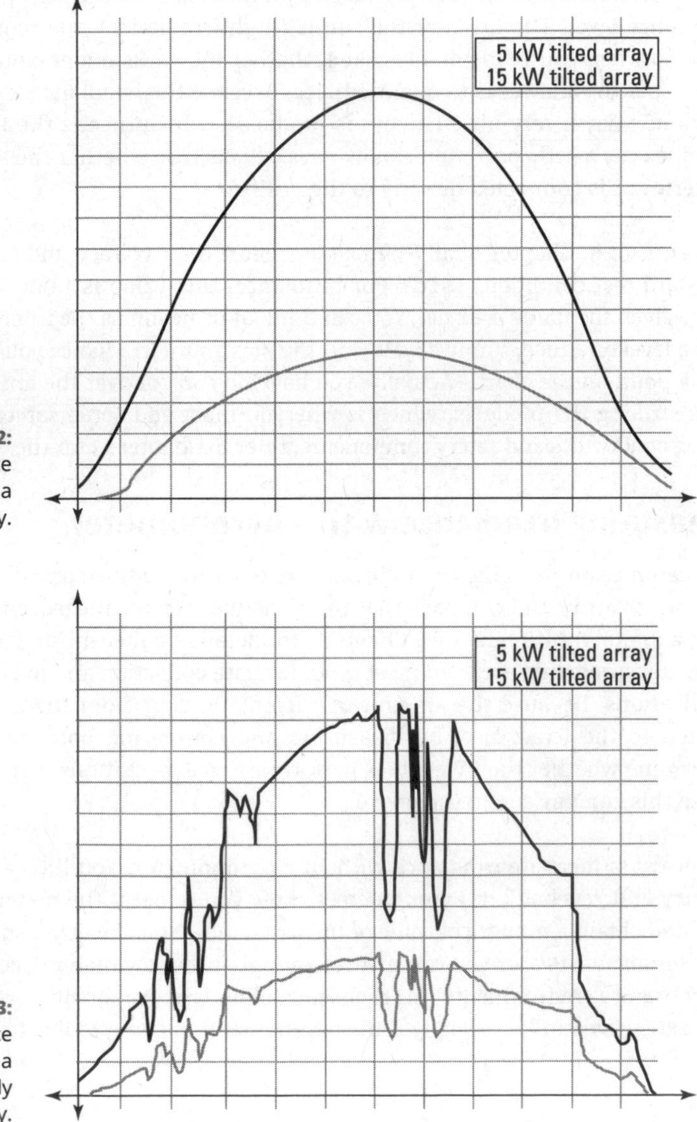


**FIGURE 4-2:** Irradiance measured on a cloudless day.

**FIGURE 4-3:** Irradiance measured on a partially cloudy day.

## Relating current and voltage to irradiance

PV modules produce *current* (the flow of electrons) and *voltage* (the pressure that makes electrons move) when exposed to sunlight and the electrons are given a path to flow in. Although voltage isn't greatly affected by irradiance, current is directly dependent on irradiance. As soon as ambient light is present (at sunrise or even during a cloudy day), the PV module will have nearly full voltage present. Current, on the other hand, will vary throughout the day as the irradiance values increase and decrease.

REMEMBER

The exact amount of current produced by a module is directly proportional to the irradiance level. The brighter the sun is (high irradiance), the more current will flow. So when you're inevitably asked whether PV works under cloudy conditions, know that the answer is (a qualified) yes. Because the irradiance levels are so low on cloudy days, very little current is produced, which means the PV module can provide only a little power on cloudy days (this is true whether the PV system uses batteries or is connected directly to the utility).

TIP

As I explain in Chapter 6, all PV modules report their voltage and current levels at standard test conditions (STC). For irradiance, this value is 1,000 $W/m^2$, and it's often given the name *peak sun.* You can think of peak sun as the intensity of the sun at sea level on a nice, sunny day. It isn't the maximum irradiance you'll ever see, but it is a point on the chart. (Actually, you have no control over the amount of irradiance striking the modules, which is why you must add some safety factors when sizing conductors and safety components; refer to Chapter 13 for the details on this.)

## Measuring irradiance with a pyronometer

You can measure irradiance on the job site to estimate the current value from the module or array and compare that to the actual current measurement you make with a digital multimeter (see Chapter 3 for details on measuring current). Irradiance measurements are used most often for data collection and in troubleshooting applications. Because the amount of current flowing from the array is directly related to the irradiance, by measuring and comparing both, you can quickly determine whether the PV array is performing to expectations. I show you how to make this comparison in Chapter 6.

REMEMBER

Never try to measure irradiance without a pyronometer. You'll do a horrible job if you try and "eyeball" it. Pyronometers allow you to point the meter in any direction and obtain a numerical value of the irradiance. You can buy a small hand-held unit for about $150, or you can invest in a high-accuracy meter (like the kind used in laboratory environments) for thousands of dollars. Personally, I recommend the less expensive kind, especially if the pyronometer is going to live in your toolbox.

To use a pyronometer, simply turn the device on, place the sensor in the same plane as the array, and read the meter. Keep in mind that the sensor consists of a small PV cell that has been calibrated at the factory. It may not be the most accurate tool, but it's a handy device for what you need.

## Calculating solar radiation energy: Irradiation

A successful PV system designer and installer not only knows how to track irradiance levels over the course of the day (I cover irradiance earlier in this chapter) but also how to calculate the energy values from the solar radiation at the site. These energy values are referred to as *irradiation* and are based directly on the irradiance levels received. You use irradiation values to help figure out how much energy a PV array will produce at a particular site.

As I explain in Chapter 3, energy is determined by multiplying power (measured in either watts or kilowatts) by a duration of time, typically an hour. Consequently, energy is typically measured in kilowatt-hours (kWh). The power received from the sun is measured in $W/m^2$, but with a quick conversion (1,000 W = 1 kW) you can transform that into $kW/m^2$. If you take that power value and multiply it by the number of hours that receive that much irradiance, the result is $kWh/m^2$.

However, as you can see if you refer to Figures 4-2 and 4-3, irradiance values aren't constant for hours at a time. In fact, they can change with the blink of an eye, which leads to the following question: If the irradiance values are only constant for seconds, or minutes at best, how am I supposed to take that information and turn it into $kWh/m^2$, which is the energy value I really need?

Love it or hate it, the answer lies in calculus. What you're really after is the area beneath the curve, and you can find it with a few quick calculations. If you're a calculus-hater, you're in luck — I don't mess with any advanced math here. That's because the good folks at the National Renewable Energy Laboratory (NREL) have done the hard work of calculating areas beneath curves and have broken down the results into a single number for easy use.

Take a look at Figure 4-4. It's actually the same irradiance curve as the one I present in Figure 4-2 for a perfectly cloudless day, with the exception of a square box drawn over the curve. This box represents the equivalent area covered by the curve. In other words, the NREL has turned a curve into a box. As I show you later in this chapter, the NREL publishes this information for a number of locations for your reference.

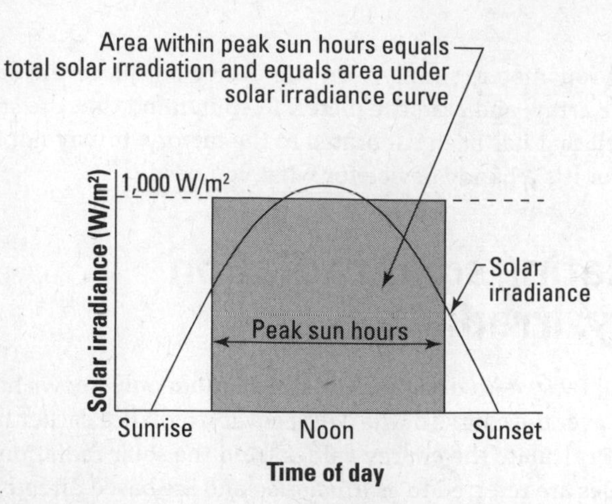


**FIGURE 4-4:** An irradiance curve with a box over it can help you calculate irradiation.

With the area beneath the curve represented as a box, you have one irradiance value to use and an easily defined number of hours. Now you can simply take the irradiance value, multiply it by the number of hours, and find the energy value (yes, that's right — irradiation). For example, if the top of the box intersects the irradiance line at 850 W/$m^2$ and the number of hours is equal to 4, then you can calculate the energy received at this location as follows:

$$850 \text{ W/m}^2 \times 4 \text{ hours} = 3{,}400 \text{ Wh/m}^2 = 3.4 \text{ kWh/m}^2$$

## Just for a day: Peak sun hours

Typically, irradiation values have an associated time period, such as the number of kWh/$m^2$ over the course of a day, month, or year. As a PV designer, plan to look at all of these periods, but focus on the daily values.

The amount of solar energy received at a particular location each day is measured in terms of the energy per unit area per day (kWh/$m^2$/day). The PV design lingo for this numerical value is the number of *peak sun hours* or *insolation*. In the sections that follow, I explain how to use peak sun hours to determine a PV array's energy output and point you toward some handy charts and maps that list peak sun hours.

TIP

The terms *peak sun hours* and *insolation* are used interchangeably because both refer to the solar energy received at a site over the course of one day. They're also both measures of irradiation (solar energy), but they have a specific time value (one day) associated with them.

### Using peak sun hours to calculate energy output

In the earlier "Relating current and voltage to irradiance" section, I tell you that the term *peak sun* refers to an irradiance value of 1,000 W/$m^2$. Well, the number of peak sun hours describes the number of hours each day that the irradiance value equals 1,000 $W/m^2$.

You probably don't even have to refer to Figure 4-4 to remember that the sun doesn't just pop up over the horizon, go to 1,000 $W/m^2$, and sit there all day. At times, the irradiance values will be very low; other times, they'll be very high. The number of peak sun hours is merely an estimation of the amount of time each day that the irradiance is equal to a peak sun. And because PV models are rated for their output under peak sun conditions, the number of peak sun hours each day indicates how many hours each day the PV array will operate at its full power output.

I think an example, complete with numbers, would be a good idea here. To start, divide your insolation value by peak sun to get a number of hours per day. Check out the example in the earlier "Calculating solar radiation energy: Irradiation" section, where the box in Figure 4-4 intersects the irradiance line at 850 $W/m^2$ and the amount of time is 4 hours. The result of the equation is 3,400 $Wh/m^2$ of energy (or 3.4 $kWh/m^2$). Because this measurement was taken during a single day, it's actually an insolation value (3.4 $kWh/m^2$/day). By dividing that value by one peak sun (1,000 $W/m^2$, which is equal to 1 $kW/m^2$), you get the number of hours each day that the PV array will operate at its rated output.

$$3.4\ kWh/m^2/day \div 1\ kW/m^2 = 3.4\ hours/day$$

You then multiply this number of hours by the power value of a given array to figure out the expected maximum energy output. So if I have a 3 kW array on my house and the number of peak sun hours today is 3.4, I can calculate the expected energy output by multiplying the two together.

$$3\ kW\ array \times 3.4\ peak\ sun\ hours = 10.2\ kWh\ of\ energy$$

*Note:* The number of peak sun hours varies with a number of factors, most notably the time of year and your position on the globe. I show you more details on these effects later in this chapter.

## Referring to handy charts and maps

After you know how to make calculations using peak sun hours, how do you know precisely what peak sun hour numbers to use for your client's area? NREL to the rescue! You can find peak sun hours data for more than 200 U.S. locations on the NREL Web site at `rredc.nrel.gov/solar/pubs/redbook`. Granted, other sources of peak sun hours data are out there and you can spend a lot of time dissecting all of them, but I show you the NREL data so you can acquire a fundamental understanding.

In Figure 4-5, you see an example of the data set NREL publishes in its *Solar Radiation Data Manual for Flat-Plate and Concentrating Collectors.* (Nicknamed "the NREL

Redbook," this resource also has a good introduction to help you better understand all the terms the Redbook uses in its charts.) Although all the values shown are based off of arrays that point true south, these tables also list multiple solar resource values based on the tilt of a PV array that's positioned on a horizontal surface rather than a vertical one. On the left-hand side of Figure 4-5, you see 0°, Latitude −15°, Latitude, Latitude +15°, and 90°. Latitude −15° means the tilt angle of the PV array off of the horizontal is equal to the local latitude minus 15°. (I get into the details of these values later in this chapter. For now, just focus on the fact that the PV array is moving from flat to vertical as you look down the far-left column.)

**Portland, OR**

WBAN NO. 24229

LATITUDE: 45.60° N
LONGITUDE: 122.60° W
ELEVATION: 12 meters
MEAN PRESSURE: 1017 millibars

STATION TYPE: Primary

**Solar Radiation for Flat-Plate Collectors Facing South at a Fixed Tile ($kWh/m^2/day$), Uncertainty ±9%**

| Tilt(°) | | Jan | Feb | Mar | Apr | May | June | July | Aug | Sept | Oct | Nov | Dec | Year |
|---|---|---|---|---|---|---|---|---|---|---|---|---|---|---|
| 0 | Average | 1.2 | 1.9 | 3.0 | 4.2 | 5.3 | 5.9 | 6.3 | 5.4 | 4.1 | 2.5 | 1.4 | 1.0 | 3.5 |
| | Min/Max | 0.9/1.5 | 1.4/2.4 | 2.4/3.8 | 3.4/4.8 | 4.3/6.2 | 4.8/6.9 | 5.0/7.2 | 4.6/6.3 | 3.2/4.9 | 1.8/3.2 | 0.9/1.7 | 0.7/1.3 | 3.1/3.8 |
| Latitude −15 | Average | 1.7 | 2.5 | 3.6 | 4.6 | 5.4 | 5.8 | 6.3 | 5.9 | 5.1 | 3.4 | 1.9 | 1.4 | 4.0 |
| | Min/Max | 1.1/2.7 | 1.4/3.6 | 2.7/5.0 | 3.5/5.4 | 4.3/6.3 | 4.7/6.8 | 4.9/7.2 | 5.0/7.0 | 3.7/6.2 | 2.3/4.6 | 1.1/2.8 | 1.0/2.5 | 3.5/4.3 |
| Latitude | Average | 1.9 | 2.6 | 3.7 | 4.5 | 5.0 | 5.3 | 5.8 | 5.6 | 5.1 | 3.6 | 2.1 | 1.6 | 3.9 |
| | Min/Max | 1.2/3.0 | 1.4/3.9 | 2.6/5.2 | 3.4/5.3 | 4.0/5.9 | 4.3/6.2 | 4.6/6.7 | 4.8/6.7 | 3.6/6.4 | 2.3/5.0 | 1.2/3.1 | 1.0/2.9 | 3.4/4.2 |
| Latitude +15 | Average | 1.9 | 2.7 | 3.5 | 4.1 | 4.4 | 4.6 | 5.1 | 5.1 | 4.9 | 3.6 | 2.1 | 1.6 | 3.6 |
| | Min/Max | 1.2/3.2 | 1.3/4.0 | 2.5/5.1 | 3.1/4.9 | 3.5/5.2 | 3.8/5.3 | 4.0/5.8 | 4.3/6.1 | 3.4/6.1 | 2.2/5.0 | 1.2/3.2 | 1.0/3.1 | 3.2/3.9 |
| 90 | Average | 1.8 | 2.3 | 2.8 | 2.9 | 2.8 | 2.7 | 3.1 | 3.4 | 3.7 | 3.0 | 1.9 | 1.5 | 2.6 |
| | Min/Max | 1.0/3.1 | 1.0/3.5 | 1.9/4.1 | 2.2/3.5 | 2.3/3.3 | 2.3/3.1 | 2.5/3.4 | 2.9/4.0 | 2.6/4.7 | 1.8/4.3 | 1.0/2.9 | 0.9/3.0 | 2.3/2.9 |

**Solar Radiation for 1-Axis Tracking Flat-Plate Collectors with a North-South Axis ($kWh/m^2/day$), Uncertainty ±9%**

| Axis Tilt (°) | | Jan | Feb | Mar | Apr | May | June | July | Aug | Sept | Oct | Nov | Dec | Year |
|---|---|---|---|---|---|---|---|---|---|---|---|---|---|---|
| 0 | Average | 1.5 | 2.4 | 3.8 | 5.2 | 6.6 | 7.4 | 8.3 | 7.3 | 5.7 | 3.4 | 1.7 | 1.2 | 4.5 |
| | Min/Max | 0.9/2.3 | 1.3/3.4 | 2.7/5.4 | 3.7/6.4 | 4.8/8.0 | 5.8/9.2 | 6.1/9.9 | 5.7/9.0 | 3.8/7.2 | 2.2/4.7 | 1.0/2.4 | 0.8/2.0 | 3.9/5.0 |
| Latitude −15 | Average | 1.9 | 2.9 | 4.3 | 5.6 | 6.8 | 7.5 | 8.4 | 7.7 | 6.4 | 4.1 | 2.1 | 1.6 | 4.9 |
| | Min/Max | 1.1/3.2 | 1.4/4.3 | 2.9/6.3 | 3.9/6.9 | 4.9/8.3 | 5.8/9.3 | 6.1/10.1 | 6.1/9.6 | 4.2/8.2 | 2.5/5.8 | 1.2/3.2 | 1.0/2.9 | 4.2/5.4 |
| Latitude | Average | 2.0 | 3.0 | 4.3 | 5.5 | 6.5 | 7.2 | 8.1 | 7.5 | 6.5 | 4.3 | 2.2 | 1.7 | 4.9 |
| | Min/Max | 1.2/3.5 | 1.4/4.6 | 2.9/6.4 | 3.8/6.8 | 4.7/8.0 | 5.6/8.9 | 5.8/9.8 | 5.9/9.5 | 4.2/8.4 | 2.5/6.1 | 1.2/3.5 | 1.0/3.3 | 4.5/5.4 |
| Latitude +15 | Average | 2.1 | 3.0 | 4.2 | 5.3 | 6.1 | 6.7 | 7.6 | 7.2 | 6.3 | 4.2 | 2.3 | 1.7 | 4.7 |
| | Min/Max | 1.2/3.6 | 1.3/4.6 | 2.8/6.3 | 3.6/6.5 | 4.4/7.5 | 5.2/8.3 | 5.5/9.2 | 5.6/9.0 | 4.0/8.2 | 2.5/6.1 | 1.2/3.6 | 1.0/3.4 | 4.0/5.2 |

**Solar Radiation for 2-Axis Tracking Flat-Plate Collectors ($kWh/m^2/day$), Uncertainty ±9%**

| Tracker | | Jan | Feb | Mar | Apr | May | June | July | Aug | Sept | Oct | Nov | Dec | Year |
|---|---|---|---|---|---|---|---|---|---|---|---|---|---|---|
| 2-Axis | Average | 2.1 | 3.0 | 4.4 | 5.6 | 6.9 | 7.7 | 8.6 | 7.7 | 6.5 | 4.3 | 2.3 | 1.8 | 5.1 |
| | Min/Max | 1.2/3.7 | 1.4/4.7 | 2.9/6.4 | 3.9/6.9 | 5.0/8.4 | 6.0/9.5 | 6.2/10.3 | 6.1/9.7 | 4.2/8.4 | 2.5/6.1 | 1.3/3.6 | 1.0/3.5 | 4.3/5.6 |

*Source: National Renewable Energy Laboratory*

**FIGURE 4-5:** An example of NREL peak sun hours data.

As you move horizontally across any of these different tilt-angle designations, you see numerical values for the number of peak sun hours for each month. The unit for these values is $kWh/m^2/day$, or peak sun hours. Think of it as the average number of hours each day of that month when your PV array will operate at its rated output.

TIP

If you keep moving to the right along any row, you see that the last value is the average peak sun hours for the year for an array at that particular location. You can quickly compare this value among all the tilt angles listed to get an idea of the best tilt angle for an array at the client's specific location. (I tell you all about tilt angles later in this chapter and about evaluating your client's site in Chapter 5.)

REMEMBER

As a general rule, the number of peak sun hours increases as you move from winter to summer and then begins to decrease again as you move back toward winter. The exact amount of change is dependent on the tilt angle of the array and the latitude.

You can get an idea of the peak sun hours available across the United States in Figure 4-6. As you can see, the greatest amount of peak sun hours occurs in the southwest, and the numbers decrease as you move north. This chart can help you generalize and compare your location to others, but you should only use it to create a rough estimate of your location. You always need to look up the specific site you're working in to make an accurate energy-production estimate.

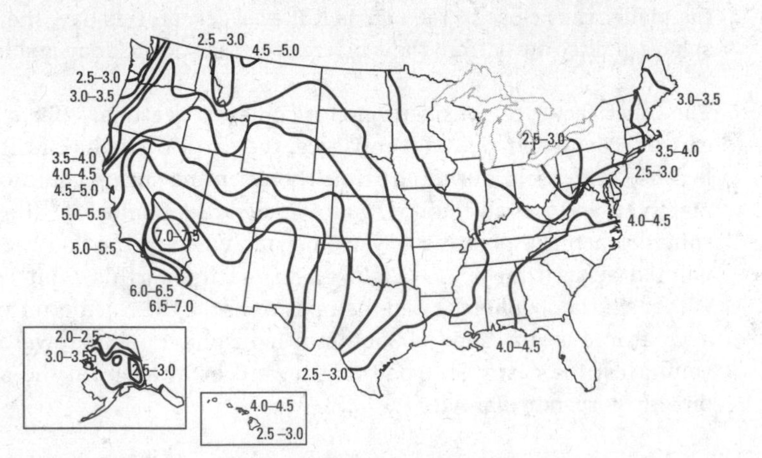


**FIGURE 4-6:** An insolation map of the United States.

# Examining the Effects of the Sun's Path on the Earth

To succeed as a PV system designer and installer, you need to have a firm grasp on the relationship of the sun and the earth, particularly in terms of how they're positioned relative to each other throughout the calendar year. For someone in northern Alaska, these changes are much more dramatic than for someone in Hawaii, but you need to understand the concepts regardless.

In the following sections, I describe general seasonal effects based on the earth's movement, the sun's relationship to your client's specific location, and solar time; I also explain how to analyze sun charts and introduce the solar window.

## Getting a grip on seasonal effects

The number of daylight hours each day has an obvious effect on a PV array's production: More sun means more solar energy. As a PV installer, you need to be able to visualize how the position of the sun changes with each season and the effect that has on the arrays you're designing and installing. In other words, you need to be able to take *seasonal effects* into account.

One important factor to consider is the motion of the earth around the sun. Earth takes an elliptical path around the sun, meaning that on the summer solstice (approximately June 21), the earth is actually at its farthest point from the sun. On this day, the Northern Hemisphere is tilted toward the sun, and that half of the world has its longest day (and shortest night) of the year. Time moves on, and the earth continues to orbit the sun. On the winter solstice (approximately December 21), the planet is as close to the sun as it'll ever get; on this day, the Northern Hemisphere is tilted away from the sun, creating the shortest day and longest night.

The other factor to consider when it comes to seasonal effects is the tilt of the earth's axis. When viewed from space, the earth's axis has a tilt of 23.5 degrees. Because of this tilt, during the times between the spring equinox (approximately March 21) and the fall equinox (approximately September 21), the Northern Hemisphere is actually pointing toward the sun, whereas the Southern Hemisphere is pointed away from it (see Figure 4-7). As the earth's orbit continues toward winter, the hemispheres swap their positions, so the Northern Hemisphere points away from the sun, and the Southern Hemisphere points toward the sun. On the equinoxes, the earth isn't pointing toward or away from the sun; instead, it's directly perpendicular with it.

Take a look at Figure 4-8 and pretend that your perspective of earth is the same as the sun's (doing so helps you visualize how the tilt of the earth affects the amount of sunlight striking any part of the earth). Notice how the tilt during the winter and summer restricts the sun's access to the hemispheres that are pointed away from the sun and how the sun has full access to the entire globe on the equinoxes.

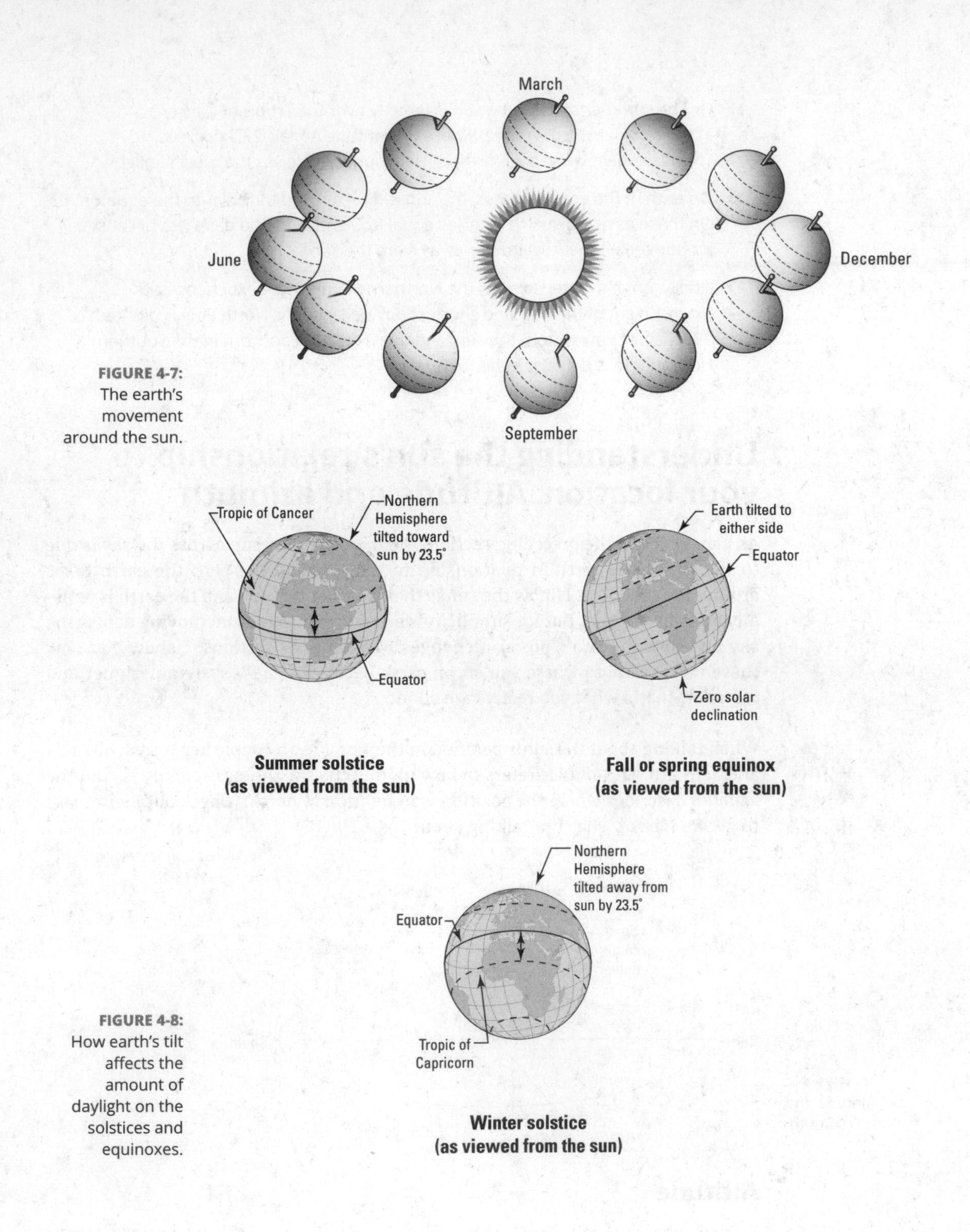


**FIGURE 4-7:** The earth's movement around the sun.

**FIGURE 4-8:** How earth's tilt affects the amount of daylight on the solstices and equinoxes.

» On the summer solstice, the sun is directly over the Tropic of Cancer in the Northern Hemisphere, where the latitude equals 23.5 degrees. This is the date when the Northern Hemisphere receives the most sunlight.

» On each of the equinox dates, the sun is directly perpendicular to the equator. On these dates, the earth receives equal hours of light and darkness because it's pointed neither toward nor away from the sun.

» Finally, on the winter solstice, the Northern Hemisphere experiences its longest night of the year and shortest day because the North Pole is pointed 23.5 degrees away from the sun, and the Tropic of Capricorn in the Southern Hemisphere is perpendicular to the sun.

## Understanding the sun's relationship to your location: Altitude and azimuth

As I show you in the preceding section, the motion of the sun across the sky is due to the tilt of the earth in relationship to the sun and the path the earth takes around the sun. (Yes, I know the sun is the stationary object, and the earth is actually orbiting the sun. But for simplicity's sake, I refer to the sun moving across the sky and how the sun's position changes.) In the next sections, I show you how these facts relate to where you sit on earth because as a PV system designer and installer, that's what you really care about.

REMEMBER

When talking about the sun's position in the sky, I use a couple key terms: altitude and azimuth. The *altitude* refers to how high in the sky the sun actually is, and the *azimuth* describes where the position is in relation to north. (Check out Figure 4-9 to get an idea of what I'm talking about.)

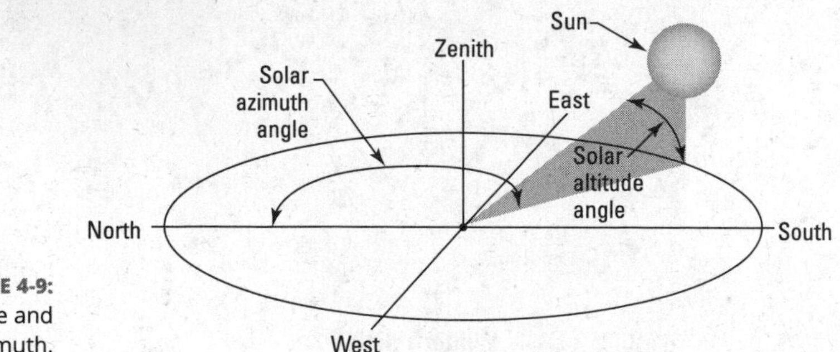


**FIGURE 4-9:** Altitude and azimuth.

### Altitude

You're probably well aware that the angle of the sun off the horizon varies throughout the year. The amount of variation is consistent across the globe, with the exact measurement dependent on the time of day and your specific latitude on the earth.

*Latitude* is defined as the number of degrees north or south of the equator. Your latitude affects where the sun is positioned in the sky throughout each day (relative to your position, that is). I explain how to visualize these solar positions with the help of sun charts later in this chapter.

To view the changes in the solar altitude, picture yourself standing on the equator for an entire year (I suggest picturing yourself with your favorite cold beverage in hand because a year can be a long time). On the spring equinox, the sun is directly overhead, or 90 degrees from your perspective. As the earth continues to orbit the sun, it moves toward the summer solstice. When it gets to June 21, the sun is no longer directly over your head; it has actually moved 23.5 degrees to the north. As the earth continues its path, it comes back around to the fall equinox, and the sun is directly overhead once again. And as you can probably figure out by now, when the earth moves to the winter solstice position, the sun is now 23.5 degrees to the south of your location on that day. By the time the earth gets back to March 21, your year is over, and the sun is again directly overhead.

You can now apply this motion and sun-to-earth relationship for any latitude on earth. The position directly over your head is known as the *zenith angle,* and it has a numerical value of 90 degrees. The sun's highest altitude on each and every equinox is the zenith angle (90 degrees) minus the latitude you're standing at. Alternatively, the sun is at a position of 23.5 degrees greater than the equinox position on the summer solstice and 23.5 degrees less than the equinox position on the winter solstice (in the Northern Hemisphere, that is; in the Southern Hemisphere, the calculations are the same except that the summer solstice down there is December 21 and the winter solstice is June 21). If you're having trouble with this concept, head to the later "Interpreting sun charts" section; the charts really help drive this point home.

## Azimuth

Just as the sun has a position in the sky based off of the horizon (altitude), it also has a position in the sky that moves from east to west. This position is known as the solar azimuth, and it has a steady movement on a daily basis.

The earth rotates around its own axis once every day, or once every 24-hour period. Because a full rotation of the earth measures 360 degrees, the motion of the sun is 360 degrees divided by 24 hours, which equals 15 degrees per hour. So over the course of an hour, the motion of the sun from east to west is 15 degrees.

Typically, north is considered the zero point, and the number of degrees the sun is from that point gives you the azimuth angle. The zero point is true north (I go over true north and south versus magnetic north and south in Chapter 5). Using this convention, when the sun is at a position directly to the east of your location, it can be described as having an azimuth of 90 degrees. If it's directly to the south of your position, it has an azimuth of 180 degrees. And when it moves due west of you, it has an azimuth of 270 degrees.

With this convention, you don't need to give a direction as well as the number of degrees because the numerical value tells the whole story. Not all sources use this convention though. Some use south as the zero point and require you to designate the direction (east or west) along with a numerical value to describe the position of the sun along the horizon. I find it easiest, however, to keep the convention with zero being north, east at 90 degrees, south at 180 degrees, and west at 270 degrees.

## Ticking off solar time

One point worth noting is the difference between solar time and clock time because the two very rarely match. When looking at your client's location, you care about locating the array based on solar time, not necessarily the time on your watch. The sun (and lucky ol' Arizona and Hawaii) doesn't have a clue about daylight saving time; it operates solely on solar time.

Jumping from clock time to solar time when you're in the field isn't a crucial task. You need to be aware of the distinction, though, especially when you're discussing the solar window (which I fill you in on later in this chapter) and how that window relates to your client's site. In Chapter 5, I explain how to appropriately apply this window.

## Interpreting sun charts

Sun charts allow you to figure out how the sun's path looks on paper. You can use them to pinpoint the sun's location at any time of day and any time of year, which is helpful when you're evaluating a particular site for potential shading issues. The sun charts in this section are the basis for the shading-analysis tools I describe in Chapter 5, so don't skip over this material.

The solar resource available at a location is affected by that location's position on the globe, the time of year, and the local climate. However, the path of the sun may look exactly the same in two very different locations. For example, from where I sit in Oregon, the path of the sun on a daily basis will appear identical to someone in Milwaukee, Wisconsin, because we're sitting at similar latitudes, but the exact number of peak sun hours (solar energy) we experience each day will be different because of the differences in our weather patterns. So I may not notice any differences in the days when I go to watch the Milwaukee Brewers play some baseball, but if I bring my PV array with me, it'll probably perform differently than it does back home in Oregon.

Figure 4-10, the sun chart for 30 degrees north latitude shows a typical sun path. The curves represent the path of the sun across the sky for various times of the year. The tallest curve is the path of the sun on the summer solstice, the middle curve represents the equinox paths (March 21 and September 21), and the lowest curve is the path on the winter solstice.

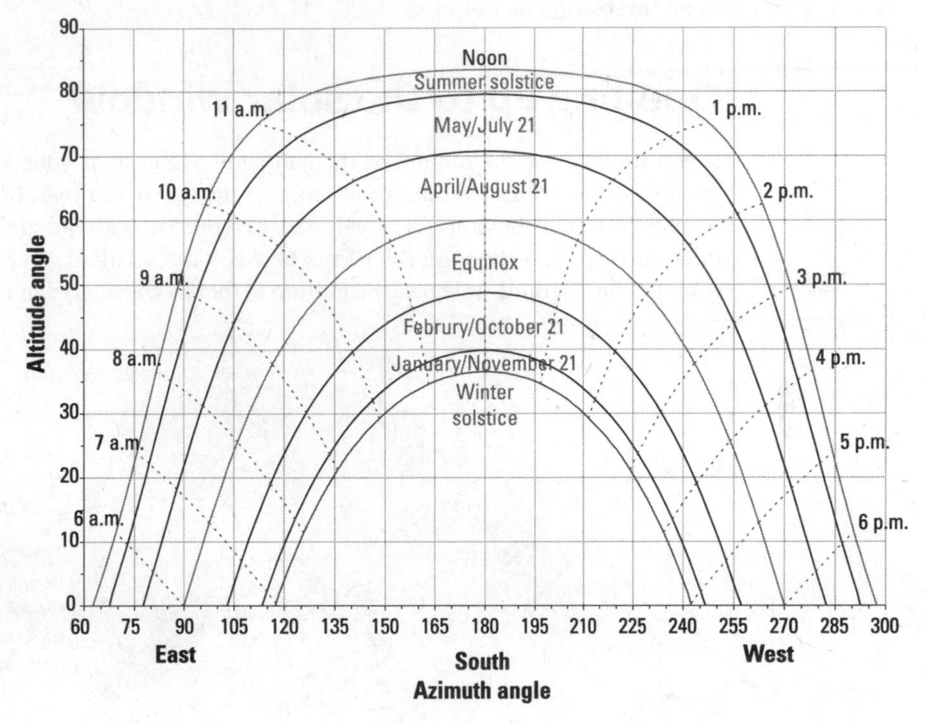


**FIGURE 4-10:** A sun chart for 30 degrees north latitude.

REMEMBER

You can use a sun chart like the one in Figure 4-10 to quickly and accurately determine the sun's altitude and azimuth. The azimuth angle is given along the *x*-axis, and the altitude angle is given along the *y*-axis. The times of day, which are based off of solar time (see the preceding section), are indicated by the dashed lines that intersect the sun paths moving east to west.

Notice that at solar noon on the equinox dates, the altitude of the sun is equal to the zenith minus the latitude (90 degrees – 30 degrees = 60 degrees). You can further evaluate the sun chart to see that the altitude difference between summer solstice and the equinox at solar noon is 23.5 degrees.

See how the sun rises north of east and sets north of west during the summer and just the opposite during the winter? All of these parts of sun charts help you understand the path of the sun and describe the solar window, which is the basis of your solar site assessments; I introduce you to the solar window in the next section.

TECHNICAL STUFF

One thing I find extremely fascinating is the fact that no matter your latitude, the sun rises due east of south and sets due west of south on the equinox dates. When you stop and think about it in terms of the earth's relationship with the sun, as I describe earlier in this chapter, it makes sense. Nonetheless, it always strikes me as an interesting fact of life.

## Opening up to the solar window

If you think about the motion of the sun across the sky at your location, you can imagine it appearing at various positions over the course of a year. In fact, this movement is very predictable and relatively easy to represent with the sun charts described in the preceding section. All the points between the solstices can be defined as the *solar window*. Figure 4-11 shows the portion of the sky considered in the solar window.

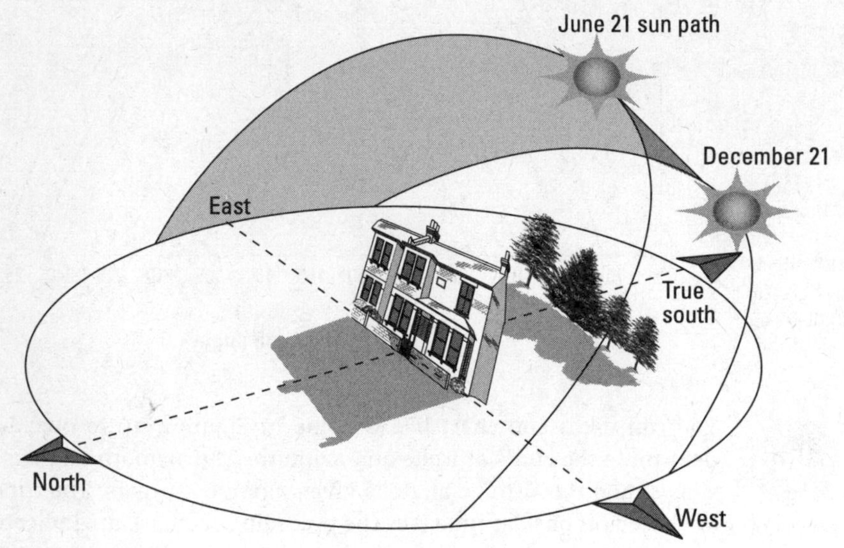


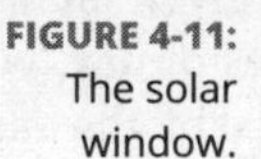

**FIGURE 4-11:** The solar window.

The path of the sun on each solstice date defines the extreme sun paths for a location. By defining these two paths, you know that the sun will always be within that "window" every day of the year. (The solar window is represented by the half gray, half clear arc in Figure 4-11.) As you design a PV system, your goal is to keep as much of that window open for the PV array as possible. If a tree or building is inside that window, the PV array will be shaded for some amount of time, reducing the energy output.

REMEMBER

The exact solar window varies based on the latitude of your client's site, but it's equally important regardless of location. Because the sun will always appear in the solar window over the course of a year, you need to keep this window clear from objects that will cast a shadow on the PV array. You don't have to go to the ends of the earth to make every minute of this window clear, but you should plan for the window to be open three hours before and three hours after solar noon every day of the year. I cover how to apply this window when selecting a site for your client's PV system in Chapter 5.

# Positioning PV Modules to Make the Most of the Solar Resource

After you know how the sun moves at your client's location, you need to relate that movement to the placement of the PV modules. The angle off of the horizon on which the modules are mounted and the direction they're pointed in reference to true north are the key factors for your installation. I explain some basics to know in the sections that follow; flip to Chapter 5 for full details on performing a site-survey analysis for a PV system.

## Introducing tilt angle

*Tilt angle* is the number of degrees the PV modules are mounted off of the horizon. It's a critical consideration in any PV system installation. Installations vary from nearly flat on some large commercial roofs (called low-slope roofs) to a very slight tilt (5 to 10 degrees) in some applications to vertical walls (which are at a 90-degree angle) and everything in between.

REMEMBER

Tilt angle is one component in the goal of pointing the PV array toward the sun. The exact location and angle you place the modules will likely be based on a combination of design considerations, including

- **Aesthetics:** Even though the PV nerd in me cringes to think of any sacrifice in energy production, the realist in me knows that if PV installations look like absolute garbage on customers' houses, the industry as a whole loses. So please keep in mind the aesthetics of the final product before you go trying to get every last kilowatt-hour out of an array. Making aesthetics a priority means you may need to mount the PV array on a roof with a tilt angle other than the optimal tilt in order to keep the array from looking like an art project gone terribly wrong.
- **The end goal:** Another consideration is what the end goal of the PV system is. If the system
  - Will be used as the primary power source in an off-grid home, then you want to do all you can to tilt that PV array at the optimum value for the client's location

- Is a grid-direct system in suburbia, you can take a hit on production by installing the array at a less-than-optimum tilt to maintain a reasonable installation because the utility will be present to help with power demands that are greater than what the PV array can produce
- Resides on a commercial rooftop, you may consider keeping a very low tilt on the modules to help reduce structural loading on the facility

Knowing about the tilt of the earth's axis (which is 23.5 degrees) relates to tilt angle. How? Refer to Figure 4-5 for a refresher on the solar radiation data sets. In particular, look at how the tilt angle of the modules is represented in terms of the number of degrees off the horizontal in relation to latitude. The tilt angles in relation to the site's latitude are affected in the following ways:

» If you adjust the tilt of the array by latitude minus 15 degrees, the PV array is perpendicular to the sun just before and just after the summer solstice. Because the sun is higher in the sky at these times of year, reducing the tilt angle of the array makes the array perpendicular to the sun and maximizes the array's energy output during the summer months.

» If the PV array is mounted with a tilt angle equal to the local latitude, it's perpendicular to the sun twice a year (on each equinox date) and very close to perpendicular for the weeks before and after the equinox; this makes the array perpendicular to the sun's position in the sky for the greatest number of hours throughout the year.

» If the desire is to have the array perpendicular to the sun in the winter, then a tilt angle of latitude plus 15 degrees is optimum. This helps the array maximize energy production during the winter months.

The idea behind varying the tilt angle is to maximize the PV array's energy production by pointing the array perpendicular to the sun as much as possible. So if your client needs to maximize energy production in the summer months (perhaps because she's running a water pump), you should mount the array so the tilt is equal to a value of latitude minus 15 degrees. On the opposite end, if the client wants more energy production in the winter, the array should have a tilt angle of latitude plus 15 degrees.

REMEMBER

The exact tilt angle that maximizes the annual energy production varies based on the local climate. For most locations, the optimum tilt angle (assuming you don't change it) is somewhere between an angle that's equal to latitude to an angle of latitude minus 15 degrees.

TIP

The NREL data for peak sun hours that I present earlier in this chapter can help you define the optimum tilt angle (see the "Referring to handy charts and maps" section). Simply compare the annual average peak sun hours for each tilt angle and see where your client's site is maximized.

The reason for the magical 15-degree value is the fact that the sun moves 23.5 degrees between the equinox date and the solstice date and then back again. By adjusting the tilt angle by 15 degrees rather than the full 23.5, you allow the array to be perpendicular to the sun for the days and weeks surrounding the time of year you're optimizing for.

*Note:* If you read almost any older textbook on the subject of PV module mounting and how best to determine tilt angle, you'll see a nearly universal answer: Tilt the PV array at an angle equal to the latitude for maximum annual energy production. Although this would be an absolutely correct statement if your client's site were always free of cloud cover and other weather variations, such as a lot of fog in the fall and winter, it may not be an absolute truth for your client's location. In Chapter 5, I show you how to use some commonly available tools to help you determine the tilt angle that'll serve your client best.

## Orienting your array to the azimuth

Another major component to consider when planning the array location to maximize energy output is the orientation with respect to true north, or the azimuth. The best way to refer to the azimuth of an array is to refer to the number of degrees the array is facing with reference to true north, exactly the same as you refer to the sun's position.

With this notation, the array has an azimuth of 90 degrees if it faces true east, 180 degrees if it faces true south, and 270 degrees if it faces true west. Bear in mind that true east, true south, and true west don't refer to compass directions; instead, they take magnetic declination into account. (*Magnetic declination* is the difference between true north and magnetic north; flip to Chapter 5 to discover how to account for magnetic declination.) For an example of what I mean, look at Figure 4-12. It shows a PV module that has an azimuth that's east of south. You could also say that this module has an azimuth of approximately 165 degrees based on how far it is from true north.

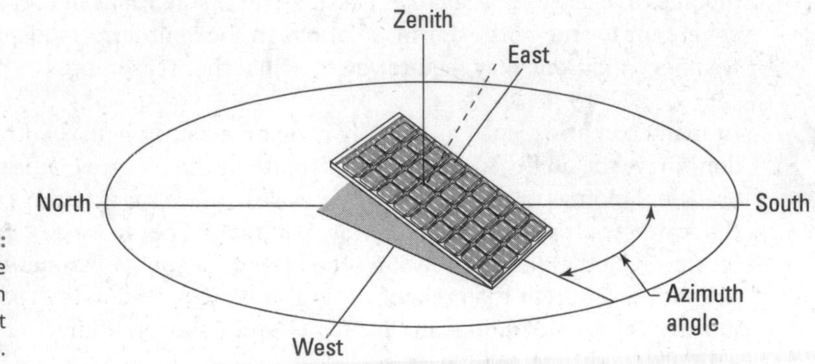


**FIGURE 4-12:** A PV module with an azimuth of about 165 degrees.

## TO TRACK OR NOT TO TRACK?

One of the most requested items in all PV systems is a *tracking system,* a mechanical assembly that holds the PV array and follows the sun throughout the day. Somewhere along the line, a majority of people decided that a tracking system is an absolute requirement for any PV array. They're definitely cool and a great idea (plus they can increase the overall energy output), but an honest evaluation must take place before you make the call that a tracking system is a necessity for a particular PV array.

First, you need to consider the fact that tracking systems contain assemblies that move and will eventually fail. One of the inherent beauties of PV arrays is the fact that they don't require moving parts — you can't say that for many power-producing systems. The addition of a tracking system takes away from that beauty somewhat. Not only have PV module prices been dropping, but placing additional PV modules on inexpensive racking allows you to gain more energy than by placing a smaller array on a tracking system.

Second, consider that tracking systems increase the total energy yield of the array (after all, they keep the modules pointing at the sun a greater number of hours per day) — but at a cost. Typically, tracking systems tend to cost three to four times as much as a similarly sized stationary racking system. In many situations, the economics of tracking systems and the maintenance they require make them less viable options.

On the other hand, some installations, such as a PV array that's driving a water pump all summer long, are a perfect fit for a tracking system. In this type of scenario, the additional hours may be a necessity, and a tracking system is the only way to achieve them. But for most folks, a simple stationary mount (like those described in Chapter 16) does just fine without adding another piece of equipment that you have to service periodically.

REMEMBER

Because irradiance directly affects the amount of current that a PV array produces (as I explain in the earlier "Relating current and voltage to irradiance" section), it stands to reason that you want a PV array perpendicular to the sun (in other words, directly facing it) as much as possible. So generally you must point the modules as close to true south as possible (for installations in the Northern Hemisphere) or to true north (for installations in the Southern Hemisphere), but local weather conditions may require you to adjust that reasoning.

TIP

In many scenarios, such as a PV array on an existing house, you as the installer don't have the ability to change the azimuth of the PV array, at least not easily. If sufficient room is available, you may choose to place the array on a racking system that can track the sun (see the nearby "To track or not to track?" sidebar for more on this subject). But often racking the system in such a manner that the array's azimuth is different than that of the house it's attached to isn't worth it. (I cover mechanical considerations in Chapter 16 and the overall effect of nonideal array azimuth conditions in Chapter 5.)

IN THIS CHAPTER

» Preparing yourself to conduct a site survey

» Identifying the basic pieces of information you need to gather about a site

» Recording data about magnetic declination, tilt angle, azimuth, and shading

» Evaluating the information obtained during your site survey

# Chapter 5

# Properly Selecting a Site for a PV System

One of, if not *the*, most exciting portions of the overall system installation process is the site survey. Why? Because you get to work with a blank slate (after all, the client shouldn't have another PV system anywhere on his building or property) and create a PV system from the ground (or roof) up. As the person who performs the site survey, you need to be able to identify any potential trouble spots and the best ways to address the issues they pose from the beginning of the project.

The site survey is generally your one chance to obtain all the required information about the site to create a proposal that works for both you and the client. It's also the only time you can really work with your client to establish his goals and expectations for his PV system before you're too far along in the process. I explain the steps and methods you need for a site survey in this chapter. I also give you an idea of what to do with all the data that you gather.

# Setting the Stage for a Site Survey

When you're at a client's house or business to perform a site survey, you must be diligent about collecting the information you need. You don't want to leave out any information that may prove critical for you to provide an estimate and a quality design for your installation crew. Return trips to gather information that should've been collected the first time around do nothing but waste your time and risk making you look less than professional. In this section, I provide some pointers on how to prepare for a site survey.

## Putting aside enough time

REMEMBER

You can't conduct a solid site survey without allowing yourself ample time and staying focused (a site-survey form can help focus your itinerary; see the next section). I suggest allocating at least an hour of on-site time. If through preliminary phone or in-person conversations with the client you get a feeling that he'll require extra time, then schedule yourself enough time to answer his questions while leaving enough time to collect the information you need.

Speaking of questions, the one you'll hear most often (aside from "What's the average cost?") is "How much money will this save me each year?" Although this can be a difficult question to answer without knowing a number of specifics, you should have an answer for it.

Use the PV Watts tool to become familiar with what a 1 kW PV array can produce in the client's area. (Figuring out what a 1 kW system can produce means you can easily do the math to adjust the values for your client's site.) This free tool, found at `www.nrel.gov/rredc/pvwatts`, takes information from you and applies weather information from collected data and returns estimated energy values for a PV system. One of the best features of this tool is the ability to vary input factors such as the direction you want your array to point or what tilt angle you want it to have. With just a few mouse clicks, you can run some scenarios and be able to determine the best solution for your site. (***Note:*** If you're new to PV Watts, I suggest using Version 1 first and then moving on to Version 2 as you become more proficient with the tool.)

By taking the time before the site survey to acquire this information, you can give your client an estimate for the array size and the amount of money the system can save him each year after you complete the site survey. Just be sure to give the disclaimer that you won't know the exact size of the array or how much energy it'll produce until the design is done.

## Creating a standard site-survey form

So that you don't forget any of the information you need to acquire while conducting a site survey, seriously consider carrying a standard site-survey form. Using a standardized site-survey form keeps you on track by reminding you what to collect and giving you a place to write all of that down. It also improves the quality of information you walk away with and can save you time when a client whom you gave a bid to a year ago suddenly calls and says that he's ready to buy. With all the necessary information recorded on your site-survey form, you can save yourself another trip out to the site.

REMEMBER

A good site-survey form can be easily replicated by others within your organization, meaning no matter who's conducting the site survey, the required information will be obtained and recorded. In a program such as Microsoft Word, you can quickly create a site-survey form that contains places for the following information:

- The site address and client's contact information (e-mail and phone number)
- The type of system desired (grid-direct; utility-interactive, battery-based; or stand-alone, battery-based)
- The utility company (along with contact information if you're not familiar with the local utility)
- The information from the client's main distribution panel (see the later "Collecting Basic Information during a Site Survey" section for the how-to on obtaining this data)
- Space for notes on the proposed inverter location
- A note section for options for wiring the array to the inverter
- The roof tilt and azimuth if the array will be roof mounted *or* the estimated array tilt and azimuth if the array will be ground mounted
- Information on the type of roofing material and its condition
- A note section for any issues with the structure
- A list of pictures to take
- Space for notes about the customer's requests or special site considerations

## Toting a site-survey bag

When you go to do site surveys, you'll be climbing on roofs, traipsing through fields, and walking through all sorts of buildings — all while carrying the tools you need to complete the survey. For these reasons, I can't stress how essential a dedicated site-survey bag is.

TIP

I prefer a tool bag with multiple pockets and a heavy-duty strap so you can carry the bag over your shoulder. The strap is key because you'll be carrying the bag up and down ladders and in and out of access hatches.

REMEMBER

The following tools should always be in your site-survey bag:

- A good-quality digital camera (see the next section for tips on using this tool during a site survey)
- Measuring tapes, including a
  - 25- or 30-foot tape
  - 100-foot tape
  - Wheel tape (this one's critical for commercial applications)
- A compass
- An angle finder (see the later "The tilt angle" section for the scoop on this tool)
- A calculator
- A clipboard and notebook with pens and pencils
- A flashlight and extra batteries (note that a headlight works great in attics while allowing you to keep your hands free)
- A shading-analysis tool (I describe several of these tools later in this chapter)
- Screwdrivers (you can get away with just one if you have a 10-in-1 screwdriver)
- A digital multimeter (see Chapter 3 for an example)
- Tinted safety glasses

REMEMBER

Although a ladder won't live in your bag, make sure you have it on hand. Very few site surveys can be completed without accessing a roof or attic.

# Picture This: Documenting Your Entire Site Survey with Digital Photos

One of the best pieces of advice I ever received regarding site surveys was to take enough of the right photos so I could re-create the site when I got back to the office. A good-quality digital camera is therefore an essential tool for any site-survey bag.

If you're in the market to buy a new digital camera, do a little research first. I suggest looking for a camera that can sustain some abuse (like being dropped and/or being exposed to water). Simple point-and-shoot cameras are sturdier (for your purposes, anyway) than the fancy digital SLRs out there, plus you can get high-enough-quality pictures from them. Compare different models within your price range using the solid reviews found at `reviews.cnet.com/digital-cameras`.

TIP

Chances are you'll find yourself doing multiple site visits in a day, so before you do anything else during a site visit, take a photograph of the front of the building. Doing so helps you establish the location of all the pictures that follow. Without establishing your location in the beginning, you can easily jumble all of your pictures together.

REMEMBER

To help yourself make sense later on of the photos you take, be consistent with your procedures. By keeping a consistent order to your site survey, you'll be more likely to remember the information needed at each step. I personally like to start at the PV array location and follow my way through the rest of the system, but I also know some installers who like to work in the opposite direction. Find the method that makes sense to you and stick with it.

Here are some tips for taking photos during your site survey:

- When you're looking at mounting the PV array on a roof or on the ground, take multiple pictures from each corner of the array location so you can see potential issues from each vantage point during the design process.
- Use reminders as to where you're taking the pictures from. For example, if you're planning to use a satellite image or a roof plan for recording measurements, then take a picture of the image/plan you're using while pointing at the location you're at. So if you're on the southwest corner of the roof, take a picture of the roof plan while you point at the southwest corner of the roof. Then take the pictures from that vantage point. You can do the same thing if the array will be mounted on the ground; just follow the same process to fully document the site.
- Repeat these steps for every location where any PV equipment will be installed. I recommend identifying alternate equipment locations too. After all, you never know whether the location you covet for the inverter is really reserved for some prize-winning rose bushes.

TIP

Another trick is to use the video feature of your digital camera and record the same vantage point you just photographed. Doing so gives you the ability to talk to the camera and remind yourself of certain points or give co-workers some insight into factors or considerations worth noting.

## Collecting Basic Information during a Site Survey

You need to gather certain pieces of information during each and every site visit; you also need to look out for oddball things that may cause problems later. In the sections that follow, I look at the most basic (yet critical) pieces of information you need to acquire.

WARNING

When performing a site survey, gathering the information is vital, but so is maintaining your safety. Any time you're climbing on roofs, walking through attics, or opening electrical boxes, you're doing real work. All of these activities have inherent dangers that you need to think through. If during your site survey you find yourself in a situation that feels unsafe or makes you uncomfortable, immediately stop what you're doing and find a different way to perform the task. Chapter 15 covers typical safety issues and how to handle them; the guidelines I present there apply to site surveys as well.

### General site information

When you show up to perform a site survey, it's easy to get so focused on all the little details you need to document that you forget to look at the big picture. Yet taking an objective look around and asking a few questions can save you from wasting your time. Following are four tidbits of general site information you should find out before you start delving into minutiae:

- **What's the shading like at the proposed location?** I explain how to perform a basic shading analysis later in this chapter, but you also need to account for the other shading-related site conditions that may cause problems. The one that catches people the most is *future shading* — objects that aren't there now but may shade the array in the future. For instance, those small trees on the south side of the building won't always be small, and the empty land to the south of the property may one day be the home of several tall buildings.

  TIP

  If the view from your client's site is clear now, you may be able to protect it from any potential shading issues. Many areas throughout the United States have solar-access laws that allow a person to protect his solar resource from shading sources on adjacent properties. These laws are typically enforced at the municipal level, so check with the local building department to see whether solar-access laws apply in your client's region.
- **Are there any restrictions for the site?** Such restrictions can include but aren't limited to homeowners' associations, city covenants, and historic districts.

REMEMBER

At the very least, your client should know whether he belongs to a homeowners' association; you can then ask the association's contact person about any association restrictions. The local building department can provide you with any other special requirements that are in place based on the building's location.

- **What agencies need to give approval for the installation?** You absolutely need to know what (if any) restrictions are in place due to local specialty codes and firefighter access. The best place to find out what's required for agency approval is the local building department (the same office that issues permits, as I explain in Chapter 14). You also need to check in with the local utility to make sure you meet its requirements (Chapter 17 has more on what you need from the utility).
- **What's your client's target budget?** You can easily establish a rough estimate of the PV system's cost based on your equipment choices and the installation specifics, but you need to know what your client's financial thresholds are. For example, if there are substantial unforeseen issues, how much money is the client willing to spend to fix an underlying problem before moving on?

## Structural and mechanical information

How much physical space is available for the installation? Typically, PV systems are installed on the roofs of buildings or on free land space. Your task during the site survey is to make sure the space available will suffice for the client's desired PV system. Your client may have an idea of where he wants the array to go, but it's your job to make sure a better alternative doesn't exist.

REMEMBER

The area you have your eyes on may be the same as someone else. Always verify that other plans don't exist for the space you want to use, such as plans for solar thermal collectors or skylights.

Here are some additional structural and mechanical questions that you should ask if the array will likely be mounted to a roof:

- **What are the dimensions and shape of the roof area available?** Taking the dimensions of the roof area you plan to install on will help you sketch out the roof later when you're ready to plan how the array will be arranged on the roof (later in this chapter, I give you tips on how to use satellite photos and three-dimensional drawing programs to help keep this sketch accurate). During the site-survey process, you also need to identify obstructions (such as plumbing vents, chimneys, and attic vents) on the roof as well as their locations.
- **What condition is the roof material in, and how old is the roof covering?** Placing an array on a roof that will need to be replaced in a few years doesn't make a lot of sense. If a reroof is in order, suggest it be done now and be sure

to work closely with the client and the roofer to coordinate phases of the project so you can continue with the PV system design and installation in a timely fashion.

TIP

As I state in Chapter 16, a possibility for dealing with a reroof and keeping your project moving is to do some of the rack installation before the roofers come in and allow them to seal the roof attachments.

- **What's the roof framing like?** The roof's framing plays an important role. Most modern homes and commercial buildings (those built since the mid-1970s) tend to have roof framing that's adequate for a PV array mounted parallel to the roof so long as a single layer of lightweight roofing material (such as composition asphalt shingles or wood shake) is used as the roof covering. Why? Because the roofs of modern homes are designed to handle multiple lightweight roof layers. As long as only one layer is present, adding the weight of a PV array will be less than the structure's limitations.

  If the building was constructed *before* the mid-1970s, plan to spend a few hundred dollars to have a structural engineer evaluate the roof for you and outline any changes you need to make to safely support the array. Make sure this consultation happens as early as possible in the system design process. ***Note:*** Many commercial buildings are built with the minimum requirements; therefore, they may not be able to support a roof-mounted PV array of any size.

- **What are the dimensions and spacing of the roof framing?** Most residential roofs have rafters or trusses that are made of dimensional lumber that's either 2-x-4 or 2-x-6 and are spaced 2 feet apart, which gives you a relatively narrow space to hit for your array's attachment points. For commercial roofs, the structures vary greatly. Some buildings have lumber, similar to residential roofs; others use very large wood support members; and some use steel supports. Consequently, you should take the time to verify the roof structure in order to properly attach the array in any system that's being installed on a commercial roof.

REMEMBER

  Always do your best to verify the roof framing composition and orientation when conducting your site survey. (See Chapter 16 for more information about mounting an array.)

WARNING

  Be sure to carefully evaluate rafters that are *overspanned* — a situation where the rafter has too much space between vertical support members. Different spans are allowed based on lumber type and roof-loading restrictions, but as a general rule, if the rafters have a span of more than 7 feet between supports, you should investigate the need for adding support by consulting a structural engineer.

After roofs, ground mounts are the most popular type of racking system. Unless your client's site has unusually loose soil (like sand), you can work with a racking company (and maybe an engineer, if necessary) to determine the best possible mounting solution for the array. Of course, before you start talking to a racking company, you need to make sure the location is suitable for mounting an array. I cover the issues you need to consider for ground-mounted PV arrays in Chapter 16.

## Electrical information

If you're installing a stand-alone, battery-based system, you don't really need to worry about examining the existing electrical service and equipment because the stand-alone system will be providing all the energy for the client moving forward. (You will, however, have a lot of components to size and to specify, as I explain in Chapters 12 and 13.) For utility-interactive systems (whether grid-direct or battery-based), you have a number of items to review while you're on-site because you'll eventually connect the PV system to the utility:

- **What are the specifications for the main distribution panel (MDP) and the main circuit breaker protecting the panel?** The ratings on the MDP and the main circuit breaker play a major role in determining a PV system's maximum size. When looking at the existing electrical service, you need to document the specifics on the MDP and any subpanels you want or need to use (I cover these panels in Chapter 2), including their physical locations. The voltage for most electrical services in residential applications is 240 VAC at various current levels; for commercial applications, the voltage is usually 208 VAC or 480 VAC.

  *Busbars* are the pieces of metal in the back of the MDP that connect the circuit breakers in the panel to the wires coming from the utility (you can't see them when the cover of the MDP is on). Every MDP has a rating for its busbars on the label attached to the inside of its cover. This rating is a value for the amount of current that can flow on the busbars inside the panel without causing any problems. Standard residential MDPs have either a 200 A or 225 A busbar rating and may simply be labeled 200 A Max (or 225 A Max); commercial busbars typically start at 200 A for small facilities and can exceed 1,200 A in large facilities, with a number of options in between.

  The other specification for the MDP (and any subpanel used) is the rating of the main circuit breaker protecting the panel. For the MDP, this is often the same size as the busbar rating. The ratings for circuit breakers in subpanels vary based on the loads located in the subpanels.

- **Are there any open breaker spaces on the main electrical panel?** Look in the MDP or subpanels to check for available space to put a breaker. Then connect the inverter's output wires to one of these panels by placing a breaker in the panel and wiring the inverter to it. (This process is similar to putting a breaker in the panel for a new set of outlets or a new load except that the electrons are running in the opposite direction.) If the panel is full, you need to either make room or replace the existing panel with a larger panel.

» **Where will the inverter and required disconnects be located?** Determining a location for the inverter and required disconnects means you have to know the specifics of the equipment you'll be using as well as what the local jurisdiction requires. Typically, the inverter needs to have disconnecting means (both AC and DC) within sight of the inverter.

REMEMBER

The majority of the inverters on the market today have integrated disconnects, but they aren't all considered AC and DC disconnects, which means you need to know for sure what type of product you're using. If the inverter doesn't have integrated disconnects, make sure it has disconnects next to the inverter and in locations that satisfy the *National Electrical Code*® (*NEC*®). Refer to Chapter 17 for exact location requirements.

In addition, some utilities require visible, lockable disconnects at the utility's meter location. If this is a requirement at your client's site, become familiar with the utility requirements *before* conducting a site survey so you aren't surprised later. Ask the utility for its net metering or PV interconnection rules. If the utility does have a requirement, make sure you reserve the proper space for the required disconnect.

REMEMBER

The electrical connections are a critical part of the whole installation (as you find out in Chapter 17). Before you get too far into the system design, however, you need to know what the licensing requirements are in the jurisdiction in which you're working. Some jurisdictions require that the people doing certain portions of the work hold specific licenses. (The most common example is the requirement that a licensed electrician make the final wire connections from the inverter to the utility.) Consequently, you may need to hire certain tradespeople to do certain portions of the work.

TIP

To find out the requirements in the state where you're working, visit the state's construction contractors board or licensing department Web sites. These sites will list the requirements for working in the construction trades.

# Measuring Information in Degrees

Some factors that are relatively constant on sites within a geographic region are the *magnetic declination* (the direction a compass points toward north versus true north), insolation data (the number of peak sun hours), and local climate conditions. Factors that differ on nearly every installation, no matter the location of the site, include the tilt angle of the array (the number of degrees it is off of the horizon), the azimuth the array faces (the number of degrees from true north), and the shading considerations.

In the following sections, I explain how to measure the three pieces of vital information that are measured in degrees: magnetic declination, the array's tilt angle, and the array's azimuth. Flip to Chapter 4 to find out how to measure peak sun hours and consider seasonal effects; head to the later "Exploring Shading-Analysis Tools" section in this chapter for pointers on analyzing shading concerns.

## Understanding magnetic declination

The term *magnetic declination* refers to the number of degrees that a compass needle differs from true north. (If you do a fair amount of navigation — flying, boating, mountaineering, and the like — you may refer to this as *magnetic variation;* that's just another term for essentially the same thing.) For locations in the Northern Hemisphere, you want a PV array to be facing true south as much as possible — something you can't make happen if you follow exactly what a compass tells you. If you were to use your compass and point the array the same direction as the "south" needle on the compass, you'd be as much as 20 degrees off from true south, depending on your location. To make sure an array points toward true south, you need to know what your magnetic declination is.

Depending on where you are in the United States, a compass's north needle points either to the east of north (the western half of the country) or to the west of north (the eastern half of the country); the Mississippi River is roughly the dividing point. The designations are referred to as *eastern (positive) declination* for locations in the western part of the country and *western (negative) declination* for locations in the eastern region of the country. These designations refer to the direction that the north needle is pointing in reference to true north.

REMEMBER

Ultimately, you need to concern yourself with true south more than true north (which is contrary to what most people are taught). Why, you ask? Well, for the Northern Hemisphere, the sun is located in the southern half of the sky, which is where you want to point your PV modules. You need to turn yourself 180 degrees and think about how the declination value affects the magnetic south and true south locations. For example:

- When you're on the west coast and using a compass to find south, the south needle is actually pointing southwest of true south (because the north needle is pointing northeast of true north). To account for this, you need to turn yourself to the east to find where true south is.
- When you're on the east coast, the opposite occurs. The south needle points southeast of true south (because the north needle points northwest of north). Turn to the west to find true south.

Figure 5-1 gives you the visual of these two examples. (Don't worry if the concepts still don't make sense; they can be difficult to grasp. I still have to stop and think the process through to make sure I get it right on a roof.)

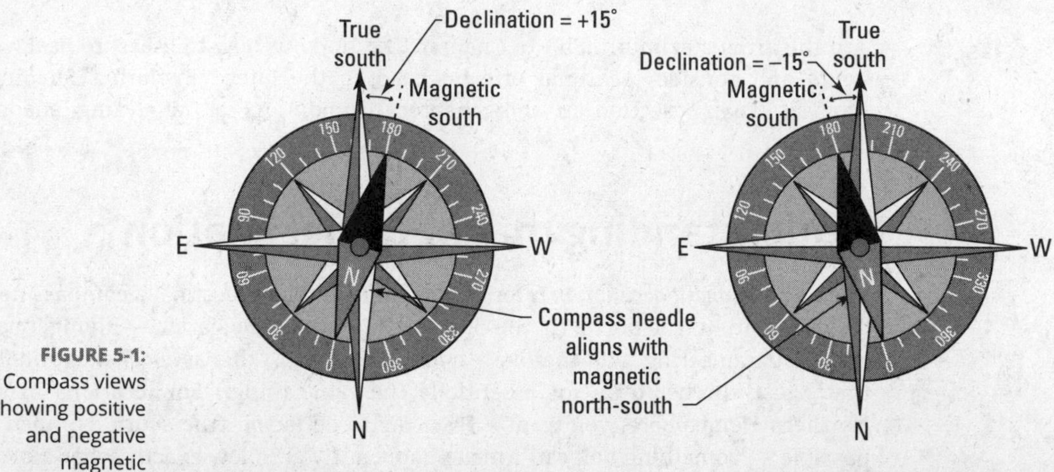


**FIGURE 5-1:** Compass views showing positive and negative magnetic declination.

To add to it all, magnetic declination is a dynamic value, meaning the number of degrees your compass lies to you is constantly changing. It doesn't change rapidly, but it does change. True south, however, never changes, so after you find it, you don't have to worry about going out to the client's PV array every five years and adjusting it a few degrees.

The National Oceanic and Atmospheric Administration (NOAA) has a great declination calculator on its Web site that can help you find the exact value for your client's location; you can find it at `www.ngdc.noaa.gov/geomagmodels/Declination.jsp`. To find the point on your compass that relates to true south, you need to know the magnetic declination for your client's area and subtract that from 180 degrees (north is 0 degrees, so south is 180 degrees on the compass):

- For locations in the western part of the United States, true south is designated on the compass as 180 degrees – declination.
- For locations in the eastern part of the United States, you still subtract the declination value, but because the declination is a negative number to begin with, the equation becomes 180 degrees + declination.

Technology can help you evaluate the declination at a site during a site survey, but even the greatest technology requires you to input the starting information correctly. If you can at least keep the general rules about magnetic declination in mind, you should see consistently accurate data — provided, of course, that you're not trying to use your compass in a location that's close to steel objects, which cause a compass needle to point in any number of directions and prevent it from finding magnetic south. If you can't find magnetic south, you'll have a tough time finding true south by subtracting the magnetic declination value.

## Calculating the array's tilt angle and azimuth

In a typical PV installation, the array is located on top of a building. In this scenario, the most cost-effective installation method places the array parallel to the roof and pointing in the same direction as the roof because the additional requirements for securing an array that isn't parallel to the roof surface can become overwhelming. Aesthetics are another consideration. I know, I know. Beauty is in the eye of the beholder, but an array that looks awkward on a roof doesn't make passersby say, "Gosh, I sure would like that on my roof!"

Knowing an array's tilt angle and azimuth can help you properly place one on a roof. I explain what you need to know to calculate these values during a site survey in the following sections. (If these concepts don't sound familiar to you, check out Chapter 4 for a quick introduction.)

*Note:* The approaches in the following sections refer to roofs and the methods used to calculate an array's tilt angle and azimuth in relation to a roof. If you're installing an array on the ground, the same approaches apply; you just don't have a roof to reference and need to use landmarks instead.

### The tilt angle

You have a couple options for figuring out the tilt angle of an array that will be both functional and eye pleasing:

- **Use an angle finder.** To use this tool, which is also called an *inclinometer,* simply place it on the face of the surface to be measured and wait until the rotating dial comes to a stop. The face of the tool has degree measurements on it in relation to a flat surface so you can record the angle off of the horizon by noting the degree location where the dial stops spinning. Because most PV arrays are mounted parallel to a roof, this angle is the same for both the roof and the array.
- **Use a little math.** If you can measure the amount of height the array changes (rise) over a certain horizontal distance (run), you can calculate the corresponding angle. After all, who doesn't love some good old-fashioned trigonometry? A common approach is to give the rise of a roof over a distance of 12 inches.

TIP

You may hear a roof slope (angle) referred to as *X:12.* What this is saying is that the roof rises *X* inches for a run of 12 inches. The smaller the *X* number is, the lower the slope is for that roof. If you're given the roof slope this way, you can quickly convert the variable to a number of degrees by using a calculator with basic trigonometry functions. The calculation is: arctan (rise ÷ run). The arctan function is represented as $\tan^{-1}$ on many calculators. Therefore, if I tell you that a roof has a slope of 6:12, you can calculate the number of degrees with this operation: arctan (rise ÷ run) = arctan (6 ÷ 12) = arctan 0.5 = 26.6 degrees.

TIP

If you aren't given the roof slope, the easiest way to figure it out is by using an angle finder. However, you can also use a bubble level and a tape measure. If you choose to go this route, hold the level with one end touching the roof and the bubble centered in the viewing glass (in other words, hold it so that the level is level). Use the tape measure to see how many inches the roof surface is from the end of the level that's above the roof. You now know that the roof rises that many inches (from the tape measure) over a run that's equal to the length of the level.

## The azimuth

Satellite images are an immense help when conducting a site survey, especially when you need to calculate the azimuth. With the help of satellite images, you can know the azimuth of the roof with magnetic declination accounted for before you even set foot on it.

These images show the site in relation to true north and true south. So if you find your array location (typically a building) on the satellite image, you can determine the roof's azimuth without having to use your compass. Why? Because the satellite images have a built-in compass that's pointing true north. If you can find your client's building on the satellite image, you can use this built-in compass to estimate the true azimuth of the building (with magnetic declination accounted for).

TIP

Another handy online tool is available at `www1.solmetric.com/tools/RoofAzimTool.htm#`. It lets you find a satellite image of a building and use your mouse to determine the building's azimuth.

WARNING

The downsides to satellite images are their resolution and the time when they were taken. A remote location may have poor image resolution. Other times, the image is a few years old, and the building you're looking for isn't in the image yet. In these situations, you have to obtain the information you need on-site. The address look-up tools on these products aren't perfect either, so you may need to confirm with your client which building is the right one before you head to the site.

Depending on the services you need, one of these options may work for you:

- Google Earth (`earth.google.com`) and Bing Maps (`www.bing.com/maps`) let you see the building you're looking for with a decent amount of detail after you type in the address. Some of these images can also be used in conjunction with free three-dimensional drawing programs (such as Google SketchUp, found at `sketchup.google.com`) to really help you evaluate and analyze your site.
- Subscription-based programs such as Pictometry (`www.pictometry.com`) can give you even more detail.

» Smartphones offer applications that can come in handy during a site survey. You can download apps to your phone that perform shading analysis, calculate tilt angles, and even help with some of the *NEC®* calculations I present in Chapter 13. These phone apps are currently available solely for the iPhone, but with the speed at which the wireless world moves, I'm sure that will all change very quickly.

If you print a couple copies of the satellite images for your client's site beforehand, you can take notes and record measurements directly on the printouts without having to re-create the building or array location with hand drawings. I recommend printing at least two or three clean copies so you have plenty of space to make notes for yourself.

## Exploring Shading-Analysis Tools

When you perform a *shading analysis* during your site survey, you look at the area surrounding the proposed PV array location and estimate the amount of sunlight that's blocked from obstacles like trees and buildings. This analysis is what allows you to give your client a realistic expectation of the energy that can be delivered by the PV system over the course of a year.

Shade on a PV array can drastically reduce the array's power output, which in turn reduces its energy production. Although a perfectly sunny spot is preferable, the reality is that shading is a simple fact of life in many locations. Some situations can be altered (a tree can be trimmed or removed), but others can't (the neighbor's house isn't going anywhere). The goal for any PV site is a location that's shade free three hours before and three hours after solar noon (I define solar noon in Chapter 4). This time frame allows the PV array access to full sun during the portion of the day when the irradiance values are at their highest. (If this time window isn't an option, like in the winter, look for a window of two hours before and two hours after solar noon as a minimum.)

By using a *shading-analysis tool*, which recognizes the shading objects in the PV array's *solar window* (all the points between the sun's lowest and highest points in the sky) and calculates the effect that shading has on the array, you can effectively determine the shading effects from objects located in the array's vicinity. Figure 5-2 illustrates how a shading-analysis tool transfers the obstructions in the site's solar window to a sun chart that you can analyze (I explain how to perform this analysis later in this chapter). You can also use shading-analysis tools to propose "what-if" scenarios, like what's the overall impact if we remove that tree in the distance?

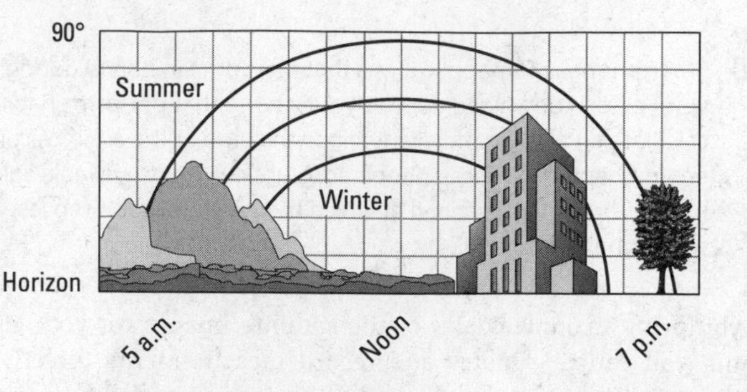


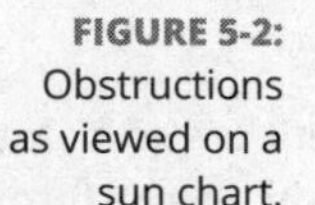
**FIGURE 5-2:** Obstructions as viewed on a sun chart.

The shading-analysis tools I outline in this section work the same way. I've used all three tools and feel that they're quality products that deliver good information when used correctly. (Because they all work off of the same principle, you may want to check out the sun-path information in Chapter 4.) Of all the tools I suggest you keep in your site-survey bag (see the earlier "Toting a site-survey bag" section), a good shading-analysis tool will be your single biggest expense (unless of course you buy the best-of-the-best digital camera). It's a tool you can't do without though, so research your options and think about how (and how often) you plan to use it before buying one.

REMEMBER

Always use a shading-analysis tool at the proposed array location. If you can't get to that, use the manufacturer's instructions to adjust the information you collect so that it describes shading at the actual array location.

TIP

One of the greatest features of the following three shading-analysis tools is their ability to evaluate a site for an entire year at once. They all use site-specific solar charts and local weather data for the site analysis, which means you can go out to the site at any time (as long as enough light is available so you can see the area surrounding you) and collect all the information you need.

- **ASSET:** The Acme Solar Site Evaluation Tool (abbreviated as ASSET and found at `www.we-llc.com/ASSET.html`) is a shading-analysis tool that uses a digital camera and software to evaluate a site. All you have to do is set the camera up in a special base that incorporates a level and compass. ASSET then takes seven pictures, starting with the camera pointing due east and rotating the camera to the west. The result is a panoramic picture of the site.

  After you have your panoramic view, you download the pictures into the ASSET software so the program can return a single picture (made of the seven individual ones) with the solar path on top of the site. ASSET also recognizes shading objects and reports the overall loss of the solar resource due to shading.

- **Solar Pathfinder:** The Solar Pathfinder (found at `www.solarpathfinder.com`) has a long history within the solar industry. It consists of a plastic base that integrates a level and compass and holds a paper sun chart. (The company makes sun charts for various latitudes, so the Solar Pathfinder can be used virtually anywhere.) The final piece is a transparent dome that sits directly over the sun chart in the base. This dome reflects the surroundings and allows you to project those objects down to the sun chart below. You can then either trace those objects directly onto the sun chart or take a digital picture of the tool and download that picture into Solar Pathfinder's optional software for further analysis.

The software option is a powerful tool that can help you better estimate the shading effects on that site. In fact, if you're purchasing a Solar Pathfinder for the first time, don't even consider the software optional; just buy it. The additional information you get from the software is worth the investment.

- **Solmetric SunEye:** One of the newest entrants in the shading-analysis-tool market is the Solmetric SunEye (found at `www.solmetric.com`). It's a fully digital tool that incorporates a fish-eye camera, compass, and level into a base with a screen showing the site complete with sun chart and obstructions. The machine evaluates the site by taking a photo at the array location and using some site-specific data, such as magnetic declination and local climate readings (note that you have to input some basic site parameters in order for the SunEye to be able to retrieve the proper information). It then returns detailed information about the shading effects.

# Interpreting the Data and Bringing It All Together

During your site survey, you go to the site and gather all the data you need (described earlier in this chapter). That's great, but what do you do with all that data afterward? As you find out in the following sections, you compile it and analyze it to determine the best PV system solution for your client. Specifically, you must evaluate the following:

- The area available for the array
- The options for mounting the array
- The electrical considerations
- The array orientation
- The shading effects

## Analyzing reports from your shading-analysis tool

Shading-analysis tools generate a report that's associated with the site. This report helps you during the design and installation portions of a job because having this information allows you to avoid shading issues that may reduce the overall energy output. Here are some important pieces of information revealed in a typical report from a shading-analysis tool (this is the information your client is most likely to care about):

REMEMBER

- **The effect of shading on an annual basis:** This is the most basic level of information you absolutely have to walk away with. It may be reported as a percentage of the ideal insolation or as a new amount of insolation. Either way, you can have your shading-analysis tool break down this information month by month so you can see where the big hits are. Ideally, you'll have very few losses due to shading. In the next section, I show you how to incorporate the shading losses and the array's orientation losses to account for all the site-specific losses (note that the combination of shade and orientation losses shouldn't reduce the site's potential by more than 25 percent).
- **Estimations of the energy production based on different scenarios:** These estimates allow you to vary the tilt and azimuth to see how such adjustments may benefit the installation.
- **Predictions of how the PV array would perform if the obstructions were removed:** You can estimate the array's production if, say, a tree weren't in the way and allow the client to choose between keeping the tree or obtaining additional energy from the array by removing the tree.

# Considering the total solar resource factor

REMEMBER

When shading-analysis tools evaluate a site, they typically look at the shading effects as well as the effects of the tilt and azimuth. The combination of these factors is defined as the *total solar resource factor* (TSRF). The individual components of the TSRF are often referred to as the *shading factor* (SF) and the *tilt and orientation factor* (TOF). (*Orientation* is another word for azimuth.) Both are reported as a percentage of the ideal situation. For shading, the ideal situation is an absence of shading; for tilt and orientation, the ideal situation is pointing a stationary array at a specific tilt and azimuth to have the array produce the maximum amount of energy annually. You need to figure out what the TSRF is so you can accurately predict the amount of energy that the PV array will produce annually.

The SF value is, obviously, totally dependent on shading. If a site were to have no shading, the SF would equal 100 percent. That's pretty difficult to achieve, but an excellent site has shading factors greater than 95 percent (meaning less than 5 percent of the solar resource over an entire year is lost due to shading).

The TOF value, given in terms of a percentage, is estimated by evaluating the PV array's tilt and azimuth compared to what the absolute ideal tilt and azimuth for a PV array in that same location would be. These values are typically represented as a graph with concentric circles (or ellipses) that represent different levels of the resource. Figure 5-3 shows one such graph. The vertical axis represents the tilt angle of the PV array, and the horizontal axis represents the array's azimuth. The innermost circle represents the ideal tilt and orientation for this location. As the array moves away from that circle, the amount of solar resource is reduced. To estimate the TOF with such a graph, simply find the point on the graph where the proposed array lands and apply that percentage for the TOF.

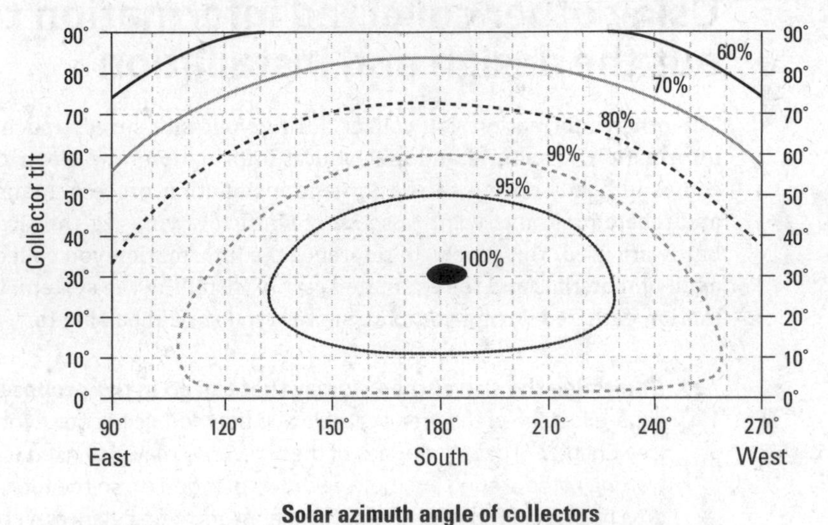


**FIGURE 5-3:** A typical tilt and orientation graph.

To calculate the TSRF, you multiply the SF by the TOF. The result is the total resource available after these inefficiencies are considered. For example, if you perform a shading analysis and find the annual reduction of the resource due to shading is 9 percent, that would equate to an SF of 91 percent. If you located the array in a location where Figure 5-3 applied and the array was tilted at a 20-degree angle and a 165-degree azimuth, the TOF value would be 97 percent. You can determine this percentage by placing an X at the intersection of the 20-degree tilt and the 165-degree azimuth on the tilt and orientation graph. The X falls between the 95-percent ring and the 100-percent ring, so you can estimate the value to be 97 percent. The resulting TSRF value would be 91 percent × 97 percent, or 88.3 percent. This TSRF value tells you that the site's output will be reduced by nearly 12 percent when compared to a PV array oriented toward the ideal location with absolutely no shading.

The TSRF should be greater than 75 percent for your PV arrays. Anything less than that is marginal at best. Many rebate programs require that the TSRF exceed a minimum value (typically 75 to 80 percent), so be sure to find out the restrictions tied to any rebate money your client may be expecting.

You can then use the TSRF to estimate the amount of energy the array will produce by multiplying it by the peak sun hours data available from sources such as the NREL Redbook (see Chapter 4). Another source for this information (as well as handy tools to estimate the TOF values) is Solmetric, the manufacturer of one of the shading-analysis tools I recommend earlier in this chapter; you don't have to use Solmetric's tool in order to use some of the features of its Web site (`www.solmetric.com`).

## Using other collected information to plan out the design and installation

The other information you collect during your site survey (such as general site information, structural and mechanical information, and electrical information) comes in handy during the design and installation process. Many municipalities and rebate programs want to see some basic drawings of the site and associated hardware used. You'll need to reference the information you collected to establish a bill of materials and the estimated cost of installing the system. Using the information gathered during your site survey, you should be able to

- **Determine the size of the PV array that can go in the proposed area.** The square footage of the area available and the module of choice for the array (see Chapter 6) dictate the size of the array. You may also need to account for walking paths around and between rows of modules, so the total available area may actually be less than what it appears at first glance. A good plan for a roof-mounted array is to allow for access around the array. If you pack a roof with modules, not only will the installation process be difficult and more dangerous, but you'll also make it nearly impossible for anyone to get on that roof in the future.

  A good number to keep in your head is 10 W per square foot of open roof area. So if you have 200 square feet of open roof area, an off-the-cuff estimation for the PV array is 2,000 W, or 2 kW. This value accounts for space that can't be fully utilized by the array. It's slightly on the conservative side, meaning you can probably fit more wattage in that area, but it's a great starting point and generally within reason.

- **Establish any potential upgrades to the structure for either the array or the other components.** If you spot a potential problem area in the roof framing information that you collected during the site survey (see the earlier "Structural and mechanical information" section), you can use this information, along with an engineer's report, to justify the addition of more bracing or supports to the existing roof.
- **Identify the racking structure used.** You can narrow down your options based on the defined array location and the array's potential size.
- **Decide on the best method to run all the conductors to and from the inverter and tie into the electrical system.** As part of the site survey, you examine the potential wire runs and the best methods for getting from the array to the inverter and then to the MDP (or subpanel).
- **Identify all the major components for installation per local requirements.** By becoming familiar with the local jurisdiction's requirements as well as the local utility's requirements, you can establish all the major components needed and their approximate locations to save yourself some time and hassle during the installation process.

# 2

# Digging into Complete System Details

## IN THIS PART . . .

Are you ready to peek behind the curtain and see how the major parts of a PV system function? I hope so because in this part, I detail the specific components used in PV systems and how they actually work.

Chapter 6 introduces you to PV modules and how they're made; it also explains how you get electricity from sunlight. Chapter 7 describes the batteries used in PV systems and helps guide you in choosing the right battery for your client's situation. Charge controllers are examined in Chapter 8, including how they act to maximize a PV array's input to the batteries. Chapter 9 covers the different inverters available, focusing on the major differences in technology and the features you need to look for when selecting one. Finally, Chapter 10 touches on the different safety components used in PV systems in order to help keep you (and your clients!) safe.

IN THIS CHAPTER

» Seeing how solar cells form the foundation for PV modules

» Understanding the differences in the various PV modules out there

» Recognizing key electrical specifications

» Examining standard test conditions and calculating the effects of environmental conditions

» Checking out graphs that link current and voltage

# Chapter 6
# PV Modules: From Sand to Electricity

Electricity does pretty powerful work: Electric vehicles move you from point A to point B, computers give you access to an unearthly amount of information, and electric heaters do a marvelous job of keeping you comfortable in the winter. When electricity is properly contained, you can't see or feel it — a fact that causes many people to consider it a magical phenomenon. Add PV modules and their ability to harness energy from the sun and convert it into electricity, and you have another layer of mysticism.

Electricity generated from PV modules can do a lot for people, but it can't do everything. Certain electrical loads (such as water heaters, electric stoves, and anything else that uses electricity to create heat) aren't good matches for solar electric systems, especially battery-based systems. Fortunately, the good folks engineering and manufacturing the devices used in PV systems have made them capable of integrating seamlessly into existing electrical systems. Grid-direct systems (see Chapter 11) operate automatically, and battery-based systems (see Chapter 12) are getting smarter all the time.

Because the PV module is the heart of any PV system, you need to know the secrets to the magic tricks behind one. In this chapter, I help you become comfortable with the terms and concepts used in describing PV modules — specifically, I explain their construction and manufacturing, note different types of modules, and walk you through their electrical specifications under standard test conditions. Knowing how PV modules are affected by the intensity of sunlight and varying temperatures helps you further understand the requirements of installing PV systems. By having a solid understanding of what PV modules can do, you can create a PV system that will, at a minimum, meet your client's expectations and requirements.

# Creating Solar Electricity: It All Starts with a Cell

Many people aren't willing to accept "magic" as the answer to the question "How does a PV module make electricity?" That's why, in the following sections, I provide you with a working knowledge of the basics behind PV construction and manufacturing at the cell level. I also explain how cell construction allows solar cells to do useful work through a process known as the photovoltaic effect.

## Getting a grip on cell construction and manufacturing

The basic building block for all PV modules is the *solar cell,* a roughly 6-x-6-inch object that starts its life as sand (actually silicon) and is then wired within a PV module to produce the voltage and current desired by the manufacturer. (*Current* is the flow of electrons, and *voltage* is the pressure that makes electrons move. Flip to Chapter 3 for an introduction to voltage, current, and other electricity basics.) Solar cells are manufactured in such a way that when they're placed in sunlight, the photons in the light excite the electrons in the cells. When the module is connected to an electrical circuit, useful work, such as turning a fan or powering a refrigerator, can be done.

In this section, I explain the two parts of the cell-manufacturing process that help cells do this useful work: doping cells and keeping electrons separated.

### Doping solar cells to create semiconductors

When a solar cell is completely manufactured, it becomes a *semiconductor,* a material that acts as both an electrical conductor and insulator. Solar cells become conductive when exposed to light, which makes them able to pass current. However, silicon — the primary ingredient of solar cells — is naturally a much better insulator than a conductor. *Insulators* inhibit the flow of electrical current, which isn't a desired feature for solar cells. In order to enable the flow of electrons (and therefore become semiconductors), the cells are doped during manufacturing. Typically two elements, boron and phosphorous, are used in the doping process.

Unlike in sports, doping is an acceptable and highly encouraged activity in the manufacturing of solar cells. Because the silicon won't readily produce an electrical current in its natural state, the addition of the dopants allows the current to flow. Typically, boron is introduced to the silicon during the first stages of cell manufacturing, and phosphorous is introduced to the silicon by diffusing a vapor directly onto the manufactured cell.

The addition of these dopants adds electrons and electron holes to each side of a solar cell. The phosphorous atoms have extra electrons within them, and the boron has extra electron holes, waiting to be filled with the electrons. The phosphorous-doped side becomes known as the *N type,* or the negative side of the cell (the side facing the sun), and the boron-doped side becomes the *P type,* or positive side (the side facing away from the sun).

## Creating a one-way electron path with a PN junction

When sunlight hits the phosphorous-doped (N type) side of a solar cell, the electrons in the cell become excited. They're so anxious to get moving that they'll gladly go to the boron-doped (P type) side of the cell if given the proper path.

That path involves a junction between the positive and negative side of the cell. This *positive-negative junction* (or *PN junction*) acts as a diode, allowing the electrons to pass from the positive (bottom) side to the negative (front) side of the cell but not in the reverse direction. This means the electrons flow from the negative side of the cell through the circuit and to the positive side of the cell. As more electrons move from the negative side to the positive side, the electrons on the positive side are pushed up through the PN junction to the negative side of the cell, and the process continues as long as sunlight is present. The PN junction ensures that the electrons move through the circuit. Figure 6-1 shows an example of a solar cell and a PN junction.

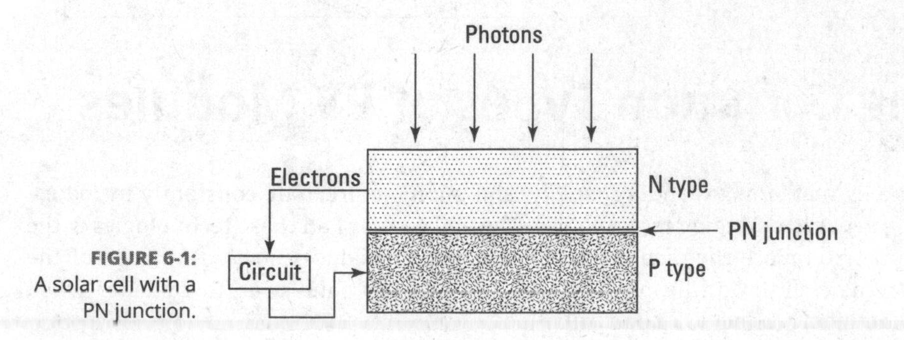


**FIGURE 6-1:** A solar cell with a PN junction.

## Connecting cell construction to the photovoltaic effect

The phrase *photovoltaic effect* describes solar cells' ability to produce voltage and current when exposed to sunlight. Here's a step-by-step breakdown of how a cell's construction allows that to happen (check out Figure 6-2 to see how electrons move through the PN junction):

1. **Energy from the sunlight's photons excites the electrons located on the solar cell's N type, giving them the potential (voltage) to move.**
2. **When the solar cells are connected to a load, the excited electrons start moving (current flow) from the N type to the P type, performing useful work along the way.**
3. **The electrons go to the cell's P type and combine with the electron holes.**
4. **As sunlight continues to strike the cell and more electrons are sent through the circuit, the electrons are forced from the P type back to the N type through the PN junction to continue the process.**

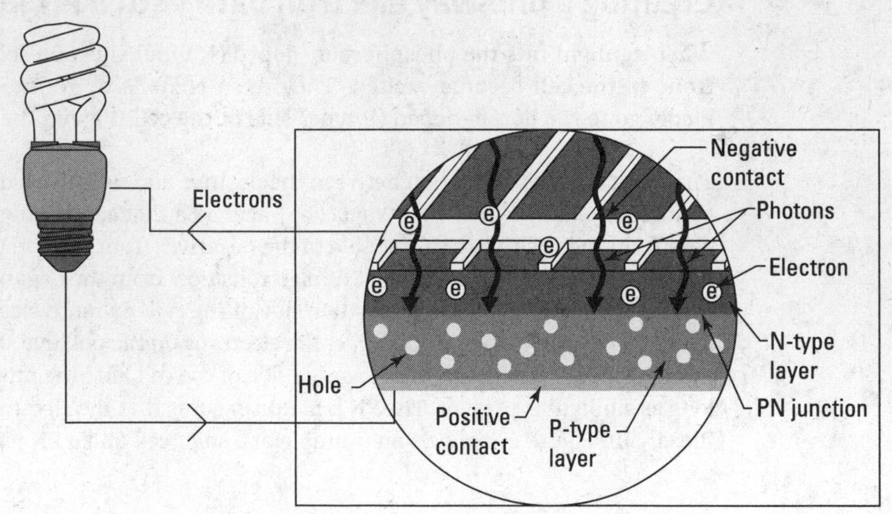


**FIGURE 6-2:** The movement of electrons though the PN junction.

# Reviewing Common Types of PV Modules

The PV market is expanding rapidly, and manufacturers are constantly introducing new and emerging technologies. The end result of all these technologies is the same: You place the module in the sun, and it produces power. But each of the commercially available products has its own pros and cons to consider when selecting a product for a particular job.

In the sections that follow, I focus on the commercially available PV modules currently used for residential and commercial systems; the ones you're most likely to run into are crystalline modules and thin film modules.

*Note:* I don't cover the technologies primarily used on industrial-scale projects. And, as with any technology field, a number of materials in research and development as of this writing may very well turn the solar industry on its head in the future.

## Checking out crystalline modules

Crystalline PV modules, which are made by grouping a number of individual solar cells together, are currently the most common module type for residential and commercial applications. One of the main reasons why crystalline modules are used so frequently is that they're more efficient than other PV technologies. Typical crystalline modules are rated at 11 to 14 watts per square foot ($W/ft^2$). Some of the higher-efficiency modules are rated in excess of 17 $W/ft^2$, which allows a consumer to generate a greater amount of energy in a limited space, like on her roof.

Manufacturers don't tend to report the number of watts per square foot (or *power density*) for their modules, but you can find out this information pretty easily. Just take the module's rated power output, as described in the later "Maximum power point" section, and divide it by the module's square footage. The watts-per-square-foot value can be more telling than any efficiency value reported because a common goal of PV systems is to place as much power as possible in a given area.

As you find out in the next sections, two main types of crystalline modules exist — those containing monocrystalline cells and those containing multicrystalline cells (see Figure 6-3). Both types of modules start from the same raw material, but the manufacturing processes differ: The monocrystalline modules have a uniform molecular structure and are more efficient, whereas the multicrystalline modules have many structures, resulting in a less efficient module.

### The monocrystalline kind

Monocrystalline modules begin as a molten vat of purified silicon that has been doped with boron to create electron holes (see the earlier "Doping solar cells to create semiconductors" section for more information on this). A starter seed, a crystal about 4 inches long and 2 inches in diameter, is introduced to the silicon-boron mixture that becomes the structure for the solar cells. During the manufacturing process, the silicon aligns itself with the starter seed and takes the exact same crystal structure of the seed.

The starter seed is then drawn out of the mixture, and a crystal grows around it, forming the beginning of the *ingot*, a 6- to 8-inch-diameter crystal. The ingot continues to be pulled from the molten vat until it reaches the desired length — about 6 feet. This ingot comes out as a cylinder, thereby giving monocrystalline cells a circular shape (at least initially).

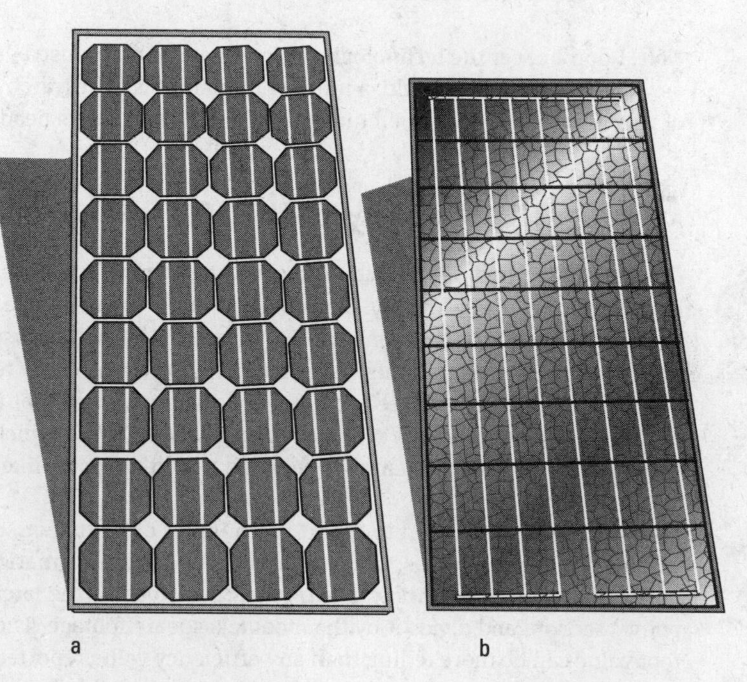


**FIGURE 6-3:** A monocrystalline cell module (a) and a multicrystalline cell module (b).

The ingot is then sliced into very thin wafers that are exposed to the diffusion process to introduce the other dopant (phosphorous). At this point, the solar cells are complete, and an electrical grid can be placed atop them (in a process that best resembles silk-screening onto a T-shirt) to effectively allow the electrons to flow. (You can think of this process like the toast you order at your favorite breakfast joint. The loaf of bread is like the ingot, the slices of bread are the wafers, the pieces of toast are the cells after doping, and the jam that makes it all worthwhile is the electrical grid.)

REMEMBER

Monocrystalline modules are typically more efficient than their multicrystalline counterparts on the cell level because the molecular structure of the ingot is uniform from top to bottom (refer to Figure 6-3). This characteristic allows the photons to move the greatest number of electrons when in sunlight because the cells are all lined up and facing the exact same direction. In a multicrystalline cell, the crystals have various shapes and point in different directions, slightly reducing the efficiency.

Monocrystalline cells are circular when they start out their lives, but because PV modules are rectangular in shape, the cells need to be squared off in order to fit into the module. Because making the circular cells into perfect rectangles would result in a high amount of waste, the manufacturers cut corners off the cells and square the edges to create octagons. The resulting octagonal cells are then capable of being packed into a module frame more densely than if they'd remained circles, thereby reducing the amount of dead space in the module. As you can see in Figure 6-3a, the octagons allow the cells to be placed closely together but not right next to each other, creating the dead space within the monocrystalline modules.

TIP

Most manufacturers use a white back sheet so you can immediately spot monocrystalline modules due to the small amount of white space at the corners of all the cells.

### The multicrystalline kind

Multicrystalline cells are manufactured differently than monocrystalline ones — the ingots are essentially brick shaped or cubes rather than cylinders. The manufacturing process begins with a vat of molten silicon-boron mixture, but instead of pulling a crystal out of this mixture, the mixture is formed in a cubic crucible, which results in the silicon cooling and forming multiple crystals. After the silicon has cooled and the ingot is sliced into thin wafers, the dopant (phosphorous) and electrical grid are added to the modules.

REMEMBER

The efficiency of multicrystalline modules is reduced due to the many crystal structures in the cubes. When the photons strike the cells, they have a more difficult time knocking the electrons free thanks to the many different surfaces present. On the plus side, the cells can be made into squares or rectangles very easily, a fact that allows the multicrystalline modules to have their cells packed one next to the other with very little space between them (refer to Figure 6-3b). The end result is that multicrystalline modules have power ratings per unit area that are similar to that of their monocrystalline counterparts even though they're less efficient on the cell level.

## Looking at thin film modules

The phrase *thin film module* is a catchall for a number of different PV technologies. Saying to a bunch of solar-technology junkies that your PV module is thin film is akin to telling dog people that you have a dog. Most folks interested in solar technology immediately want to know what kind of thin film product was used in the module, just like most dog lovers want to know what kind of dog you have.

Thin film technologies vary in their raw materials and exact manufacturing processes, and I show you some of the technologies in this section. However, at the most basic level, all thin film technologies deposit a material that can produce the photovoltaic effect onto a backing material (called a *substrate*). This substrate can be a sheet of glass, PVC roofing material, or even a foil sheet. The name *thin film* implies only that the material on the substrate is extremely thin, ranging from just nanometers to a few micrometers thick. As a point of reference, a human hair is approximately 100 micrometers thick, and crystalline solar cells are approximately 250 micrometers thick — the equivalent of two and a half human hairs.

In the following sections, I focus on the thin film modules you'll see most often. (One interesting up-and-coming module type is the hetero-junction module, which uses both crystalline and thin film technologies; see the nearby "The best of both worlds: Hetero-junction modules" sidebar for details.)

## Amorphous silicon

*Amorphous silicon* (aSi) is an extremely prevalent type of thin film module. It's based on a silicon technology that involves depositing silane gas on a substrate. One of the advantages of aSi is its ability to be incorporated on nonrigid substrates (such as a flexible vinyl sheet or PVC roofing materials), which allows aSi PV modules to become part of the roofing material. Using aSi has become a popular method for large commercial flat roofs, despite the fact that aSi has a reduced power-per-unit-area value (approximately 50 percent to 60 percent less than crystalline modules). But by incorporating it into the roofing material and installing it on very large roof areas, aSi can become an attractive option, both in terms of aesthetics and overall cost.

REMEMBER

An aSi module can be manufactured in a variety of forms to help accommodate the specific application. For example, they can be made to stick right on a metal roof or become part of the roofing material used on commercial roofs. Even though aSi modules aren't as power dense as the crystalline kind, they're able to use sunlight when it's at lower light levels, like in the early morning and late afternoon, which means aSi modules have an increased energy output compared to crystalline modules.

## Cadmium telluride

*Cadmium telluride* (CdTe) modules are another currently available type of thin film technology. To construct CdTe modules, a very thin layer (I'm talking mere micrometers thick) of CdTe is deposited on the substrate.

REMEMBER

One method is to deposit the CdTe directly onto a glass substrate. To protect the cells, a second glass layer is adhered to the first. This glass-on-glass process means the PV module can be used in place of conventional windows, allowing some light to penetrate the building while still producing electricity.

One advantage of CdTe modules is that the costs of the raw materials used are relatively low, a fact that allows the technology to be price competitive. Yet one of those raw materials, tellurium, is a rare earth element. If CdTe technology takes off, the availability of tellurium may become problematic.

## THE BEST OF BOTH WORLDS: HETERO-JUNCTION MODULES

One manufacturer is using crystalline technology in conjunction with amorphous silicon (aSi) thin films. The result is a hetero-junction module that provides a greater power output simply by taking advantage of the strengths of the individual technologies. The aSi surrounds the crystalline cell in ultrathin layers and helps increase the power output.

Hetero-junction modules, which are typically used in high-visibility applications such as patio covers or solar awnings, where aesthetics are a primary concern, use reflected light around the module to help boost the power output. They're a premium product, and their price reflects that.

### Copper indium gallium diselinide

Another type of thin film module, *copper indium gallium diselinide* (CIGS), uses four different raw materials. The substrates used can be either flexible (like those used for aSi modules) or rigid (like glass). CIGS technology has been used in the past and has recently regained popularity because it can be manufactured with nanometers (0.000000001 meters) of material compared to some of the other thin films that require micrometers of material (.000001 meters). This extreme difference allows CIGS modules to use far less raw material, helping reduce manufacturing costs.

REMEMBER

CIGS technology has been incorporated into a relatively new manufacturing process for commercial rooftop installations: The cells have been manufactured into cylinders and placed in tubes rather than the traditional rectangular modules. The advantage of this tubular format is that the cells can be perpendicular to the sun a greater number of hours per day, thereby increasing the energy production.

# Pointing Out Electrical Specifications on PV Modules

PV modules have the ability to produce power, and each module has its own specific current and voltage characteristics. You can find all of these specifications listed on the backside of every module you buy as well as on every specification (spec) sheet from the manufacturers. The power value is important because it lets you know the wattage the module can produce. From this power value, you can calculate the energy output for the system. The current is important so you know how many amps are running through the *conductors* (wires), and the voltage is important because it tells you of the module's potential to push the current.

In the following sections, I show you the different current and voltage specifications associated with PV modules. I also get you acquainted with some other key PV specifications that are worth knowing.

## Current specifications

The *current* that a module can produce is represented by the ampere, or amp, and is the number of electrons moving per second. As I explain in Chapter 3, amps are actually a rate, even though you don't see the time value in the units. The notation for current is typically I (think intensity), but you may sometimes see A used as well.

REMEMBER

The amount of current flow is directly proportional to the size of the individual solar cells within the modules, so the larger the cells, the more current you get. The amount of current flow is also dependent on the intensity of the sunlight. As the intensity increases, the electrons move faster, but the modules can only ever deliver the amount of current that the sun allows, which means they're inherently *current limited*.

Two current values are reported on PV modules and spec sheets: short circuit current ($I_{sc}$) and maximum power current ($I_{mp}$). I explain what these values mean in the sections that follow. Later in the chapter, I graph these values, and in Part 3, I show you how to use the module-specific data to properly size the wires connecting the system.

### $I_{sc}$

*Short circuit current*, abbreviated $I_{sc}$, is the value achieved if the positive and negative wires on a PV module come into direct contact with each other and there's essentially no resistance between the positive and negative sides on the module. The definition of this condition is that voltage equals zero because no potential between the two sides of the cell exists; the electrons are just trying to do what's natural to them, which is to go to the electron holes. The $I_{sc}$ values you see reported by manufacturers all occur at standard test conditions (STC; I describe these later in this chapter).

WARNING

*Shorting* the conductors (connecting the positive to the negative) won't harm the module while the current is flowing, but it could potentially harm you and the equipment if you don't disconnect the conductors properly or if you try to short more than one module at a time. (I present safety concerns and how to properly protect yourself in Chapter 15.)

REMEMBER

You use the $I_{sc}$ value whenever you need to calculate the minimum size of the conductors connected to your system (see Chapter 13). You must size conductors so as to satisfy the electrical codes, mainly the *National Electrical Code®* (*NEC®*), but not necessarily to ensure maximum system performance. The $I_{sc}$ value also dictates the ratings of other components you connect to the PV modules and array, such as overcurrent protection devices and disconnects.

### $I_{mp}$

*Maximum power current* ($I_{mp}$) is very similar to the maximum power voltage (see the "$V_{mp}$" section later in this chapter). It represents the current value when the module is producing the maximum amount of power possible. The value listed by the manufacturer is at the STC that I describe later in this chapter and is variable off of that. The current value varies greatly based on the intensity of the sunlight striking the module.

Use the $I_{mp}$ value when you need help maximizing the performance of the system. Note that this current value is often used in conjunction with $V_{mp}$ to size the conductors in the system to maximize the power output from the modules and array. (The calculations and methodologies are in Chapter 13.)

The $I_{mp}$ value is represented in a number of different ways, including $I_{pm}$, $I_{pp}$, and $I_{mpp}$. Typically, a manufacturer determines the notation it wants to use for the maximum power values and applies that notation to both the current and voltage maximum power values.

## Voltage specifications

The *voltage* from a PV module is the measurement of how much push or potential is available to move the electrons. Voltage by itself doesn't achieve any useful work — you need current present for that — but it's necessary so PV modules can have the ability to push electrons somewhere.

Traditionally, voltage is represented in equations such as Ohm's Law as E because the volt was used as the unit of measurement for electromotive force. Nowadays you see voltage represented as V, but you may also see some European companies list voltage with the designation U.

PV modules produce voltage as soon as they're placed in the sun. Each solar cell contributes some voltage — for example, a mono- or multicrystalline cell produces approximately 0.5 maximum power voltage ($V_{mp}$) when it's in the sun. In addition, it takes very little sunlight to achieve full voltage from a module. A module that isn't in the sun still has voltage present, but it can't produce power because current doesn't flow (meaning electrons don't move) unless the module is in the sun.

Two voltage values are on PV modules' listing labels and in spec sheets: open circuit voltage ($V_{oc}$) and maximum power voltage ($V_{mp}$). Later in this chapter, I show you where these values are in relation to each other on a graph. In Part 3, I show you how to apply these values when designing PV systems.

### $V_{oc}$

*Open circuit voltage*, abbreviated $V_{oc}$, is the voltage value in the absence of current flow. Resistance between the positive and negative wires on the PV module is infinite because the circuit is open. (*Resistance,* as I explain in Chapter 3, refers to resisting the flow of current.) The $V_{oc}$ value listed on a PV module is at STC; it varies based on environmental conditions (I explain STC and environmental conditions later in this chapter).

REMEMBER

When calculating the number of modules to use in conjunction with specific equipment, you must use the $V_{oc}$ value to determine the maximum circuit voltage.

### $V_{mp}$

*Maximum power voltage* ($V_{mp}$) is, not surprisingly, the value where the PV module produces the greatest amount of power. Just like the $V_{oc}$, this value moves, particularly in response to the temperature of the cells. $V_{mp}$ is often referred to as the *operating voltage* of the modules, although the true operating voltage is generally lower than this value due to the loss of voltage as temperature increases. The $V_{mp}$ value you see is always reported at STC.

REMEMBER

The $V_{mp}$ value is important because this voltage is associated with the current flowing from the PV modules. When thinking about $V_{mp}$, you need to keep two things in mind: Voltage drops as temperatures rise, and PV modules always need to have enough voltage (push) to keep the current flowing. These facts mean you must adjust the $V_{mp}$ value for high temperatures to make sure the modules are arranged such that they always have enough voltage present to push the current, regardless of the temperature.

TIP

One important point to note about the maximum power voltage is the inconsistency in the terminology. Somehow the PV manufacturers of the world can agree on testing conditions, but they can't agree on the terms used to report the data they collect. Maximum power voltage is commonly represented as $V_{mp}$, but you'll also see it listed as $V_{pm}$, $V_{pp}$, $V_{mpp}$, and $V_p$. All of these variations really say the same thing; they just use different notations to say it. (Thankfully, I've never seen a spec sheet list the open circuit with anything other than the $V_{oc}$ notation, so regardless of what the various manufacturers use for the maximum power voltage notation, you have one constant when checking a module's electrical specifications.)

## Maximum power point

The *maximum power point* (MPP) is the product of $V_{mp} \times I_{mp}$ (refer to the power equation in Chapter 3), and the units associated with it are watts (W). MPP is listed by every manufacturer at the STC I describe later in this chapter and is typically the numerical value in a module's name. The MPP of an individual module is often referred to as the *rated power output* because that's the amount of power the module can produce at the environmental conditions at which the modules are tested.

REMEMBER

The MPP is important because it's used to determine the following (I show you these relationships in Part 3):

- **The rated power output of the entire array and the components you're going to connect to:** The array's rated power output is the number of modules used in that array multiplied by the individual module's MPP value. You then use this value in relation to the inverters you connect to the array. For example, if you have ten modules that are each rated at 200 W, the array's rated power input is 10 × 200 = 2,000 W, or 2 kW.
- **The expected energy production of the PV array:** You start with the MPP value for all the modules and apply system losses and local environmental conditions to help compute the number of kilowatt-hours the PV array will generate.

Because the MPP is a product of voltage multiplied by current, it represents a variable value. The value listed by the manufacturer will rarely, if ever, be seen on operating meters because the module's voltage changes based on temperature and because the current changes based on irradiance. The likelihood of both the current and voltage being at STC when the modules are operating in the field is slim. Nonetheless, MPP is an important value because it provides a reference point when you're comparing multiple modules.

## Voltage temperature coefficient

As I note in the earlier "$V_{mp}$" section, a PV module's voltage is related to the temperature of the cells within the module. This relationship is considered to be inversely proportional because as the temperature increases, the voltage decreases. Likewise, as the temperature decreases, the voltage increases. This is an important consideration because PV modules are exposed to some of the most extreme temperatures, and their voltages react based on those temperatures.

REMEMBER

The change in voltage due to temperature is a linear relationship, meaning the change happens in the exact same increments regardless of the temperature. The amount of change is referred to as the *voltage temperature coefficient.* This value is often reported in terms of a percentage per degrees Celsius (%/°C), and it tells you that for every degree change in Celsius, the module's voltage changes by a corresponding percentage. I provide some examples of voltage coefficient values in the later "Environmental effects on standard test conditions" section.

Ideally, the modules you're working with have a small voltage temperature coefficient, meaning the voltage changes very little with changes in temperature. You can't do anything to change this value, so you need to consider it when designing your systems and specifying modules.

REMEMBER

Most crystalline-based modules (mono- and multi-) have very similar voltage temperature coefficients with most manufacturers reporting values that are comparable to their competition's. This is due to the nature of the cells and how they react to temperature. Thin film modules, on the other hand, typically have smaller voltage temperature coefficients than their crystalline counterparts, so they don't lose as much voltage in high-heat conditions.

## Power tolerance

All module manufacturers share what the power output of their modules are under STC (which I describe later in this chapter). They also tell you what the *power tolerance* of that module is, which is basically how close they guarantee to come to hitting that value. The manufacturers monitor the production process so they can accurately predict each module's power output. Because the manufacturing process inevitably creates some inconsistencies, the manufacturers guarantee that their modules' output will be within a certain percentage of the rated power output.

Many manufacturers offer power tolerances that are 0% to +3%. This says to the consumer, for example, that a module rated at 100 W will produce 100 W to 103 W. Again, this is at STC, so it's highly unlikely you'll ever see that 100 W thanks to system losses (I cover such losses in Part 3), but at least the starting point is accurately represented.

## Series fuse rating

The last major specification you should be aware of is the *series fuse rating*; it represents the largest overcurrent protection device (the fuse or circuit breaker) that can be placed on any *series string* (which is made by connecting the positive wire from one PV module to the negative wire of the next PV module). The series fuse rating for PV modules is typically 10 A or 15 A (occasionally it's even 20 A). The exact value has an effect on the wiring methods and overcurrent protection devices used. I touch on series strings in Chapter 3 and reveal how to properly size overcurrent devices in Chapter 13.

# Surveying Test Conditions for PV Modules

PV modules have to undergo some very stringent tests and certifications before they ever reach the consumer (or the installer, for that matter). An Underwriters Laboratory (UL) test protocol, UL1703, covers PV modules and requires testing of both the electrical and mechanical portions of each and every PV module.

REMEMBER

Any module you want to consider purchasing or installing should have UL1703 certification.

As part of the UL certification process, the PV manufacturer has to provide prescribed information on the module itself by applying a label on the back. Plan to use the information on the electrical characteristics most often. Other high-value information, such as mechanical data, isn't always found on the listing label, but you can usually track it down on the module's spec sheet.

After you're familiar with the different electrical specifications (which I cover earlier in this chapter), you can appreciate the consistent baselines that allow you to more easily compare the performance characteristics of different modules. I cover these baselines, as well as the effects of various environmental conditions on them, in the sections that follow.

## Standard test conditions

REMEMBER

All PV module manufacturers test their modules under *standard test conditions* (STC). The three main elements to the STC are: cell temperature, irradiance, and air mass — all of which are variable conditions that the PV modules will be exposed to after they're installed. Because these conditions affect the modules' power output, PV manufacturers had to establish a value for each of these elements that everyone could test to and report their results. Those standard values are as follows:

- **Cell temperature:** The STC for cell temperature is 25 degrees Celsius or 77 degrees Fahrenheit. (Note that it's *cell* temperature, not *air* temperature.) When a PV module is operating in the sun, it typically gets much hotter than 25 degrees Celsius. Depending on the location and the way the module is mounted, cell temperatures of 75 degrees Celsius aren't uncommon when the modules are in full sun.
- **Irradiance:** *Irradiance,* simply stated, is the intensity of the solar radiation striking the earth. The STC value for irradiance is 1,000 watts per square meter ($W/m^2$). Irradiance values vary from 0 $W/m^2$ to 1,250 $W/m^2$. The 1,000 $W/m^2$ value represents full sun, or *peak sun,* which is common to many terrestrial locations. (Flip to Chapter 4 for an introduction to irradiance.)
- **Air mass:** Air mass is a representation of how much atmosphere sunlight must pass through to strike the earth. The STC value for air mass is 1.5 (AM 1.5). Actual air mass values vary widely depending on one's location on the globe, the time of year, and the time of day.

Of these three elements, you should concern yourself with compensating for differences in cell temperature and irradiance values because these two variables directly and measurably affect a PV module's voltage and current in the following ways:

- **Voltage:** The higher the cell temperature, the lower the module voltage — and vice versa.
- **Current:** The higher the irradiance value, the more current is pushed through the module.

REMEMBER

You maximize a PV module's electrical output by keeping the module as cool as possible and pointing it directly at the sun as much as possible.

## Environmental effects on standard test conditions

When you're in the field and need to verify a module's power output, you must be able to apply voltage temperature coefficients (which I present earlier in this chapter) and calculate the module's current output based on the irradiance received at that time; the following sections introduce the info you need to know. (I show you how to apply these concepts when sizing a module to an inverter and charge controller in Part 3.)

### Temperature

WARNING

You need to be able to calculate the change in voltage due to cell temperature because every electrical component connected to a PV array has input voltage requirements that must be met. Regardless of the type of PV system used, if you apply a voltage in excess of what the equipment is rated for, you may damage it beyond repair; if you don't send enough voltage, the electronics will shut down. The excessive voltage is considered a safety concern, whereas the lack of voltage is considered a performance concern.

The 2008 version of the *NEC®* (which is the most recently updated version) states that if the manufacturer supplies the voltage coefficient data, you must use that information to calculate the temperature-adjusted voltage. If the manufacturer doesn't supply this information, you can use a table, which is included in Article 690.7 of the *NEC®*, to calculate it. The *NEC®* offers no provision for a loss of voltage due to high temperatures, but this consideration is still important to keep in mind. One of the worst calls you can get from a client is the one where she tells you her system shut down on the sunniest day of the year because the voltage loss was too excessive. In Chapter 11, I walk you through the steps you need to take to account for voltage loss in high-temperature conditions.

REMEMBER

There are two distinct voltage coefficient values, one for $V_{oc}$ and one for $V_{mp}$. The coefficient value for $V_{oc}$ is always less than the coefficient for $V_{mp}$. After all, there's a greater change in the voltage when the module is operating ($V_{mp}$) than when it isn't doing any useful work ($V_{oc}$).

When working with a crystalline PV module, check the module's spec sheet for the $V_{oc}$ and $V_{mp}$ values. If they're not there, the manufacturer should be able to supply them because they're known quantities; you just have to ask either the person you're buying the modules from or the manufacturer for this information. If the information isn't readily available, you can use the following values and be within reason (unless of course you're working with thin film PV; in that case, you have to obtain the data from the manufacturer and can't use these values):

- Temperature coefficient for $V_{oc}$ = –0.35%/°C
- Temperature coefficient for $V_{mp}$ = –0.5%/°C

TIP

Notice the negative number in front of the coefficient; it represents the inverse relationship. So as the temperature increases (a positive change in temperature), the voltage decreases. As the temperature goes down (a negative change in temperature), the voltage increases.

REMEMBER

Your reference temperature is always going to be the STC temperature of 25 degrees Celsius. So when you calculate the change in voltage, you have to compare the cell temperature in that scenario to 25 degrees Celsius. For example, if you want to find out what percentage change the module's $V_{oc}$ will have when the cell temperature is 15 degrees Celsius, you need to subtract the STC temperature from the cell temperature:

Cell temperature – STC temperature = Difference in temperature

15°C – 25°C = –10°C

You now know that the cell temperature is 10 degrees Celsius less than the STC temperature, which means you can use this information in conjunction with the given temperature coefficient to determine the percentage change.

Difference in temperature × Voltage temperature coefficient = Percentage change in voltage

–10°C × –0.35%/°C = +3.5% change in rated $V_{oc}$

I go into more detail on this in Chapter 11, so jump there to see how to use this calculation when sizing a complete PV system.

### Irradiance

The change in irradiance directly affects a module's current output. You may need to verify the performance of a single module or an entire array. By measuring the irradiance value and the current output of the module (or array), you can compare the two to see whether the modules are operating as expected. Without knowing the exact irradiance measurement, you can't accurately determine whether the current values are reasonable.

To calculate this change (which is an important task when *commissioning,* or turning on, an array; see Chapter 18), you first need to know what the irradiance value is. You can figure this out by using an irradiance sensor that points in the same direction as the module (or array). If you're using a hand-held sensor, place the bottom of the sensor on the frame of the module and read the digital display to see the irradiance value at that moment.

Because the module's output is directly proportional to the irradiance and the STC value for irradiance is 1,000 W/m², you can estimate the percentage change of current by measuring the irradiance and dividing that by 1,000 W/m². For example, if the measured irradiance is 650 W/m² and you want to know what the effect is on the current value, then you'd perform the following calculation:

$$650\ W/m^2 \div 1{,}000\ W/m^2 = 0.65$$

The current is therefore 65 percent of the module's STC value. The irradiance sensor reports the values in terms of W/m², and because the STC is 1,000 W/m², this makes the conversion easy on you. When you're in the field, you can expect to see irradiance values at or below 200 W/m² and up to 1,250 W/m². Regardless of the value you see, you can simply multiply the current value at STC by the percentage you calculate (as I show you in this section) to determine the estimated current value under those irradiance conditions.

## Relating Current and Voltage in IV Curves

As a PV system designer and installer, a graph that you'll soon come to know and love is the current-versus-voltage curve, also known as the *IV curve.* (I stands for current, and V stands for voltage.) The IV curve graphically represents the relationship of a module's current and voltage. See Figure 6-4 for a sample of a typical IV curve at STC, which is the most common way manufacturers represent the curve (I cover STC in the earlier "Standard test conditions" section).

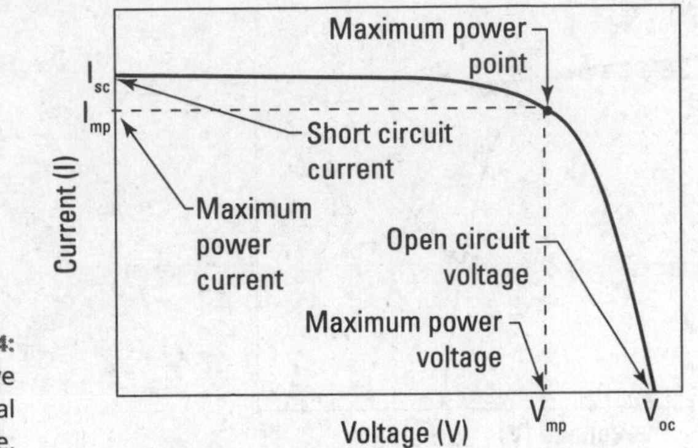


**FIGURE 6-4:** A typical IV curve for an individual PV module.

Following are the five major electrical characteristics graphed in Figure 6-4:

- **MPP:** Where current multiplied by voltage results in the highest power output
- $V_{oc}$: Where the voltage is at its maximum for a particular temperature and current equals zero
- $V_{mp}$: Where the voltage value intersects the MPP
- $I_{sc}$: The point on the curve that represents a voltage value of zero and the amount of current that would flow if the positive and negative wires were placed in direct contact with each other
- $I_{mp}$: The point where the current value intersects the MPP

You can also use an IV curve to show how environmental conditions such as the temperature of the solar cells and the intensity of the sunlight affect a PV module's current and voltage output, as I explain in the next sections.

## An IV curve with varying temperature

As I mention in the earlier "Temperature" section, a module's voltage decreases as the cell temperature increases. This is a very linear and measurable effect. In Figure 6-5, I show how a typical PV module's IV curve varies with changing temperatures. Notice how the curve moves left and right along the horizontal voltage axis but moves relatively little along the vertical current axis. This movement in the curve results in the MPP moving along with the voltage, allowing you to see that the temperature changes directly affect the voltage output, which in turn directly affects the power output.

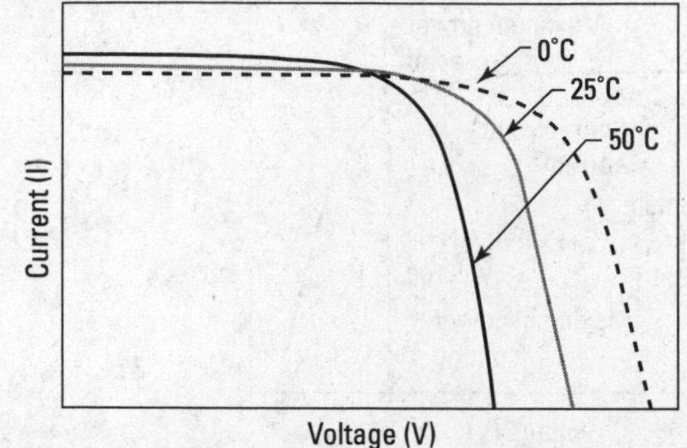


**FIGURE 6-5:** IV curves of the same module at varying temperatures.

TECHNICAL STUFF

The amount of current actually increases as the temperature increases, but the amount of that change is so much less than the change in voltage that you don't really need to consider it during the design process.

## An IV curve with varying irradiance

Irradiance varies greatly each day, starting at 0 W/m² just before daybreak and climbing higher than 1,000 W/m² at any given time. The amount of irradiance striking a PV module directly affects how much current the module can put out. Just like the relationship of voltage and temperature, this is a linear relationship that you can calculate over various conditions.

Figure 6-6 shows how the IV curve varies with changes in irradiance. Notice how the curves move in the vertical direction much more than they move horizontally. This movement indicates that the module's voltage isn't affected very much due to changes in irradiance even though the current is affected dramatically. The MPP also moves a lot due to the changes in irradiance.

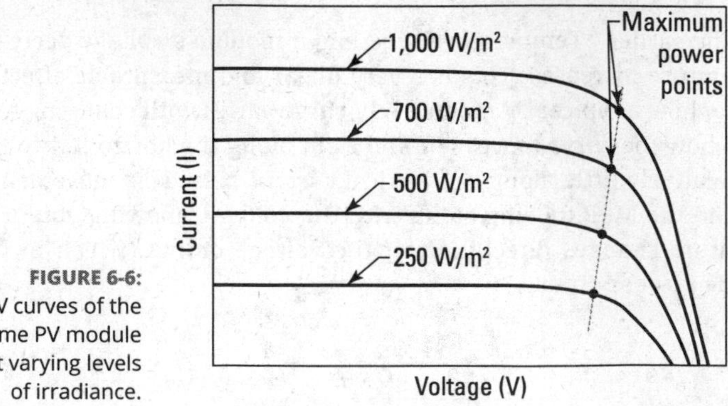


**FIGURE 6-6:** IV curves of the same PV module at varying levels of irradiance.

PV system designers generally don't consider the effect of irradiance on voltage when designing a system. Many consider the module voltage to be at full $V_{oc}$ as soon as the sun comes up and simply adjust this voltage based on temperature (I show you how to perform this calculation in the earlier "Temperature" section and in more detail in Chapter 11). This is a conservative approach that will serve you well in your own designs. It's conservative because as the sun comes up, a module's voltage jumps quickly, as much as 90 percent of $V_{oc}$ with only 200 W/m$^2$, but it isn't at full $V_{oc}$ until the intensity increases. Using this conservative method allows you to build a safety buffer into your system design and installation.

IN THIS CHAPTER

» Taking a peek at the guts of a battery and discovering what makes one tick

» Getting to know the various battery technologies

» Becoming familiar with the concept of capacity

» Knowing what to consider when selecting batteries for a client

# Chapter 7
# The Basics of Batteries

So your clients want to store energy? For PV systems, the need for energy storage leads most people to battery banks. With the exception of petroleum-based fuels, batteries are currently the most practical and cost-effective way to store enough energy for human use. Fortunately, you can incorporate batteries into virtually any PV system.

- If the utility grid is present, you can design a utility-interactive, battery-based PV system, also known as a battery-backup PV system. In this system, batteries are used only when the utility grid fails.
- In the absence of the utility grid, you can design a stand-alone (or off-grid) system. Batteries in this system are used when the PV array can't provide all the power draw for the loads at any specific time (cloudy weather, nighttime, or if a number of loads are running during the day). Stand-alone systems can actually go a step further and incorporate multiple battery-charging sources in addition to the PV array (think wind- or water-powered generators and fuel-based AC generators).

In this chapter, you get to see the types of batteries being used in PV systems. Here's your chance to explore their basic functions and features and get to know the fundamental criteria used to choose battery banks for PV systems.

TIP

Even if you think you'll never install a battery throughout the course of your PV career, you should still know all about the design and installation components regarding batteries. Offering this service will increase demand for your work because clients will see you as a true expert.

# The Fundamentals of Battery Anatomy and Operation

When I talk about batteries, I like to refer to them as buckets of energy. Batteries are a convenient way to store electrical energy for people to use at will. More accurately, batteries are a group of cells that store electrical energy and, through a chemical reaction, can deliver power to loads. In the following sections, I introduce you to the typical components and operation of the batteries used in PV systems.

## Constructing a battery, from cell to bank

The fundamental construction of individual batteries and how they're grouped together is very similar to the solar cells, PV modules, and arrays I describe in Chapter 6. For batteries, the most basic portion is the battery cell. Battery manufacturers take these battery cells and make batteries of various electrical and physical characteristics. As an installer, you take these individual batteries and make a battery bank to serve the needs of your client.

### Going cellular

Just as solar cells are the basic building blocks of PV modules and arrays (as I explain in Chapter 6), battery cells are the basic building blocks of battery banks. Each battery cell has a specific voltage potential; this voltage value is dependent on the battery technology but independent of the cell's size. The amount of energy stored is directly proportional to the cell's size, though; in other words, the bigger the cell, the greater the amount of stored energy or *capacity* (I cover capacity in greater detail later in this chapter).

Battery cells designed for PV systems are configured by the battery manufacturer to produce a desired voltage and store a desired capacity. Every cell has either a single plate with positive and negative sides or multiple positive plates and negative plates together within an individual compartment. The negative plates are separated from the positive plates by a special electrically insulating material. Each plate holds the *active material,* which is the material that reacts in the chemical process I describe later in this chapter. These plates are then immersed in an electrolyte, either a liquid or a gel, that allows the charged particles to flow between the positive and negative plates when given an electrical path in which to move, such as a DC load connected to the batteries or an inverter that's running AC loads.

If multiple plates are used in a cell, all the negative plates are connected together with a metallic bar, and all the positive plates are connected together with a similar metallic bar, creating parallel connections within the cell. Depending on the desired effect, many thin plates or a few thick plates may be together in a cell.

REMEMBER

- Multiple thin plates allow for an increased amount of surface area of the active material in contact with the electrolyte, creating a greater amount of current flow in a short period. These batteries work well as starting batteries (like the one found in your car).
- A few thick plates allow a battery to move from charged to discharged many times without damage (unlike batteries with multiple thin plates). If your system design calls for a battery to run electrical loads continuously, you need a *deep-cycle battery* (also known as a *deep-discharge battery*). A deep-cycle battery's thick plates allow it to deliver nearly all the energy stored and then be recharged fully. All the batteries used in PV systems are deep-cycle batteries because they're expected to deliver a majority of their capacity many times in their lives.

You can see multiple plates in a battery cell in Figure 7-1.

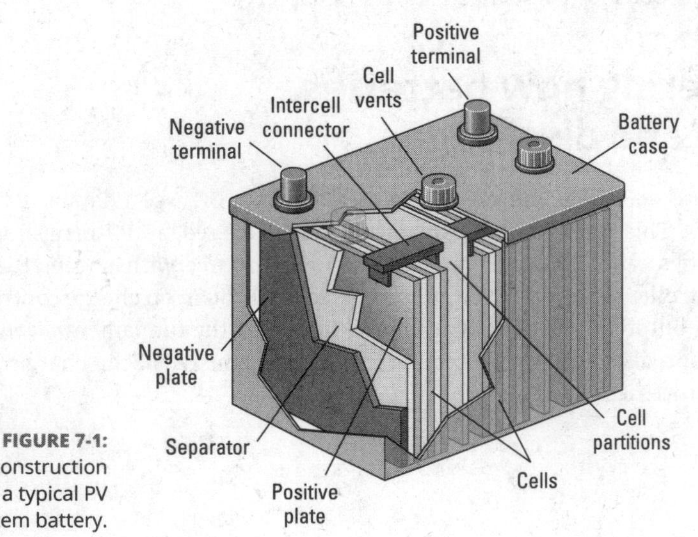


**FIGURE 7-1:** The construction of a typical PV system battery.

### Making a connection to create a battery

To create a battery, multiple battery cells are connected together within a case. The exact number of cells connected together varies by manufacturer, but the most common configurations result in 6 V and 12 V nominal batteries. The *nominal* designation indicates that the batteries are close to that voltage but can vary (I tell you about nominal voltages in Chapter 3).

The exact voltage and capacity of a battery depend on the technology used, but you can gain some quick knowledge just by looking at the top of a battery case. Each cell has its own vent, so if you look at a battery and see three vents, you know that three individual cells are in that battery.

### Taking it to the bank

Typically, individual batteries don't have the required voltage and capacity values necessary for an entire PV system, which is why you must connect them to create a *battery bank* (the total number of batteries used). By wiring batteries in series, you create a *string of batteries,* which allows the voltage to increase while the capacity remains constant. This concept is identical to the PV series string connections I show in Chapter 3.

If you require additional capacity, you can create multiple strings of batteries and place them parallel to each other. Doing so allows you to keep the voltage constant while increasing the overall capacity of the battery bank. (I explain how to properly size a battery bank later in this chapter.) This practice is the same as the parallel connections for PV strings I show in Chapter 3.

## Discovering how batteries charge and discharge

Batteries store energy in the form of direct current (DC; see Chapter 3 for an introduction). This is the same form of current delivered by PV arrays, which means batteries can be charged directly from a PV source with minimal losses. A charge controller makes this charging possible. I fill you in on charge controllers in Chapter 8, but in the sections that follow, I describe the fundamental concepts of charging and discharging and explain the basic chemistry of the charging and recharging processes.

## Getting a grip on the concepts of charging and discharging

REMEMBER

Deep-cycle batteries (described in the earlier "Going cellular" section) are designed to deliver nearly all of their stored energy to loads in a PV system, but limitations exist. The phrase *depth of discharge* (DOD) describes the amount of energy drawn from the battery. For most deep-cycle batteries, the maximum DOD you want to experience is 80 percent, meaning you remove 80 percent of the stored capacity from the battery. Although deep-cycle batteries are designed to deliver as much of their stored energy as possible, if the DOD frequently reaches a high value, such as 80 percent, the overall battery life will be reduced. Therefore, increasing a battery bank's overall life while delivering as much energy for the loads as possible becomes a balancing act that you, as the PV system designer and installer, must master.

TIP

The opposite term for DOD is *state of charge* (SOC), and it reflects the amount of energy still stored in a battery. So if I say a battery has an SOC of 20 percent, then I'm telling you that the battery has delivered 80 percent of its energy and has 20 percent of it left. This is another way of saying the battery has an 80-percent DOD. Because these terms are the opposite of each other and PV folks refer to them most often in percentages, a handy trick for remembering the relationship is that SOC + DOD = 100%.

All batteries designed for use in a PV system are rated by their voltage and capacity as well as by how many cycles they can deliver that energy. A *cycle* is when the battery goes from being fully charged to fully discharged and then recharged again. The exact number of cycles a battery can undergo during its life is affected by a number of factors, including temperature and how fast the battery is being discharged (these factors also affect the battery's capacity, as I explain later in this chapter). Typically though, battery manufacturers relate the number of cycles a battery can withstand against the DOD. So, logically, the greater the DOD a battery experiences, the fewer cycles it'll experience.

## Understanding basic charge and discharge chemistry

Batteries are able to deliver electrical current to loads in a PV system through a chemical reaction that's contained within that battery. The exact chemical reaction is dependent on the type of battery used and the materials used to push the electrons around. You don't need to get too hung up on the chemical reaction I show in Figure 7-2 (unless you do chemistry in your spare time for fun). All you need to take away is how the electrons move from one side to the other depending on whether the battery is charged or discharged.

**FIGURE 7-2:** The chemical reaction that occurs when discharging and charging a battery.

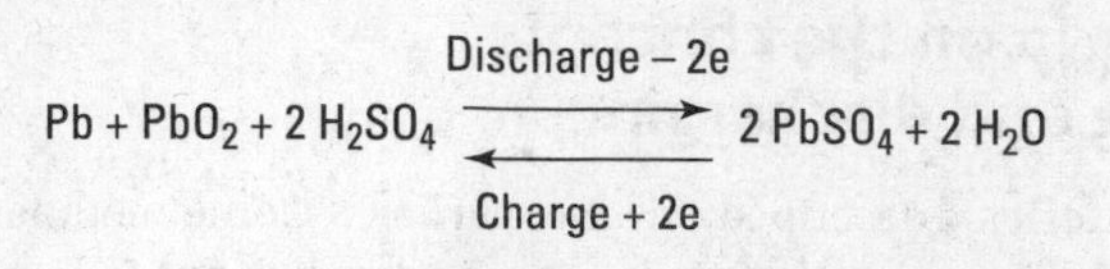

$$Pb + PbO_2 + 2\,H_2SO_4 \underset{\text{Charge} + 2e}{\overset{\text{Discharge} - 2e}{\rightleftarrows}} 2\,PbSO_4 + 2\,H_2O$$

On the left side of Figure 7-2 is a chemical equation for a fully charged battery composed of lead, lead oxide, and sulfuric acid (this type of battery is a *lead-acid battery;* see the later related section for details). As the battery is discharged, the lead and lead oxide become lead sulfate, and the sulfuric acid becomes water. So a fully discharged battery contains lead sulfate and a very watery acid solution (see the right side of Figure 7-2). The electrons don't go away; they merely move to a different form and are stored in the lead sulfate.

During the charging cycle, the lead sulfate breaks up, lead oxide re-forms, and the water becomes a strong acid again. Hydrogen gas is also released into the atmosphere, although the amount of gas released varies greatly among battery technologies. As for the electrons, they move similarly to the way electrons move in solar cells (see Chapter 6). During the charging process, current is applied into the batteries to force the electrons back. When the electrons are back where they started, they look for a way to return to the other side. You provide this path with a load, and the whole process repeats.

In theory, the recharging portion of the battery cycle should remove all the lead sulfate that has accumulated on the plates and return it to the sulfuric acid solution. However, that's not the reality. In real life, the plates become corroded as batteries age. In the case of lead-acid batteries, that corrosion is in the form of sulfation, and it reduces the batteries' ability to store as much energy as they did when they were new. As a result, lead-acid batteries become discharged quicker and appear to recharge quicker, but they're really not performing at their optimal effectiveness.

REMEMBER

The best way to prevent corrosion in a battery is to follow the charging cycles specified by the manufacturer when you set the charging levels for the different battery-charging sources (PV array, utility, and generator).

WARNING

Batteries aren't smart creatures by any means, and they can't differentiate between wires connected to their terminals and a load or wrench that was accidentally dropped across their terminals. If you ever drop a wrench across a battery's terminals, the battery will short-circuit and discharge very quickly because the electrons start flowing through the wrench from the negative plates directly back into the positive plates. Ideally the only damage will be to the wrench and the battery, but if you get caught in the path, you're in serious danger. To avoid this scenario, only use tools that are specifically made for use on batteries. You can buy tools that have specially insulated handles to protect you from connecting the positive and negative terminals.

### Charging correctly

As an installer, you need to make sure the pieces of equipment used to charge the battery bank — the charge controller and inverter — are set up so that the charging voltages and current values are correct. These values depend on the battery technology, overall battery bank size, and the battery temperatures. Consult the battery manufacturer's requirements and the charge controller and inverter installation manuals to set the charging points correctly, and flip to Chapter 8 to see the steps a battery goes through when being recharged.

# Comparing Different Types of Batteries

Even though a number of battery technologies are available commercially, relatively few are found in PV systems. The next sections get you familiar with the types of batteries that are regularly used in PV systems and help you differentiate among them to make the selection process a little easier.

## Lead-acid batteries

Far and away the most common type of battery used in PV systems, the lead-acid (LA) battery has been around since the 1860s and commercially available since the turn of the 20th century. In LA batteries, the active material used on the negative and positive plates is lead and lead oxide (sometimes with small amounts of antimony added); the electrolyte that allows charges to flow in LA batteries is sulfuric acid. Each cell is approximately 2.12 V when fully charged, so the cells in LA batteries are referred to as 2 V nominal.

LA batteries are so popular in PV systems due to their relatively low cost, robust design, and ability to achieve a high depth of discharge. LA batteries are available in either flooded or valve-regulated versions; both types have their pros and cons, as you find out in the following sections.

### Flooded lead-acid batteries

As the name implies, flooded lead-acid (FLA) batteries have a liquid electrolyte inside them that floods the cells. These batteries have removable caps on each cell and require the person maintaining the batteries (typically the system owner) to check the electrolyte level regularly and add distilled water when the fluid level begins to get low (I explain how to perform this task in Chapter 18). FLA batteries work well in stand-alone systems in which the batteries are cycled on a regular basis; they hold up better than their sealed counterparts in this environment of constant cycling because they're designed for more regular cycling and can accept maintenance charging (as I describe in this section). FLAs are much more cost effective for stand-alone clients; the additional maintenance is more than offset by the increased battery life they're able to achieve in these environments.

REMEMBER

FLA batteries require a high level of client interaction and maintenance if you expect them to last as long as possible and perform to their specifications. Why? As I describe in the earlier "Understanding basic charge and discharge chemistry" section, the chemical reaction that occurs during the charging process creates hydrogen gas and reduces the amount of water in the battery. Because flooded cells need to remain immersed in the electrolyte solution (when they're not, they become damaged beyond repair), the client must add distilled water to the cells on a regular basis. When properly maintained, an FLA battery bank should perform for ten years or more.

FLA batteries also require a periodic intentional overcharge, or *equalization charge*. This process involves intentionally increasing the charging voltage into the batteries to stir up the electrolyte solution and "scrub" the *sulfation* (corrosion) that accumulates on the plates. During this process, the batteries produce large amounts of hydrogen gas, and you smell rotten eggs. Every battery manufacturer has its own recommendation for the time between equalization charges and voltage levels.

REMEMBER

An equalization charge must always be performed in a controlled and regulated manner. If the system owner is properly trained and feels comfortable with this task, he can take it on. Otherwise, you, as the system installer, should make arrangements to provide this service on a regular basis.

WARNING

I explain how to equalize flooded batteries in Chapter 18, but you should know that before you even begin the equalizing process, you need to remove the battery cell covers to allow the extra hydrogen gas to escape easily and check that the batteries are venting to the outside (preferably through the use of a brushless fan) to reduce the possibility of fires — after all, hydrogen is a very flammable gas. Also, make sure you place the proper signs in the battery bank's location to warn others about the potential fire and explosion hazards.

## Valve-regulated lead-acid batteries

The other type of lead-acid battery used in PV systems is the valve-regulated lead-acid (VRLA) battery. Also known as a sealed lead-acid battery, VRLAs are sealed from the environment and possess no internal components that the client must maintain. The valves on the individual cells keep the batteries sealed but also allow gas to escape when necessary. As a VRLA charges, the chemical process produces small amounts of hydrogen gas and increases the pressure within the battery. As soon as the pressure reaches a certain level, the valve is pushed up, the hydrogen gas "burps" out, and the valve resets.

VRLA batteries are quite popular in utility-interactive, battery-based systems. They generally aren't cycled on a regular basis because, in most locations, the grid is relatively stable and power outages are the exception, which means the batteries are sitting in a constant state of full charge. (*Note:* VRLA batteries *should* be cycled every three to six months to help increase their life span, but they aren't hurt by sitting fully charged a majority of the time.)

Other positive reasons for using VRLA batteries in utility-interactive, battery-based systems are as follows:

- **Reduced maintenance:** VRLA batteries need less maintenance than FLA batteries because they don't require watering. Considering the fact that most homeowners have a difficult enough time remembering to change the filter in their furnace, not having to worry about adding water to a PV system's batteries is a major plus.

  The reduced maintenance aspect of VRLAs has also led to the misnomer "maintenance-free batteries." Yes, it's true that you don't have to add water to the cells, but VRLAs still need monitoring and evaluation on a regular basis if your client expects them to last within his PV system. Hearing the words *maintenance-free batteries* makes people feel as if they never have to think of the batteries again, so avoid using these words and stick with the term *valve-restricted* or *sealed.*

  To help a client with VRLAs stay on top of their performance, have him cycle the battery bank once or twice every six months and monitor the energy delivered by the system. This exercise helps the client establish the health of his battery bank and gives him an idea of the bank's capacity for the times when the batteries really need to perform.

- **Less hydrogen gas:** VRLAs don't produce as much hydrogen gas as FLAs, which can make placing them an easier task because the venting requirements are less restrictive.

- **Stackability:** VRLAs are stackable, a characteristic that allows you to use vertical space (along a wall, for example) rather than a large amount of floor space.

WARNING

With all the benefits of VRLAs, you may wonder why anyone *wouldn't* want to use them. Well, VRLAs have a few downsides to consider:

- **Excessive charging is a no-no.** VRLA batteries can't accept excessive charging voltages (unlike FLA batteries, which can). Because a VRLA battery may produce too much gas when exposed to higher voltages, its valve may not be able to effectively release the buildup and properly protect it. VRLAs therefore require careful charging per the manufacturer's specifications, as well as monitoring to make sure the proper voltage levels are maintained.
- **The expected life span is pretty short.** A properly maintained and used VRLA battery should last from five to seven years. However, I've seen VRLA battery banks go bad in less than one year. Even though FLA batteries can also be ruined, they generally have the ability to withstand more abuse and come back to life to charge another day.
- **The cost is greater.** VRLA batteries cost considerably more than their FLA equivalents, which makes them a greater expenditure now and in the (relatively) near future.

## Lead-calcium batteries

Lead-calcium batteries, which add calcium to the active material of the battery rather than antimony, are another type of battery used in PV systems, and they come in both flooded and valve-regulated varieties (see the previous sections). Lead-calcium batteries are 6 V and 12 V nominal, just like lead-acid batteries, and the capacities are dependent on the manufacturer and the physical dimensions of the battery. The addition of calcium to the active material reduces water consumption, makes the plates more resistant to corrosion, and decreases the battery's self-discharge rate.

All of these advantages come at a cost, though: Lead-calcium batteries are more expensive than their lead-acid counterparts. Also, few companies manufacture deep-cycle lead-calcium batteries at this time, which makes them less popular than the traditional lead-acid batteries.

## Nickel-cadmium batteries

Nickel-cadmium batteries (which are often referred to as NiCd batteries) have been used as rechargeable batteries for many years now. Nickel oxide hydroxide and metallic cadmium are the base materials for the positive and negative plates, respectively, and potassium hydroxide is the electrolyte in the cells. These batteries are manufactured in both valve-regulated and flooded varieties (see the earlier related sections), and they come in 6 V and 12 V nominal configurations, like the other batteries I cover in this chapter. However, they provide less voltage per cell, which means you need more cells in series in order to achieve this voltage. A 12 V nominal NiCd battery therefore has to have ten cells in series.

NiCd batteries aren't widely used in PV systems, but they do have an advantage over other battery technologies in that they can be excessively discharged with less of an effect on the overall life of the battery. This characteristic means NiCd batteries can have an increased number of cycles during their life spans.

NiCds are more expensive than other types of batteries, which is a major drawback for many folks. In addition, they react worse to cold temperatures than lead-acid batteries do, meaning they need a temperature-controlled environment to perform at their best.

If you design a system that uses NiCd batteries, be sure to adjust all the charge set-points for the chargers due to the difference in charging voltages associated with NiCds. I cover the different charge set-points in Chapter 8.

# Comprehending Battery Capacity

A battery's *capacity*, the amount of energy stored in it, plays a large role in the overall design of PV systems and the use of batteries within them. In order to select the right battery for your client's PV system, you need to have a solid understanding of how batteries' capacity values are determined and reported and how to apply those ratings to the system you're designing (I introduce you to this process of specifying batteries later in this chapter).

Batteries are rated according to the number of amp-hours (Ah) stored in them. Although this seems like an easy concept to grasp, multiple factors change this value. In the sections that follow, I describe a capacity rating called the C rate, and I explain the various factors that affect capacity.

## Considering the C rate for capacity

In order to sell batteries, manufacturers have to put labels on the sides of their batteries that list the voltage and capacity of each unit. (The voltage depends on the type of battery and the cell count inside the battery; see the earlier "Comparing Different Types of Batteries" section for more.) They can't very well say the capacity of their batteries is "around *X* Ah, depending on a number of variables," but in reality, that's precisely the case. Just like PV modules (which I cover in

Chapter 6), the characteristics of a battery vary with the environmental conditions affecting it. And just like PV modules, battery manufacturers have to pick a point at which to rate their batteries.

REMEMBER

Batteries used in PV systems are typically rated at a temperature of 25 degrees Celsius and at a discharge rate over a time period of 20 hours. This rate is referred to as the *C rate* and is notated as C/20 (C over 20). The C represents the capacity value, and the 20 represents the number of hours of discharge time. Many battery manufacturers use a C/20 rate, but some manufacturers use the C/24 value (for 24 hours) or something close to that.

Here's a numerical example to help you make sense of the C rate: If a battery manufacturer lists its battery as a 100 Ah battery at the C/20 rate, you can do the math to figure out what kind of load that represents:

$$100 \text{ Ah} \div 20 \text{ hrs} = 5 \text{ A}$$

This means that if a 5 A load were placed on that battery, in 20 hours the battery would deliver all of its stored energy and would no longer have any capacity. In reality, getting all 100 Ah out of that battery isn't realistic because the battery can't really deliver all of its capacity.

REMEMBER

As I explain in Chapter 3, amp-hours aren't really an energy measurement; that's watt-hours (Wh). *Amp-hours* are actually the quantity of electrons that flow through a circuit in a given amount of time; as you discover in the earlier "Understanding basic charge and discharge chemistry" section, the electrons are always there — they just temporarily reside in different sections. You can gauge the capacity of a battery in watt-hours by multiplying the amp-hour value by the voltage, but this isn't how battery manufacturers rate their batteries, so you have to make the conversions yourself.

TIP

The C rate can also describe the charging half of the cycle, but in this case you're talking about replacing the capacity rather than drawing it out. For example: If I say that my battery charger has a C/10 rate, I'm saying that I'll replace the entire capacity in a matter of ten hours. The best rate of recharge varies with battery technologies and even manufacturers. Finding the correct value is very important when sizing your charging sources, as I reveal in the later "Thinking about the rate and depth of discharge" section.

## Recognizing factors that affect capacity

A battery is somewhat at the mercy of its surroundings when it comes to energy delivery. As you find out in the following sections, environmental conditions dictate a battery's ability to deliver energy, as do the age and discharge rate of the battery. By understanding these factors, you can avoid costly mistakes that will require a battery replacement sooner than necessary.

## Temperature

Temperature has a very significant effect on a battery's ability to deliver and accept current.

- » A cold battery has a reduced capacity due to the fact that the chemical reaction inside the battery is slowed down and isn't as efficient.
- » A hot battery has the ability to deliver a greater number of amp-hours because the heat actually aids the chemical process. However, increased temperature ultimately reduces a battery's life, so intentionally keeping a battery hot to increase its capacity in the short term will only have the effect of reducing its life in the long term.

In a sense, batteries are a lot like humans. We're finicky creatures who want things just the way we like them, and we're pretty happy when we're in an environment that's around 25 degrees Celsius (77 degrees Fahrenheit — a pretty comfortable room temperature any time of the year). So spread the word to your clients! This is also the reason manufacturers rate their batteries at 25 degrees Celsius (as I cover earlier in "Considering the C rate for capacity"); it makes for a more impressive rating.

Ideally, the battery bank should be located in a temperature-controlled climate that's around 25 degrees Celsius. In reality, this task is difficult. When helping your client decide where to place the battery bank, aim for a location that doesn't experience extreme temperatures. A location such as an insulated garage that isn't heated or cooled is a good choice.

## Age

As a battery ages, the chemical reactions don't work as well as they did when the battery was young and fresh from the factory. The active material on the batteries becomes layered with parts of the electrolyte that didn't return into acid solution during the recharging process.

This aging process is evident in both the charging and discharging parts of a battery's cycle. Aged batteries discharge quickly due to their reduced capacity. They also seem to finish recharging quickly due to the reduced amount of active material that's actually available to the battery, allowing the battery to appear fully charged when it really isn't.

A common question asked about battery banks and aging batteries is this: Can you install a new battery into a bank of older batteries? Adding a new battery to an existing battery bank isn't a great idea because the old batteries drag the new battery down to their level in a short period, but depending on your client's reality, it may be workable. So the question becomes this: What's the state of the battery bank? If the bank is relatively young and well-maintained, the idea of adding a new battery to replace a failed unit may seem reasonable. If, on the other hand, the entire bank is on the verge of replacement in a short period, adding a new battery may not make any sense.

A bad battery makes itself known relatively quickly by dragging down the whole bank almost instantly. An aging battery bank, on the other hand, has a slow decline. When an aged battery bank can't hold a charge for a required amount of time (as dictated by what your client considers acceptable), the whole bank needs to be replaced.

## Discharge rate

A battery's capacity is directly affected by the rate at which the battery is discharged. This rate is the theoretical number of amp-hours a battery could deliver to a load. So, for example, a 100 Ah battery could, theoretically, deliver 1 A for 100 hours (1 A × 100 hours = 100 Ah) or 100 A for 1 hour before it was completely discharged.

Because a battery delivers power via a chemical reaction, the slower the reaction, the greater the battery's capacity. The opposite is also true: The faster the reaction, the smaller the capacity.

Figure 7-3 shows a typical graph of battery voltage versus energy delivered from the battery. The battery voltage is a good indicator of the battery's state of charge (high voltage means there's a high state of charge), so this graph shows the battery's capacity (how full it is) versus how much energy has been delivered. Figure 7-3 shows that a slow discharge rate (C over the number of hours) results in a greater number of amp-hours delivered from the same battery than a fast discharge rate. So in reality, that 100 Ah battery delivers more than 100 Ah if the load is only 1 A; similarly, it delivers less than 100 Ah if the load is 100 A.

As I mention in the earlier "Considering the C rate for capacity" section, battery manufacturers typically rate their batteries at the C/20 rate (or something very close to that). They then generally report the capacity values for their batteries at other C rates for your information so you can establish the discharge rate that most closely reflects your client's load profiles and use the C rate in your designs accordingly. See Chapter 12 for full details on sizing and designing a battery-based PV system.

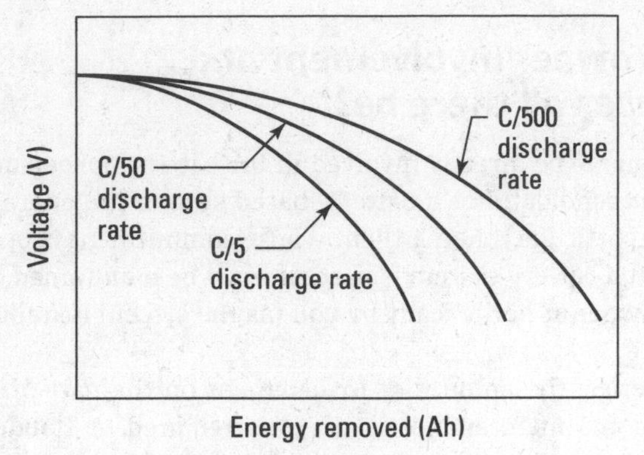


**FIGURE 7-3:** Battery capacity versus discharge rates.

# Specifying Batteries

When *specifying* (selecting) the batteries for a client's PV system, you get to take all of your battery-related knowledge and sift through the available batteries to pick the type of battery that best meets your client's needs. The sections that follow help you sort through this decision-making process.

## Specifying the type of battery to use

The first question to ask when specifying batteries is, "What's the right battery technology for my client's system?" As I show you in the earlier "Comparing Different Types of Batteries" section, the vast majority of PV systems currently being installed use lead-acid batteries. However, a lead-acid battery isn't your only option. In order to choose the right type of battery, you need to ask the key questions posed in the next sections.

### What kind of PV system will your client use?

REMEMBER

The type of system your client wants installed helps dictate the type of battery you need to use. In general, a utility-interactive system benefits more from sealed batteries than from flooded batteries because of the reduced maintenance and better ability to sit in a full state consistently. A stand-alone system tends to be a better candidate for flooded batteries (even though these require additional maintenance) because they last longer and provide more delivered energy in this environment.

## How much owner involvement and maintenance will there be?

A client who plans to be directly involved in the battery upkeep and maintenance is a much better candidate for a battery-based system in general (and a flooded battery bank in particular) than a client whose commitment is questionable from the beginning. If a battery system is used, it must be maintained, either regularly by the system owner or periodically by you (as the system installer).

For sealed batteries, the amount of involvement on the part of the owner isn't huge, assuming the maintenance is done as required. A flooded battery bank requires more attention and therefore more direct involvement from the client.

TIP

You may want to set up an ongoing service contract with clients installing a utility-interactive system. Such a contract provides for you to come in on a predetermined schedule to review the system and perform the necessary maintenance. An ongoing service contract is truly a win-win for you *and* your client because you receive a steady revenue stream for several years, and your client receives the peace of mind that his investment is being maintained as necessary. (See Chapter 18 for all the details on PV system maintenance.)

# Specifying the battery bank size

Whenever you're dealing with a battery-based PV system, one of the easiest ways to guarantee you wind up with a happy client is to make sure you size the battery bank correctly. Specifying the size of a battery bank (both voltage and capacity) is as much art as it is science because you have to take a number of considerations into account (sometimes you even have to bust out a crystal ball and gaze into your client's future). If you undersize the battery bank, your client won't be happy because the batteries will constantly be running out of power before their system can recharge the bank. If you oversize the bank, your clients may not realize it for a while, but they'll spend more money initially and may not be able to properly maintain the bank because the charging sources can't ever get the battery bank 100-percent full.

This section introduces what you should think about and why when it comes to sizing battery banks. For specific sizing instructions, head to Chapter 12.

## Calculating the required energy delivery

Your chief concern when sizing a battery bank is determining the desired amount of energy consumption. This can be a difficult, time-consuming process, but for anyone willing to make the investment in a battery-based PV system (either utility-interactive or stand-alone), the time is well spent.

WARNING

If you try to design a battery bank for a PV system without thinking about the energy consumption first, you're headed for trouble because you won't be able to accurately determine the amount of energy to store in the battery bank or the amount to produce from the PV array. It'd be like trying to buy one year's worth of gasoline at once. To do that, you'd have to think about the number of miles you drive and what the rate of consumption is based on the type of car you drive.

REMEMBER

With a battery-based PV system, you're buying all your energy at once and storing it in a battery bank. (Yes, you can recharge the battery bank, but I think you get my point.) This fact means you have to evaluate the electrical loads the client wants to connect to the battery bank. If the battery bank is for

- A stand-alone system, you need to evaluate all the electrical loads
- A utility-interactive system, you only need to evaluate the loads on the backup load center (I cover this in Chapter 2)

## Sizing to the charging source

Another consideration when sizing a battery bank is the amount of charging that's available from the main charging source, which is typically the PV system. A battery bank that's too small in comparison to the PV array won't allow the PV array to operate at its full potential, and a battery bank that's too large in comparison to the PV array will never become fully charged and will suffer in the long run.

TIP

When evaluating the PV array and battery bank together, look at the charge rate of the PV array in comparison to the overall capacity of the battery bank. Think of the PV array as your kitchen faucet and the battery bank as a glass you're trying to fill from it. If the PV array has a lot of ability to deliver current (the faucet is on at full blast) and the battery bank is small (you're trying to fill a shot glass), the PV array will quickly send a lot of current to a small battery bank (the glass has a hard time filling up effectively without spilling a bunch of water). Conversely, if the array can deliver only a minimal amount of current to an overly large battery bank, the current will only trickle in to the large battery bank. (It's like when you barely have a drip coming from the faucet and you're trying to fill a coffeepot — you're stuck waiting a long time.) Sizing the PV and battery bank properly is just like choosing the right water flow from the faucet to match the container you're trying to fill.

## Considering other charging sources

All homes and commercial properties using solar technology need to have some type of charging source in addition to the PV system. What that source is affects what you choose for your client:

- If the secondary charging source is the utility grid, the only recharging limitation is the battery charger built into the inverter. The utility grid can supply a huge amount of current, but the charger built into the inverter can allow only a limited amount of current to the batteries.
- If the charging source is another renewable energy source, such as a residential-scale wind system or a microhydroelectric system that produces power from a small creek or stream, the recharging limits are based on these resources and their availability.
- If the secondary charging source is an engine generator, recharging of the battery bank is limited to the size of the generator.

## Delving into days of autonomy

The number of days a client wants to go without needing his battery bank recharged is known as *days of autonomy.* It's a number that's defined by the client and his desires. Many people are comfortable with two to four days of autonomy. However, if the site has some critical loads that can't be without power under any circumstance, the number of days of autonomy may need to be increased into multiple weeks. The number of days of autonomy has a direct effect on the battery bank's size — the greater the number of days, the larger the battery bank needs to be.

REMEMBER

Even though the owner ultimately sets the number-of-days-of-autonomy requirement, it's your responsibility to guide him toward a number that's realistic. For most residential customers who're using a PV system in a stand-alone capacity, a few days of autonomy is adequate. Typically, a generator is included in such a system, and it can be started to help augment the lack of solar resource. For clients with utility-interactive systems, don't suggest they request too many days of autonomy. The utility typically isn't out for too long, and by building in extra days of autonomy, the battery bank may get too big and expensive.

## Thinking about the rate and depth of discharge

When designing a PV system's battery bank, you need to consider both the rate at which the battery bank is discharged and how deeply it'll be discharged. These values affect the decision-making process for battery size and should be considered independently.

As I explain in the earlier "Discharge rate" section, the rate at which a battery is discharged affects the overall capacity that the battery can deliver. Most batteries' rated capacity values are their 20-hour rate. For most residential battery-based systems, this is an appropriate value to use because residential applications tend to use a small amount of power on a continuous basis. Sure, people have brief times when they experience increased power needs, which results in an increased discharge rate, but you aren't going to put huge electrical loads such as electric water heaters and electric clothes dryers on such a system. If you do, expect a significantly higher rate of discharge on a consistent basis and plan to evaluate the battery bank at a rate other than the 20-hour rate provided, specifically a value that's less than the C/20 rate (because batteries have decreased capacities at greater discharge rates).

The depth of discharge (DOD) is the other consideration. Again, you have to look at the whole picture to help define the DOD value to use. All battery manufacturers publish charts that tell you the number of cycles expected from their batteries versus the DOD. The greater the DOD chosen, the smaller the battery bank can be because you're drawing more energy from the bank. The choice of DOD level does have its trade-offs, though. A lower DOD value (low DOD means little energy is removed from the battery) results in a greater number of overall cycles, whereas a higher DOD value results in fewer cycles.

You shouldn't necessarily design systems for an extremely low DOD. As a battery manufacturer pointed out to me once, a low DOD may increase the overall number of cycles (battery life), but it may end up reducing the amount of energy delivered. So you really need to evaluate the energy delivered over the life of the battery to get an idea of the ideal DOD to design to. For many batteries, this thinking results in a designed battery DOD of between 50 percent and 75 percent.

## Surveying the number of strings

One of your goals when sizing a battery bank is to keep the overall number of battery strings in parallel to a minimum so that all the batteries charge and discharge equally. If you use a single string of batteries, all the electrons have to flow through each and every battery in order to complete the circuit, which is great in the sense that all the batteries are charged and discharged exactly the same way. However, if you add a second string of batteries in parallel to the first, now the electrons begin to have choices along their path.

If you keep all the wires connecting two parallel strings equal in length and resistance, you won't see too many problems. In fact, two strings in parallel can have the benefit of allowing your client to limp along on half a battery bank if one battery or even one cell goes bad.

The trouble starts when you begin to get too many strings in parallel. When you have three or more strings in parallel, properly managing the connections and maintaining equal resistances becomes incredibly difficult. Therefore, you really need to keep the number of strings in parallel to a minimum.

The lower a battery's voltage, the higher its capacity. This isn't a universal truth, but it's a good guideline. Consider using a battery with a smaller voltage and a higher capacity to keep the overall number of strings in parallel to a minimum.

## Examining environmental conditions

The conditions where the batteries will be stored play a role in the battery size. If the batteries will be in a non-temperature-controlled space, such as a garage or free-standing shelter, you should evaluate (and try to minimize) the temperature swings. A cold battery can't deliver the same amount of capacity as a room-temperature or hot battery, and a fully discharged battery in frigid conditions may freeze, which can potentially ruin the battery. A battery stored where temperatures are consistently high experiences a shortened life span.

Temperature also plays a role when batteries are recharging. Fortunately, you can connect a temperature-compensation meter to charge controllers to alleviate this concern (I cover charge controllers in depth in Chapter 8). This meter tells the charge controller what the battery temperature is, thereby allowing the charge controller to vary the charging voltage and current to meet the requirements of the battery at the specific temperature.

IN THIS CHAPTER

» Looking at basic charge controller functions

» Scoping out the two main controller types: MPPT and PWM

» Selecting the option that meets your client's needs

# Chapter 8

# Keeping Current and Voltage in Check: Charge Controllers

The process of naming equipment in the PV world (and the electrical world in general) can be classified as unimaginative at best. This is unfortunate in the sense that the components don't sound like some space-age, whiz-bang item. But it's actually fortunate in that the names are generally very descriptive, allowing you to determine an item's function quickly.

Charge controllers are a prime example. When introducing charge controllers in my classes, I always give my students one guess to name their primary function. Needless to say they always guess "to control charging." Even though you may not know exactly what charge is being controlled from the name of the item, you immediately get a sense that charge controllers are somehow manipulating the current coming from a power source.

The charge controllers that are commercially available come in a variety of sizes and have an assortment of features. Small charge controllers can be used in very small systems with one or two PV modules charging a small battery bank. Larger charge controllers are designed for use with multiple-kilowatt arrays and large battery banks.

In this chapter, I tell you all about the main functions, special features, and types of charge controllers. I also explain how to *specify* (select) a controller for the particular system you're designing.

# The Essentials of Charge Controllers

Simply stated, in order to properly maintain a battery bank that's being recharged by a PV array, you must include a charge controller in the system design. In the sections that follow, I introduce the basic functions of a charge controller and describe the features offered on some models.

## Seeing how a charge controller works in stages

REMEMBER

A charge controller's main role in a PV system is to properly control the charge from the PV array into the battery bank by controlling the current and voltage from the array into the batteries. Without a charge controller, like the ones pictured in Figure 8-1, the PV array would be able to send all of its current into the battery bank without any regard for the batteries' needs. The batteries, in turn, would become overcharged and eventually ruined.

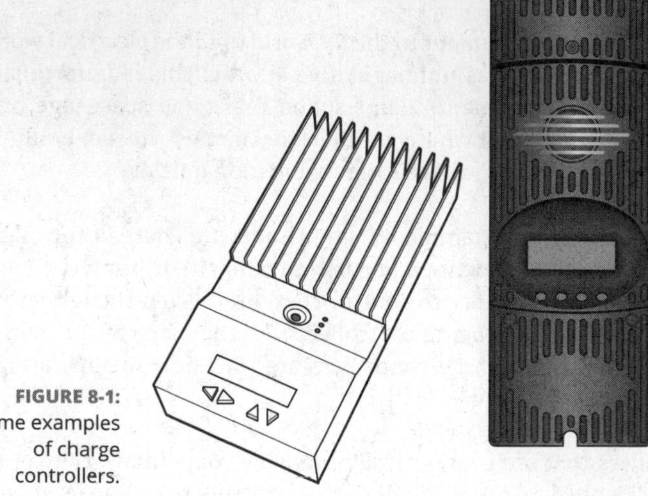

**FIGURE 8-1:** Some examples of charge controllers.

Charge controllers regulate the voltage and current sent to the batteries during the charging process. Each charge controller has multiple stages for which it regulates different voltage and current levels; in Figure 8-2, you can see three such stages. This figure shows how both the voltage and current vary over time based on the *charge set-points,* which are the voltage levels that you want to charge the batteries to; each battery manufacturer publishes its own charge set-points that you should use if you want to maximize the batteries' life span. (***Note:*** During the installation process and before commissioning the system, you must adjust the charge set-points as necessary.)

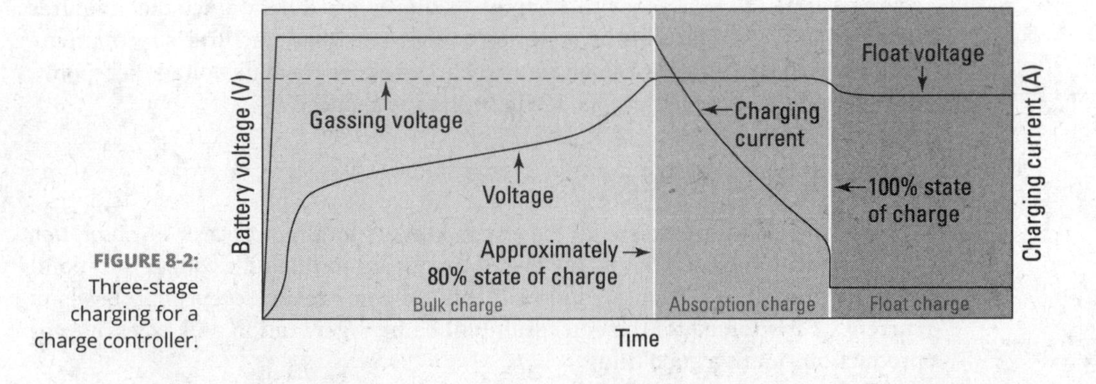


**FIGURE 8-2:** Three-stage charging for a charge controller.

In the following sections, I describe a charge controller's role during the three charging stages.

## Bulk charging

The first charging stage is *bulk charging.* It happens first thing in the morning after the batteries' voltage and capacity have been drained down since the sun set the previous day. Bulk charging pushes as many amps as possible back into the battery bank from the PV array and gets the voltage up in the process.

TIP

To better understand bulk charging, it may help to equate it with trying to fill a large glass of water from your faucet. The voltage is the water level in the glass. If the voltage in the battery (or the water level in the glass) is relatively low, then you need to allow as much current from the PV array as possible into the battery (in other words, you need to open that faucet all the way). In this analogy, the charge controller is the faucet that controls how much water (current) can flow into the glass. Both the PV array and the faucet are limited in the flow they can provide to their respective containers, but as long as the level of energy (or water) in the container is low, the container (the battery or the glass) will gladly and readily accept the flow. As the current (or water) continues to flow, the battery voltage (the water level in the glass) continues to rise. This charging process continues until the battery voltage reaches a predetermined level known as the *bulk voltage set-point*. At this point, the current needs to slow down. If it doesn't, the battery can't effectively accept the charge, and the current becomes heat (or, in the case of the water glass, the water simply starts spilling over the edge).

The exact bulk voltage set-point is determined by the battery manufacturer. Some batteries are able to take a higher charging voltage than others, so you need to make sure your controller is set correctly for the batteries you're using, as indicated in the owner's manuals provided by the controller and battery manufacturers. The voltage is taken above this set-point on purpose during the equalization charge that I tell you about in Chapter 7; this is a deliberate act that requires monitoring and should only be done according to the manufacturer's recommendations. If the batteries are consistently charged above the bulk voltage set-point, the overall life of the battery bank will be greatly reduced.

## Absorption charging

The second charging stage in the three-stage charging process is *absorption charging*. After a battery bank has been brought up to the bulk voltage set-point (see the preceding section for more on this), it can't really accept high levels of current. If it's forced to, the end result will be heat generation and excessive gas production — not a good thing.

When a battery reaches its manufacturer's bulk voltage set-point, it's really only about 80-percent full. The point of the absorption charge is to top off the battery. Think about a glass of water under a faucet. If you stop the flow just before the water spills over the edge, when you turn the faucet off, the glass isn't 100-percent full because the force of the water was pushing it over the top.

During the absorption-charging stage, the charge controller holds the battery voltage constant and reduces the amount of current sent into the battery bank. (It's like reducing the flow from the faucet to top off the water level in the glass.) When this process is done, the bank is fully recharged.

Typically, a full battery-charging cycle (bulk and absorption) is a multiple-hour event; the exact length of time required depends on the size of the battery bank and the PV array. The charge controller automatically starts and stops the charging stages, but if the charging source is the PV array, there's always the chance that the sun will disappear before the controller has the opportunity to finish its work. In this scenario, the battery bank doesn't reach 100-percent capacity and needs an additional charging source, such as the utility grid or an engine generator, to help it get there.

### Float charging

The final charging stage is *float charging*, and it's designed to keep the battery in a full state of charge after the absorption-charging stage has topped off the battery bank. Typically, a PV array spends only a small amount of time float-charging the battery bank due to the limited number of hours it has each day to recharge the bank. A charge controller enters into a float-charging stage only after the first two charging stages have been completed and when there's enough power from the array to send a float charge into the batteries.

When the number of peak sun hours is very limited (like during the winter), a PV array may not be able to get the battery bank to the float voltage at all because the lack of sun doesn't allow for a full charging cycle and because the bank may be drained relatively low due to greater use. In the summer, an array may be able to recharge the battery bank in a short amount of time, allowing it to spend a fair portion of the day in the float-charging stage.

## Surveying special effects provided by some charge controllers

Some of the small charge controllers used in PV systems have only one feature: the ability to regulate the charge entering the battery from the PV array. Others, like those designed to work with larger systems (greater than 500 W), may include a variety of additional features to complement the main battery-charging feature. The need for and use of these features, which I explain in the next sections, vary among PV systems, but all of these features are available in every type of charge controller.

### Load control

In systems that support direct current (DC) loads (namely, stand-alone, battery-based systems, although DC loads can be supported anytime batteries are present), some charge controllers employ a *load-control feature* to make sure the batteries don't become excessively discharged. This feature works by pulling electricity directly from the battery bank and sending it to the loads through the charge controller. As the loads continue to run, the battery bank's capacity is reduced and monitored by the charge controller. If the loads run long enough, the charge controller senses the batteries' reduced capacity and cuts off the flow to the loads, which ensures the connected loads don't drain the batteries too low and cause them harm. The load-control feature doesn't allow the loads to receive power again until the battery bank has been recharged to a certain point, eliminating the possibility of the loads being reconnected to the bank before sufficient capacity is restored in the batteries.

REMEMBER

DC lighting loads are some of the most common devices used in conjunction with the load-control feature, although most any electrical appliance that can run off of DC electricity can be controlled.

## Auxiliary load control

In certain situations, there may be a need (or desire) to run loads only when the battery bank is being charged excessively or when the battery bank is running low and needs attention. These *auxiliary loads* (additional loads on top of what the building is using) are used to enhance the safety or performance of the entire PV system. Fortunately, some charge controllers include relays that can close an electrical switch when the battery reaches a certain level and send power to an auxiliary load.

REMEMBER

One common auxiliary load is a fan connected to a flooded battery bank's enclosure. (I introduce flooded batteries in Chapter 7.) When the PV array (or other charging source) brings the battery voltage up to a certain charge level, the flooded batteries begin to release hydrogen gas. The auxiliary-load-control feature of a charge controller is activated when the batteries reach the predetermined voltage and power is sent to the fan (which in this case is the auxiliary load). Other auxiliary loads such as warning lights or alarms can be connected to warn your client about a battery's excessive or reduced voltage levels.

## Status meters

Some charge controllers feature status meters that can either be integrated into the face of the controller or be run remotely for a client to see in a convenient location (such as the kitchen or other living space). *Status meters* allow PV system owners to evaluate the battery and PV array voltage levels of their system with a quick glance.

TIP

Other, more sophisticated status meters also track the energy values into and out of the battery bank. These amp-hour meters are especially useful to owners of stand-alone, battery-based systems who rely on their battery banks for the majority of their power. By using a status meter that includes an amp-hour meter, your client can accurately know the status of her battery bank and know when she needs a secondary charging source, such as an engine generator, to bring the battery capacity back up. This level of monitoring is actually quite important for stand-alone systems because the batteries are the user's main power source; as such, she should know their status at all times.

# Maximum Power Point Tracking Technology

The technology that allows a PV array to deliver the maximum amount of energy to a battery bank is known as *maximum power point tracking* (MPPT). MPPT charge controllers gained popularity in the early 2000s when manufacturers released highly reliable and accurate versions that allowed users to maximize the charging ability of their PV array and, in some cases, reduce the required PV array size for battery charging compared to some of the older technology.

In the following sections, I explain the magic behind MPPT controllers and outline their pros and cons so you can evaluate whether this technology is the right solution for your clients.

*Note:* All commercially available grid-direct inverters also use MPPT technology; see Chapter 9 for an introduction to inverters.

## How MPPT works

An MPPT charge controller uses the three charging stages presented earlier in this chapter to allow a PV array to operate at its maximum power point (abbreviated MPP) regardless of the voltage of the battery bank connected to the controller. Other charge controller technologies, such as pulse-width modulation (covered later in this chapter), can't fully use a PV array's MPP.

In Chapter 6, I explain that the MPP is defined as the point on the IV curve where the current multiplied by the voltage yields the highest power value. (In other words, it's the product of the maximum power voltage, $V_{mp}$, and the maximum power current, or $I_{mp}$.) For a typical 12 V nominal panel, the voltage associated with the MPP is somewhere around 17 V. PV manufacturers realized early on that this was the voltage value required to effectively charge a 12 V nominal battery bank in nearly all worldwide geographic locations. (Keep in mind that module voltage decreases when the module temperature rises, so the extra voltage is necessary to push the electrons into the battery bank when the module's temperature is elevated.)

The maximum power voltage of 17 V doesn't always equate directly to the required voltage needed to charge a battery bank, though. Depending on the technology and the charge set-point, the voltage necessary for charging a 12 V nominal battery bank can range anywhere from 13 V to 15 V. Therefore, a PV module can produce more voltage than a battery bank can fully use. Enter the MPPT controller.

REMEMBER

MPPT controllers take the power from a PV array at the MPP, regardless of the required battery voltage, and deliver that same amount of power (minus efficiency losses, of course) to the battery bank because they're able to reduce the voltage from the array to the battery's required level. And because power is the product of voltage and current, if the voltage is decreased, the current is increased in order to keep the same power level. MPPT controllers boost current into the battery bank in relation to the current received from the array.

I think this concept is best illustrated in Figure 8-3, which depicts the power curve for a typical 12 V nominal PV module. The peak of the curve represents the *maximum power value*, which is the level that the PV module can produce. The graph also shows the location of a typical battery charge set-point. If you move straight over to the right from that point, you'll see the power level associated at the battery-charging voltage. The difference in the MPP and the power level associated with the battery-charging voltage represents the increased power output due to the use of the MPPT technology. The PV array's power levels move throughout the day depending on the environmental conditions, and MPPT controllers adjust right along with them.

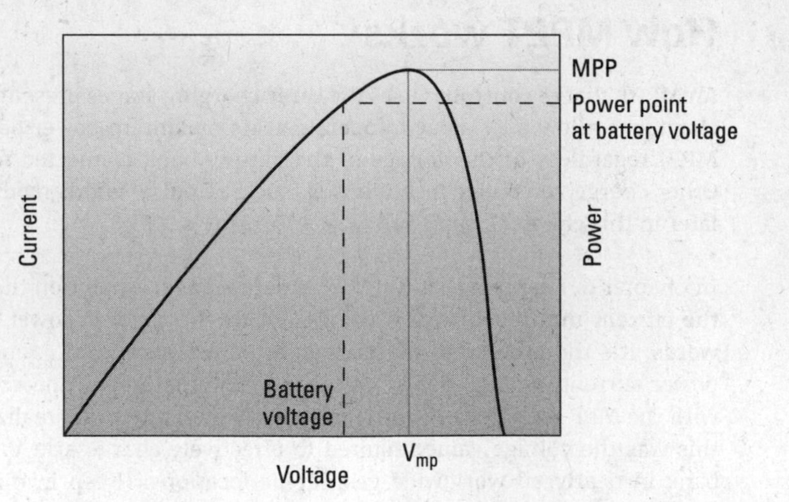


**FIGURE 8-3:** The power gained through the use of MPPT controllers.

## The pros and cons of MPPT controllers

REMEMBER

MPPT controllers have become the most popular charge controllers for larger battery-based PV systems (both stand-alone and utility-interactive) thanks to their ability to fully use the power produced by a PV array. Another good thing MPPTs have going for them is their ability to take a PV array wired for a higher voltage and still charge a low-voltage battery bank. For example, with the help of an MPPT controller, you can take a PV array that's wired in a series configuration up to 150VDC and still charge a battery bank all the way down at 12 V nominal. Having a higher-voltage array allows your client to locate the PV array farther from the battery bank and not have to take out a second mortgage for the length of wire connecting the two. Finally, MPPT manufacturers are constantly adding features and increasing efficiencies. These improvements help you, the designer and installer, by increasing the flexibility in your design.

A major drawback to MPPT controllers is the cost. This technology comes at a price (about $800 for a standard unit as of early 2010). Justifying the extra expense for very small systems that don't fully realize all the benefits can be difficult, which is why MPPT controllers are often used only in larger arrays.

# Pulse-Width Modulation Technology

Although not as sleek and sophisticated as MPPT (which I fill you in on earlier in this chapter), pulse-width modulation (PWM) charge controllers are very effective in charging battery banks and will likely be a popular technology used in PV systems for years to come. In the sections that follow, I describe the workings of PWM technology and note the pros and cons of using it so you can decide what's best for your clients.

## How PWM works

Just like MPPT controllers, PWM controllers regulate battery charging via multiple set-points. However, unlike MPPT controllers, PWM controllers can only use the voltage from the array that equals the voltage required by the batteries. (For example, if the battery bank needs 14 V to charge and the array can supply 17 V, the controller can only accept the 14 V.) This characteristic inherently reduces the overall power available from the PV array because the battery-charging voltage rarely matches the array's maximum power voltage. Because the battery bank dictates the voltage, the amount of current sent into the battery from the array is also limited (so the current value from the array that's associated with the battery-charging voltage is different from the maximum power point current).

As the battery bank gets full, the PWM controller regulates the charge into it by *pulsing the charge* (turning the power on and off) from the array into the bank many times each second. Because the pulsing of the power happens so fast, the batteries "see" the current flow from the array as a slowly declining line, as shown in the graph. This pulsing of the current, where the controller starts and stops the current flow for various amounts of time, allows the battery to accept the charge and become fully recharged.

TIP

I like to think of the way a PWM controller works as standing with your hand on a water faucet and rapidly turning the water on and off as your glass begins to fill up. By stopping the flow for brief periods, the glass can accept all the water coming into it without losing any of it.

### The pros and cons of PWM controllers

REMEMBER

PWM controllers may not be as technologically advanced as MPPT controllers, but they're a proven technology that works well in many applications because they can be used with all battery technologies — even with small PV systems that have just a few PV modules charging a few batteries. Also, they're a lower-cost option compared to MPPT controllers, and they come in sizes to match very small PV applications (of course, they can also support multiple-kilowatt installations). They can even serve as effective load controllers for wind and microhydroelectric systems if your client needs that.

WARNING

The main drawback of PWM controllers? Because they aren't as efficient as MPPT controllers in transferring the power generated by a PV array into a battery bank, you may need more PV modules in an array to get the same charge as you'd get with the help of an MPPT controller — a fact that ultimately costs your client more money.

## Specifying a Charge Controller

When it comes time to specify the charge controller in a client's system, you need to look at the system as a whole and how the charge controller will fit into it. Chapter 12 outlines the methods used to size a charge controller based on the system design and electrical requirements in relation to both the PV array and the battery bank.

REMEMBER

Make sure you always consider the voltage and current values during the charge controller selection process. As I note earlier in this chapter, MPPT and PWM controllers come in a variety of sizes. Each size of controller is rated according to its maximum and minimum voltage levels, but the current level a controller can handle is actually the more critical specification. Every charge controller is limited in the amount of current it can process due to its type (MPPT or PWM) and size (small or large). Consequently, you need to evaluate the amount of current the PV array will produce to specify the correct charge controller for your client.

TIP

If the controller needs only to handle the battery charging and maybe control a single load, a basic PWM charge controller should suffice. However, if the application requires advanced metering and the ability to run auxiliary loads, a more advanced MPPT controller is generally your best bet (although you may also need to suggest multiple MPPT controllers to efficiently address the client's needs).

IN THIS CHAPTER

» Getting the scoop on grid-direct inverters

» Exploring the ins and outs of battery-based inverters

» Selecting the right inverter for the system you're designing

# Chapter 9

# Inverters: AC (From) DC

If PV modules (see Chapter 6) are the heart of any PV system, inverters are the brains. These devices take the direct current (DC) power produced by the modules (and stored in batteries) and turn that electricity into the alternating current (AC) electricity that people use in their homes and businesses. Not only that, but inverters are smart enough to realize when the utility is gone and when the batteries need some extra attention. Impressive, huh?

As I reveal in Chapter 2, you use one of two inverter types in any PV system:

- **Utility-interactive inverters** are classified into two subcategories:
  - Grid-direct (for use in grid-direct PV systems)
  - Battery-based (for use in utility-interactive, battery-based PV systems)
- **Stand-alone inverters** (for use in stand-alone, battery-based PV systems) are all battery-based.

In this chapter, I describe the operation and features of both grid-direct and battery-based inverters. I also help you discover what the considerations are when specifying the inverter for any PV system you design and install.

## Getting the Goods on Grid-Direct Inverters

The majority of PV systems installed today feature inverters that connect directly to the PV array on one side and to the utility on the other side. These inverters don't use any method of energy storage and are most often referred to as grid-direct inverters. They're the most widely installed inverters due to their increased efficiencies and relatively simple installation, and they're used only in grid-direct (battery-less) PV systems (I describe these in Chapter 2).

All grid-direct inverters are considered utility-interactive because they require the presence of the utility in order to operate; they work in parallel with the utility to supply power to common loads (including the lights in your home, your television, and anything else that uses electricity). They even have the ability to send power back into the utility's grid.

In this section, I describe the basic operation of grid-direct inverters as well as their standard features, power output sizes, and technological differences.

### Basic operation

Although the actual electronics inside a grid-direct inverter aren't simple, the basic process of how they work is.

1. **First, DC power is delivered from the PV array to the inverter.**

   With very few exceptions, all types of grid-direct inverters work well with high-voltage PV arrays. The voltages typically fall between 150 V and 600 V, with 600 V being the absolute limit for residential and commercial grid-tied systems due to electrical code limitations.

2. **The inverter then takes this DC power and turns it to AC power through various methods.**

   The exact method for turning the DC to AC depends on the manufacturer and its choice of technology. I present the major technologies later in this chapter.

3. **Usually, the AC output from the inverter is then connected directly to a load center that's also connected to the utility grid.**

   The load center is usually the main distribution panel (MDP; see Chapter 2).

4. **The AC power then either flows into loads that are connected to the load center, such as a refrigerator or lights, or goes back into the utility's power lines and runs the meter backward.**

This utility interconnection requires you to work with the client to let the utility know that the PV system will be connected to its utility grid and will have the potential of sending power back to the grid. (I present the different ways that system owners can connect to the utility and receive full credit for their PV systems in Chapter 2.)

Grid-direct inverters don't employ any form of energy storage, so if the utility doesn't provide them with a stable, consistent power source, they can't run. In fact, grid-direct inverters are so aware of small changes in the utility's power that they shut down on occasion to meet the necessary safety standards. This characteristic can be seen as a limitation, but it's a deliberate feature of the inverters that allows them to operate safely with the utility grid (see the later "Safety features" section for more information).

## Standard features

All grid-direct inverters have some basic standard traits, regardless of their size or technology. They all incorporate some basic safety features, use maximum power point tracking, and possess some type of user interface.

### Safety features

One of the most important safety features in all grid-direct inverters (as well as utility-interactive, battery-based inverters, which I cover later in this chapter) is the ability to detect when the grid is suddenly disconnected. Grid-direct inverters have very sophisticated monitoring equipment that can detect the absence of the grid in fractions of a second and turn off the inverter automatically in response. The name given to this process is *anti-islanding*, and it's a requirement for all grid-direct inverters connected to a utility grid.

The term *islanding* refers to a situation in which the utility grid is out and alternate power sources from people's homes are still connected to the grid and sending power back into those "dead" lines. Islanding presents a safety issue for utility workers who may be working on those lines; they may think the lines are safe to touch when in fact electricity is present from a different source. That's why grid-direct inverters are required to recognize utility disturbances and stop producing power immediately. The inverters look at two parts of the utility power: the electrical frequency and the voltage. If either part goes out of the specified allowable range for the inverter's operation, the anti-islanding feature activates, and the inverter turns off.

In the U.S. market, all grid-direct inverters require testing to the same standard to meet the anti-islanding requirement. The test is known as UL1741, and the testing procedure covers several safety issues for many of the electrical components used in PV systems. Many companies perform this test, but they all follow the same rules when doing so.

After the inverter turns off for anti-islanding, it monitors the grid to verify that the voltage and frequency values are within the specified ranges. When these two parameters are met for five continuous minutes, the inverter can then turn back on and resume producing power. This entire process is automatic within the inverter and doesn't require any interaction from the user. (That's good news for your client!)

Grid-direct inverters have to react to the frequency and voltage parameters they see from the grid. Therefore, it may be possible for the grid to experience a brown-out situation where power isn't entirely lost but the inverter still shuts down. Another situation that may cause a grid-direct inverter to shut down even though the grid is still up is when there are excessive voltage drops in the wires connecting the inverter to the utility.

Another important and standard safety feature in residential grid-direct inverters is ground fault protection (GFP). GFP is integrated into the inverters and is there as a fire-safety device. GFP is different from the ground fault circuit interrupter (GFCI) that most folks are familiar with. GFCI outlets, the ones with two buttons in the center, are used in homes and are required in locations such as bathrooms. The difference between GFP and GFCI is that GFCI outlets are installed to protect people from receiving a shock, whereas GFP devices are meant to prevent faults from starting fires. (I cover ground faults and GFP in depth in Chapter 10.)

## Maximum power point tracking

All grid-direct inverters employ maximum power point tracking (MPPT) to produce as much power as possible from the PV array. MPPT allows the inverter to harvest the maximum amount of power and deliver it to the load center, just like some charge controllers (see Chapter 8) use it to deliver the maximum amount of power to a battery bank.

*Note:* Only one MPPT tracker in a grid-direct inverter looks at a system's PV array. See the nearby "Multiplicity: Inverters with more than one MPPT" sidebar for when to use inverters with multiple MPPT inputs.

## User interface

Because you can't simply look at a PV array and know whether it's working, inverter manufacturers have decided that some sort of display or interface is necessary to indicate how the array is functioning. Consequently, today's inverters generally include options for the user to see what his entire PV system is doing. With very few exceptions, grid-direct inverters now come standard with a display built into the unit, a feature that allows your client to obtain critical information such as voltage, power output, and total energy production from the inverter itself.

In addition, inverter manufacturers offer a wide variety of ways to collect, store, and display the data gathered by the inverter. Many have methods for connecting the inverter to the Internet and allowing the user to visit a Web site and see real-time information about his system. Some even make it possible to connect the inverter directly to a computer, a feature that allows the system owner to collect and store data locally so he can review it periodically to identify potential issues if they occur.

### MULTIPLICITY: INVERTERS WITH MORE THAN ONE MPPT

The majority of grid-direct inverters available for the PV industry use a single maximum power point tracker (MPPT), which means if you have an array that needs multiple orientations, you should consider an inverter that can track the maximum power point (MPP) of arrays facing multiple directions.

Say you have two strings of PV modules on the same house but they're facing different directions. The IV curves for the two strings are therefore different because the strings aren't receiving the same amount of irradiance from the sun (Chapter 6 introduces you to IV curves). If you use an inverter that has only one MPPT and both strings are connected to the same inverter, the inverter will have a hard time locking onto the true MPP for the array. In fact, there really isn't a true MPP for the array but rather two separate MPPs for the strings. The severity of the difference is dependent on how different the two orientations are. The solution? If you use an inverter with multiple MPPT devices, each string can be connected to its own MPPT, and then the inverter can lock onto the individual MPPs and be able to deliver the maximum amount of power for the whole array. Currently your options for inverters with multiple MPPT inputs are minimal, but you can find them; contact the company you buy your equipment from for help.

## Power output sizes

REMEMBER

Even though all grid-direct inverters operate on the same principles, they come in a variety of *power output sizes* (which are very often referred to simply as the inverters' *size*). Following are the three power output sizes you can expect to run into:

- **Microinverters:** These are inverters that connect to a single PV module rather than a string of modules. PV folk like these inverters because their ability to turn the individual modules' power to AC reduces total system losses from factors such as shading. Microinverters are typically less than 250 W each, which matches well with the commonly used PV modules.
- **String inverters:** Although the true definition of a *string inverter* is an inverter that's connected to a single series string (see Chapter 3 for details on series strings), the term has become synonymous with small (less than 15 kW) inverters that attach strings of PV modules for power outputs ranging from 1 kW to 15 kW.
- **Central inverters:** Associated with larger commercial projects, *central inverters* range in size from 15 kW to 1 MW. Central inverters operate a lot like the string inverters used in residential applications, just on a much larger scale (so they're kind of like string inverters on steroids).

Regardless of their power output size, grid-direct inverters usually look pretty similar; Figure 9-1 shows you a typical one.

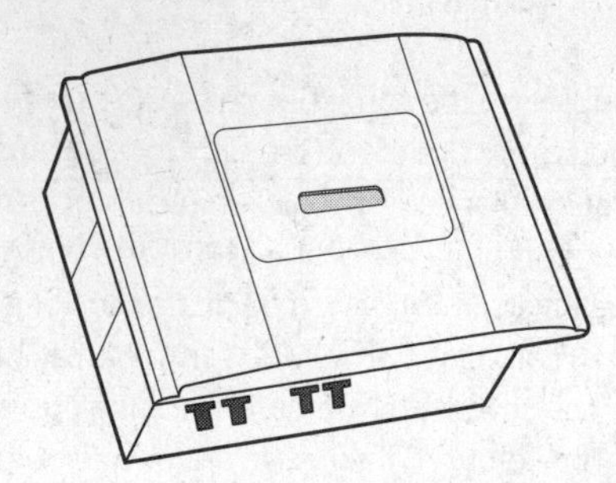

**FIGURE 9-1:** Your average grid-direct inverter.

## The importance of transformers

Just like all cars perform the same basic function — transporting people from here to there — all grid-direct inverters have the same purpose: to take DC power from the PV array and turn it into AC power for the load center or the utility. But just like cars, a grid-direct inverter can have a number of different forms. The biggest difference between one inverter and the next is transformer related: Different manufacturers may use different transformers in their inverters, and some manufacturers may not use transformers at all. I present the main transformer-related differences among grid-direct inverters in the sections that follow.

## Large transformers and low-frequency technology

Low-frequency grid-direct inverters are currently pretty common in PV systems. They take the high DC voltage from the PV array and, through a series of switches, turn that DC into AC with the help of a transformer that keeps the AC and DC isolated by inducing the switched DC voltage across the transformer and creating the AC current on the other side of it. Low-frequency inverters get their name because the frequency at which these switches are operating is relatively low in comparison to high-frequency inverters (described in the next section).

REMEMBER

The advantages of low-frequency inverters include a very robust design that allows manufacturers to reduce the number of parts required to make AC from DC and keep overall costs low. However, because these inverters use a large transformer, power losses occur across that transformer. (Also, the transformer adds a lot of weight to the unit.)

## Small transformers and high-frequency technology

High-frequency inverters use a switching technology that's similar to that of their low-frequency counterparts. The difference is that high-frequency inverters do their switching much faster, which means they can use small, lightweight transformers rather than large, heavy ones.

Some manufacturers who work with this technology are able to use the high-frequency transformers to create multiple small inverters inside the same box. What's the advantage of that, you ask? Well, if the PV array is producing low levels of power, one inverter can be off, letting the second inverter operate at a higher efficiency level. After the power level gets high enough, the second inverter turns on, preventing the first inverter from being overworked.

## Transformer-less technology

Some inverters on the market don't use any transformers at all. These inverters keep the DC and AC separated electronically and prevent any DC injection into the AC line by using *firmware* (small electronic programs) specialized for the inverter.

REMEMBER

Transformer-less inverters are commonly referred to as *ungrounded inverters* because a connection to the grounding electrode system isn't required due to the lack of a transformer. However, transformer-less inverters still need to include an equipment ground in order to reduce shock hazards. Turn to Chapter 17 for information on grounding requirements in PV systems.

REMEMBER

The largest positive feature for transformer-less technology is the increase in overall inverter efficiency. Without a transformer present, the inverter has one less step to make and can turn the DC power from the array into AC power more efficiently. The major downside involves all the additional requirements you have to go through when installing a transformer-less inverter in order to be compliant with the *NEC*®. These extras can translate into higher costs during the design and installation process.

# Investigating Battery-Based Inverters

When energy storage is a requirement for your client (calling for a utility-interactive, battery-based or stand-alone, battery-based PV system), a battery-based inverter is your go-to choice. Actually, battery-based inverters are better described as inverter/chargers because they have the ability (when needed) to accept an AC power source, such as the utility grid or a generator, and then turn the AC electricity into DC electricity for battery charging. Figure 9-2 illustrates your average battery-based inverter.

Utility-interactive, battery-based inverters operate almost identically to stand-alone, battery-based inverters, but a few major differences exist. In this section, you find need to know about the workings of both types of inverters and describe the features that come standard on any battery-based inverter.

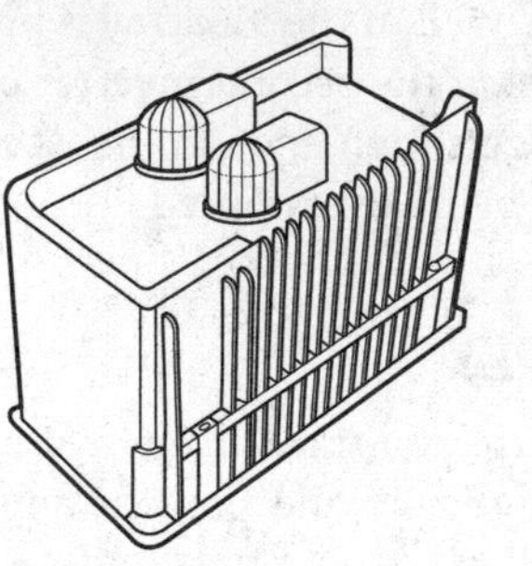

**FIGURE 9-2:** A typical battery-based inverter.

## Utility-interactive inverter operation

A utility-interactive, battery-based inverter works in different ways depending on whether the utility is up and running or down and out. I explain the differences in operation in the following sections.

### When the utility is working

When the utility is up and running and the battery bank is full, power moves through a utility-interactive, battery-based inverter in the following route:

1. **The PV array starts producing DC power in the morning and sends it to the battery bank.**
2. **If the battery bank is full, the charge controller "talks" to the inverter and sends the DC power toward it.**

   The battery bank will always be full unless there has been a power outage and the client has run loads using the battery bank.
3. **The inverter accepts the power from the PV array, changes it into AC power, and passes it though to the backup load center.**

   REMEMBER

   If the PV array is producing more power than the backup load center is consuming, the inverter takes the excess power, turns it into AC power, and sends it to the main distribution panel (MDP).
4. **The MDP then disburses power to the loads connected to it.**

   If the PV array is producing more power than the MDP and the backup load center combined, the inverter can push the AC power back into the grid, running the meter backward.
5. **As the PV array slows down and eventually stops producing power, the inverter stops sending power back toward the utility; the utility power then begins to flow into the MDP, just like it would if no PV system were present.**
6. **The utility power continues through the inverter into the backup load center so that the loads always have power available and the batteries remain full.**

REMEMBER

The connection between the utility and a utility-interactive, battery-based inverter allows current to flow both ways (think of it as a two-way street). Stand-alone, battery-based inverters can't do this; they only allow current to flow from the power source to the loads. *Note:* In both types of inverters, the AC output connection goes directly to a load center, so no power source is available to send power back toward the inverter (consequently, the AC output connection acts as a one-way street).

## When the utility is down

No utility can stay live all the time, which is why utility-interactive, battery-based inverters are ready for the occasions when one goes out. They're able to recognize the outage and automatically disconnect themselves from the utility connection, eliminating any possibility of the inverters islanding and sending power back to the utility. (I explain islanding in the earlier "Safety features" section.)

REMEMBER

Because utility-interactive, battery-based inverters are connected to the utility, they have to conform to the same anti-islanding standards that grid-direct inverters do, which means they have to monitor the utility and disconnect themselves when the voltage and frequency levels are out of the specified parameters.

At the same time that it disconnects from the utility, a utility-interactive, battery-based inverter immediately begins drawing DC power from the battery bank and sends AC power to the backup load center. It then continues powering the loads from the battery bank until either the utility power returns or the batteries discharge and can't support the loads.

REMEMBER

If the power outage is extensive, the PV array can and will continue to charge the battery bank through the charge controller (I outline this process in Chapter 8), giving the battery bank an extended run time. When the utility power returns, the inverter reconnects to it, allowing the battery bank to recharge directly from the utility. This way the battery bank can be ready to supply power if another outage occurs. (*Note:* Most utility-interactive, battery-based inverters allow you to defeat this function if you'd rather have the PV array charge the batteries.)

# Stand-alone inverter operation

REMEMBER

During normal operation in a stand-alone system, the battery-based inverter accepts DC power from the battery bank, turns it into AC power, and delivers that AC power to the loads connected in the main distribution panel. Simple, right? As long as the capacity in the batteries remains high enough, the inverter is able to continue delivering power to the loads.

After a battery bank discharges to the level you've designed for, the inverter can either alert the system owner to manually start a generator or it can automatically start one in order to make sure the batteries don't discharge more than he wants them to and cause damage (you can read all about the effects discharging has on a battery in Chapter 7). As soon as the generator connects to the inverter, the latter stops drawing DC power from the battery to make AC and starts passing the generator's power through to the loads and charging the batteries with any remaining available power.

The charger built into a stand-alone inverter performs a multistage charging cycle similar to the one I describe in Chapter 8. Most systems are set up to allow the batteries a deep discharge before calling for help from the generator, so this cycle generally takes several hours to fully charge the battery bank. After the batteries are full again, the generator can be turned off (manually or automatically), and the inverter can return to its regular job of powering loads through the battery bank.

## Standard features for all battery-based inverters

Battery-based inverters are the workhorses of the inverter world. They're capable of handling a variety of environments and delivering high-quality, reliable power. In the sections that follow, I describe some standard features found in all battery-based inverters.

REMEMBER

Most, but not all, of the battery-based inverters used in PV systems are actually inverter/chargers. On the other hand, some battery-based inverters are exactly that — inverters without the ability to charge a battery bank from an external AC source (such as the utility grid). With battery-based inverters, a number of features vary from manufacturer to manufacturer, so if you need a specific function, verify that the inverter can deliver what you want it to before you get too far into your design.

### Considering safety

The safety features built into battery-based inverters tend to focus on the safety of the system. These inverters keep batteries from becoming too deeply discharged by alerting the system owner of a low-battery situation, starting a generator, or even shutting down; when they act as chargers, they make sure batteries are charged correctly by using multistage charging. And of course, when battery-based inverters are connected to the utility, they isolate themselves from the utility during power outages (just like grid-direct inverters do; see the earlier "Safety features" section for more information).

WARNING

Unlike grid-direct inverters, ground fault protection (GFP) isn't a standard feature of battery-based inverters, although it can be added to battery-based systems elsewhere. Other safety features, such as disconnects and overcurrent protection, are installed in external boxes located adjacent to battery-based inverters. See Chapter 10 for more details on these types of protection.

### Interacting with users

So many variables require programming in battery-based inverters that some level of interface with the system user is necessary. Depending on the inverter's manufacturer, this interface can take the form of a screen and touchpad built into the inverter or a hand-held controller wired directly to the inverter. Typically, the interface not only allows for inverter programming but also gives the user some basic system information (think battery voltage and current levels).

## Sizes of battery-based inverters

Battery-based inverters come in a large range of power outputs (sizes). You can buy a small 100 W inverter that connects to the DC plugs in your car all the way up to 6 kW units. The most common types used in PV systems start at 1 kW of AC output and range up to 6 kW.

For battery-based systems that require more power than a single inverter can provide, multiple units can be *stacked*, or connected together in such a way that they can provide more power to the loads. Depending on the manufacturer, you can stack individual inverters together to provide up to 36 kW. Of course, having multiple inverters means the inverters need to talk with each other. This communication is typically handled by connecting the inverters together via a communications cable (similar to the cable you use to connect your computer to the Internet).

## Low-frequency transformer technology

All battery-based inverters use inverter technology that's similar to the low-frequency, transformer-based, grid-direct inverters. However, the battery-based inverters are limited to a 48 VDC nominal input.

# Specifying Any Inverter

To *specify* an inverter is to decide which kind of inverter to use, either grid-direct or battery-based, depending on the system you're designing. The following sections note the big-picture considerations you should have in mind as you work to pick the make and model of inverter that's a good fit for your client's needs. (When you're ready to size the inverter you've chosen, head to Chapter 11 if you're working with a grid-direct one or Chapter 12 if you're working with a battery-based one.)

TIP

After designing and installing a few systems, most PV pros have one or two brands of inverters that they prefer, and they tend to stick with these brands. But just because you like a certain brand doesn't mean you should stop checking out the market every now and again. Because the PV industry is growing rapidly, manufacturers are constantly releasing new and improved products that may serve future clients even better than your current favorite. You may wind up discovering your new favorite inverter simply by giving a new one a shot.

# Grid-direct

Grid-direct inverters are designed to operate within all the AC voltages offered to residential and commercial buildings from the utilities in the United States. Consequently, you need to consider the operating range for the DC voltages from the PV array and the overall power output of the inverter when specifying an inverter for a grid-direct PV system. Of course, you also want to ensure that the inverter you choose works well for your client's needs (as well as your own). The next sections are here to help you out.

## Matching the inverter to the PV array

REMEMBER

The *DC voltage window* (the range of allowable voltages for an inverter) and the AC power output value are your two big concerns when matching an inverter to a PV array. Following are some specifics about each one:

- **DC voltage window:** The manufacturer defines this window and whatever type of technology the inverter uses to turn DC power from the PV array to AC (this technology can be low frequency, high frequency, or transformer-less; I describe all three earlier in this chapter). The typical DC voltage window for grid-direct inverters used in residential systems is from 250 VDC (minimum) to 600 VDC (maximum). At first glance, this window seems extremely wide and relatively limitless, but as you discover in Chapter 11, this window narrows very quickly when you start evaluating real-world operating conditions.
- **AC power output value:** For grid-direct inverters, the AC power output value is directly tied to the DC input power value. The inverter's power rating is therefore evaluated in relation to the PV array's rated power values. As you find out in Chapter 6, PV modules vary in power output based on environmental conditions. PV systems also have inefficiencies all along the way, so grid-direct inverters are typically matched up with PV arrays that have larger power output ratings to make up for the energy losses. The exact amount of "extra" power from the array to the inverter is affected by a number of variables, but most grid-direct inverter manufacturers recommend between 15 percent to 25 percent more PV input power than the inverter's power output rating. For example, if you have an inverter that's rated to produce a

maximum of 3,000 W (3 kW), the maximum recommended PV array (in terms of power output) would be between 3,450 W (3,000 W × 115%) and 3,750 W (3,000 W × 1.25%). ***Note:*** Generally, you already have a desired PV array size you want to install, so use this relationship to define the inverter size you need to match your chosen array.

### Supplying the right features for you and your clients

During the specification process, you also need to evaluate the feature set associated with the grid-direct inverters available for your system requirements. Some features may be very important to you as the installer; others will be there mainly for the user to enjoy over the years.

One of the features that's of chief interest to you as the system installer is that many grid-direct inverters now come standard with disconnects that integrate to the inverter, allowing for fewer components during installation. This feature helps the installation process because the *NEC®* requires the installation of disconnects and specifies their location to the equipment they serve. By incorporating the disconnect into the inverter, you have less equipment to buy and install, which makes the system less expensive for your client and faster for you to install. (I cover safety components in Chapter 10).

REMEMBER

From the client's perspective, an important feature is the ability to access as much information about the system's performance as he's interested in. Some PV system owners go out to their inverters daily to check the energy-production values. Others, like me, know that if the information isn't readily available (and in their faces), life will likely prevent them from checking the system on a daily basis. Find out what your client's monitoring preference is and then seek out the user interface that makes the most sense for him based on those preferences, whether that's a wireless display set up in a noticeable spot in his home or a data-transfer arrangement that puts the PV system's data online or in a text message or e-mail.

## Battery-based

REMEMBER

Your first consideration when specifying a battery-based inverter should be this: What's the AC power source — the utility grid or a generator? Answering "grid" means you need a utility-interactive, battery-based inverter; answering "generator" means you need a stand-alone, battery-based inverter. From there, the questions and considerations (including what special features are desired) are very similar to the kinds you'd have if you were specifying a grid-direct inverter. (Note that the considerations that follow are all unique to battery-based inverters.)

TIP

In some cases, you'll want to have both the grid and a generator for multiple power sources. In this situation, you need to confirm that your chosen utility-interactive inverter can accept multiple power sources (not all of them can) before getting too far into the specification process.

After establishing the AC power source, you must determine the amount of power the inverter needs to supply continuously. As I show in Chapter 12, this means determining all the loads that the user plans to run simultaneously and comparing the amount of power needed to operate them all to the inverter's continuous power output rating.

TIP

For stand-alone, battery-based systems, the inverter's power output rating doesn't need to account for the PV array's power output because the battery bank is between the inverter and the PV array, acting as a buffer. When you're installing a utility-interactive, battery-based system, however, you need to look at the relationship between the two. If the batteries are full and no loads exist, the inverter must have the ability to send all the PV power into the utility grid. In this situation, if the PV array can supply more power than the inverter can process into AC power and send into the grid, the additional power won't be used, and the PV array won't be operating at its maximum efficiency. Consequently, the client won't send as much energy into the grid as he could have (which won't make him too happy with you).

Another power-related item to consider is the surge rating. Any electrical load with a motor creates an electrical surge when started. The battery-based inverter you select for your client's PV system must be able to handle this surge; if it can't, the loads won't start and could possibly cause the inverter to crash, stopping the power flow to all the loads. An inverter's surge rating is listed on its spec sheet; this rating can show up as either watts or amps, so you may need to do some conversions to accurately compare the loads to the inverter (see Chapter 3 for help making the conversion).

REMEMBER

Just as grid-direct inverters need a steady power source from the utility grid, battery-based inverters need a steady power supply from batteries and require a narrow supply voltage window from the batteries for operation. These inverters are designed to work in conjunction with a low-voltage battery bank — typically 12 V nominal, 24 V nominal, or 48 V nominal.

Finally, you need to make sure the inverter can deliver the correct voltages needed for the loads. Most deliver 120 VAC but can be configured for other voltages as necessary. For instance, some battery-based inverters come standard as 120/240 VAC for convenience, and a few manufacturers offer the ability to configure three inverters together for a 120/208 VAC system.

IN THIS CHAPTER

» Figuring out circuits

» Examining conductors and conduit

» Taking disconnects and overcurrent protection into account

» Protecting your systems from ground faults

» Labeling for the benefit of safety personnel

# Chapter 10
# Staying Secure: Wiring and Safety Components

When you install a PV system, you must make safety your top priority. That means giving the wiring and safety components of your PV designs special consideration, particularly during installation. In this chapter, I describe some common wiring and safety components used in PV systems, specifically conductors, conduit, disconnects, overcurrent protection devices, ground fault protection, and labels. (See Chapter 13 for how to size these items.)

REMEMBER

Every utility and *jurisdiction*, the government entity (typically a city or county office) that dictates the requirements for building and electrical work, you deal with will likely have slightly different wiring and safety component requirements that you must follow. I make it a point to show you the requirements as laid out by the *National Electrical Code*® (*NEC*®), but when it comes down to it, a local jurisdiction can make its own rules or have its own interpretations of the *NEC*®. If you ever have doubts as to what the requirements are in a particular area, talk with the local inspectors to find out what they want to see. (Of course, you should also purchase of copy of the *NEC*® and research it so you can become as familiar with it as possible. See Chapter 1 for how to obtain a copy of the *NEC*® *Handbook*.)

# Defining the Circuits in a PV System

A *circuit* is the path needed to complete the circle the electrons must run in to do useful work. (For instance, in a grid-direct PV system, the DC circuit is the path the electrons take from the PV modules, through the inverter, and back to the modules.) Different *conductors*, or wires, throughout a PV system connect circuits, and these conductors are commonly placed inside *conduit*, or pipes, that protect them as they run from one point to the next.

REMEMBER

The size of the conductors needed and the specific safety devices used are determined by the circuit type and location. Figure 10-1 shows a basic schematic of a PV system with the following circuits:

- **PV source circuit:** The circuit that comes from the individual PV strings is the *PV source circuit*. These circuits are often run along the back of the strings during installations; they lead to a *junction box* (a box where you can transition from outdoor wiring to indoor wiring) or a *combiner box* (a box where you can transition your wiring and also place strings in parallel; typically these boxes also have fuses inside).
- **PV output circuit:** The *PV output circuit* consists of the wires between the junction box or combiner box and the DC disconnect(s), which are often referred to as the *DC disconnecting means*, indicating that this is the method to disconnect the DC (PV) conductors from the rest of the system.

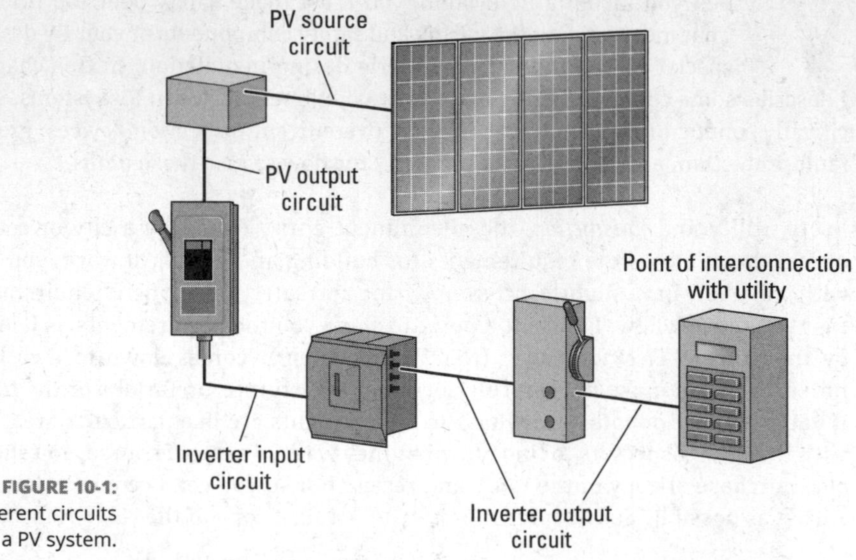


**FIGURE 10-1:** Different circuits in a PV system.

» **Inverter input circuit:** The conductors from the DC disconnect(s) to the inverter make up the *inverter input circuit.*

» **Inverter output circuit:** The conductors on the AC side of the system on the output side of the inverter comprise the *inverter output circuit.*

*Note:* The *NEC*® defines conductors based on their role in circuits. For example, the *NEC*® makes references to grounded current-carrying conductors (for both AC and DC circuits) and grounded conductors (for both types of circuits). I suggest you become familiar with the Code language and terms for your PV installations by digging into the *NEC*® and picking up a book on understanding the *NEC*®.

# Checking Out Types of Conductors

For the components of a PV system (meaning the array, batteries, charge controller, and inverter) to work together and produce energy, something needs to join them together — enter wiring. The wires used to connect all the individual pieces of equipment in a PV system are known as *conductors.* The PV side of the system has positive and negative conductors, and each one is connected to its respective part of the PV modules.

The most common material for the conductors used in PV systems is copper because it has high *conductivity* (the ability to pass current) and is compatible with the conductor *terminations* (the place where the conductors are attached to the disconnects, circuit breakers, inverters, and charge controllers).

WARNING

Less expensive aluminum conductors can be an option, but aluminum is often viewed as an inferior wiring method. If you choose to use aluminum, the conductors generally need to be larger because aluminum doesn't possess the same conductivity as copper, and all the conductor terminations in a PV system would require explicit aluminum ratings. (Flip to Chapter 13 for full details on sizing conductors and safety devices.) The terminations portion of the requirement can be the most difficult part to meet. If aluminum conductors (or copper conductors, for that matter) are connected to terminals that aren't rated for that material, the connections can eventually fail, increasing the risk of fire. My verdict: In comparison to the overall system costs, the savings associated with aluminum conductors in residential and commercial PV installations just isn't worth it.

Selecting the proper conductor isn't difficult, so there's no excuse for ever installing a conductor just because it was in the truck. Numerous conductor types are available, and each one has its proper uses and limitations, as you find out in the following sections.

REMEMBER

A few things to note regarding conductor classification:

- Conductors are always designated by an acronym that describes the conductor's properties. For example, USE is an underground service entrance cable.
- Conductors typically have multiple acronyms listed on them. As long as one of the properties noted by the acronyms meets the requirements for your installation, you can use that conductor.
- Some conductors have the designation *-2* at the end. This designation indicates that the conductors are rated for 90 degrees Celsius (194 degrees Fahrenheit) in wet or dry conditions. (I explain the significance of this temperature rating in Chapter 13.)

## USE-2

Standard PV modules come with two copper conductors preinstalled with quick-connect plugs for connection to adjacent modules and to help make the installation process easier. These factory-installed conductors, which are often *underground service entrance* (USE-2) cables, connect from module to module (or directly to a microinverter, which I describe in Chapter 9) and are secured to the backside of the PV array using specially designed clips and wire ties, as I describe in Chapter 17.

Because USE-2 is often installed along the backside of a PV array as part of the PV source circuit wiring, it gets exposed to some extreme conditions, including full sunlight and temperatures exceeding 60 degrees Celsius (140 degrees Fahrenheit). USE-2 conductors work in these extremes because they're inherently sunlight-resistant as well as heat- and moisture-resistant.

TIP

USE-2 conductors are readily available in a variety of sizes and can even be ordered in various colors (which can help when color-coding the wiring). For the portion of the PV source circuit that runs from the modules to the combiner box, you can buy rolls of USE-2 cable and install the quick-connect plugs yourself, or you can purchase USE-2 conductors that have the plugs preinstalled on the ends.

WARNING

Don't run USE-2 inside conduit and then inside buildings. It doesn't have the proper fire retardants in the insulation and isn't rated for these locations. For wires that go through buildings, you need to transition to building wiring (which I describe later in this chapter).

## PV wire

A double-insulated conductor known as *PV wire* uses cross-linked polyethylene for the insulation. This durable conductor was born out of the need for a conductor that could be used with a transformer-less inverter (see Chapter 9) because the *NEC*® requires additional protection in the form of conduit or multiple layers of insulation for the conductors on the PV side of these inverters. The two layers of insulation in PV wire help protect the copper conductors more than the single-insulated USE-2 conductor.

REMEMBER

PV wire is usually used inside PV source circuit wirings, but you can also use it as PV output circuit wiring. However, because it's more expensive than building wiring, most PV pros don't opt for this application.

TIP

PV wire isn't used in PV systems as often as the USE-2 conductor, but it has begun to gain popularity, and I recommend using it rather than USE-2 when you can. You can use it regardless of the inverter technology associated with the PV array, thereby reducing the different types of conductor you have to buy and manage on-site. Although PV wire is more expensive than USE-2, it's also more durable. Just like USE-2, for the conductors running from the modules to the combiner box or junction box, you can purchase PV wire in plug-free rolls or with preinstalled plugs.

## Building wiring

Conductors that leave the junction box or combiner box on their way to the inverter or charge controller are usually transitioned into a standard type of building wiring, such as heat-resistant thermoplastic (THHN) or moisture and heat-resistant thermoplastic (THWN-2). They can be aluminum or copper.

Building wiring is usually used for PV output circuits, inverter input circuits, and inverter output circuits. You can use it on both the DC and AC sides of a PV system, but you need to make sure you protect the conductors from damage by running them inside conduit (see the later related section). Note that building wiring can't be run in exposed locations because it doesn't have the sunlight-resistant characteristics of USE-2 and PV wire.

TIP

I recommend always using the THWN-2 conductor because it has additional temperature ratings. The -2 at the end indicates that the conductor is rated for 90 degrees Celsius (194 degrees Fahrenheit) in wet and dry conditions. Because you may have these conductors running in conduit along a roof, inside an attic, or down the side of a building, they may be exposed to some of the most extreme temperatures imaginable. Sticking with THWN-2 lets you use the same conductors in a variety of locations without too much concern.

WARNING

You can buy rolls of wire that are rated as both THHN and THWN-2, but this isn't considered standard. So if you go to your electric supply house and order THHN without specifying the THWN-2 requirement, you probably won't get what you want.

## Battery wiring

The most widely available conductors for use from the battery bank to the inverter in battery-based systems are moisture-resistant thermoset (RHW), moisture- and heat-resistant thermoplastic (THW), and underground service entrance (USE). All of these conductors are commonly made with copper.

TIP

Another battery wiring alternative is a highly flexible cable, which can have thousands of small, finely stranded copper wires wrapped together to make a large cable. People like to use these cables because tight spaces require a minimal bending radius, and the flexible wires are much easier to work with than the relatively stiff RHW, THW, and USE types.

WARNING

If you opt for flexible cables, make sure the connectors you use with them are properly listed for use with flexible stranded cables. In the past, flexible cables were used with terminals and crimp lugs that weren't properly rated. The result? The conductors failed and put the systems at risk of starting fires.

Any conductors used for battery wiring are in corrosive environments, so you need to verify with the manufacturer and the *NEC*® that the conductor type you want to use is appropriate for the location. Chapter 3 in the *NEC*® covers wiring methods and materials, so refer to the Code for more info.

## Ground wiring

All the exposed metal parts of PV systems need a connection to ground, which means you need to connect the PV module frames, racking, and metal boxes to a conductor that's connected to a large conductor that's in contact with the earth. (I get into grounding in Chapter 17.) The conductors used to make the ground connections between PV modules are almost always bare copper. After the PV system's conductors are transitioned into conduit, either bare copper or insulated ground conductors can be used.

Ground wiring can be run with all the circuits mentioned earlier in this chapter: PV source, PV output, inverter input, and inverter output. Using bare ground wire to ground the PV modules is preferable because it's properly rated for that environment and readily accessible. Using building wiring, like THWN-2, for ground wire in conduit is great; it's easier to pull and has the appropriate color-coding on the insulation. (I explain the importance of color-coding in Chapter 17.)

## Considering Kinds of Conduit

With the exception of the PV source circuit wiring, the conductors used in PV systems need proper protection from potential damage; this protection is referred to as *conduit.* (PV source circuit wiring doesn't require conduit because *NEC*® allows you to use USE-2 or PV wire in these locations.) You can see some examples of conduit found in PV systems in Figure 10-2. Conduit can come in a wide range of sizes, as you discover in Chapter 13, and a variety of materials. Selecting the right conduit for the system you're designing depends on the location where the conduit will be used and any additional *NEC*® requirements. (Chapter 17 outlines some of the specific requirements surrounding conduit installation in PV systems.)

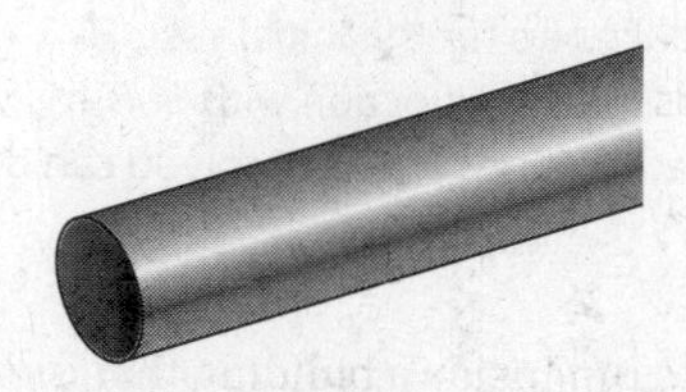

Electrical metallic tubing

Flexible metallic tubing

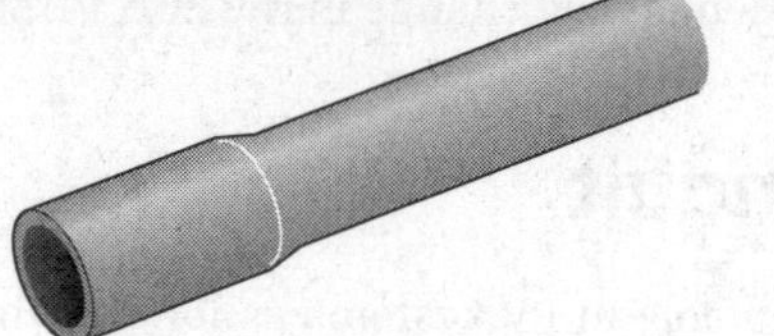

Rigid polyvinyl chloride conduit

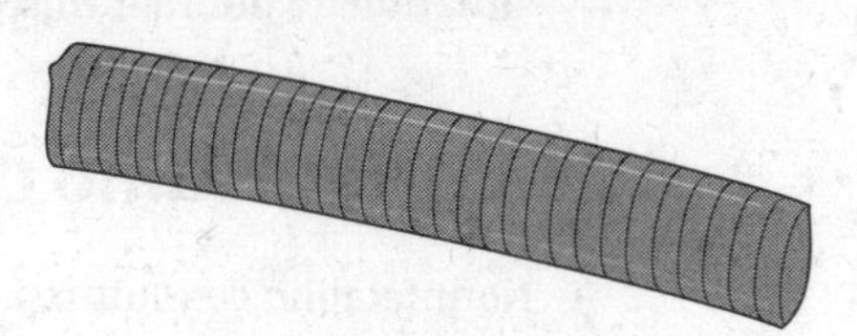

Liquid-tight flexible nonmetallic conduit

**FIGURE 10-2:** Conduit used in PV systems.

You have two main options for conduit: metallic and nonmetallic. Both provide protection for the conductors they house, but they have different installation and usage requirements.

### Metallic conduit

According to the *NEC*®, you can install PV circuits (I explain what these are in the earlier "Defining the Circuits in a PV System" section) for the PV output circuits inside a building prior to a disconnecting means as long as they remain in metallic conduit. A variety of metallic conduit options are available for use in PV installations, and each one offers excellent physical protection. Two of the more popular metallic conduit types are as follows:

- **Electrical metallic tubing (EMT):** A thin-walled, stiff pipe, EMT is used for the majority of the PV output circuit. It's an inexpensive choice that you can use both indoors and outside. You can bend it (with the help of a tool) when you need to change the conduit's direction, and you can change the conduit run using a wide variety of fittings. EMT requires more labor to install than flexible metallic tubing, but when you get the hang of it, working with EMT is easy.
- **Flexible metallic tubing (FMT):** As the name indicates, this is a flexible metallic conduit, which makes it great when you're working in a confined space. It's more expensive than EMT, but sometimes the time savings in using FMT versus EMT more than makes up for the initial cost. FMT is typically used for short runs, and it can be used in conjunction with EMT to make your job easier. FMT isn't allowed in wet locations, though, so you can only use it inside buildings.

WARNING

One wiring method for PV circuits run inside a building that *isn't* explicitly allowed by the 2008 *NEC*® is the use of metal-clad cables. Metal-clad cables are a *cable assembly* (conductors bundled together and wrapped with an outer protective covering); consequently, they don't comply with the wording of the *NEC*® in this situation. (Note that this guideline may change in the 2011 version of the Code.)

## Nonmetallic conduit

Nonmetalllic conduit is acceptable in PV systems as long as the conduit has the proper ratings (such as UV protection) and is listed for installation in a particular location (such as exposed wet locations). You can use nonmetallic conduit for any of the circuits in PV systems. Common types of nonmetallic conduit include the following:

- **Rigid polyvinyl chloride (PVC):** An inexpensive and easy conduit to work with, PVC is similar to EMT (see the preceding section) in the sense that the material is stiff and can be bent or attached to specific fittings. One feature to be aware of is that PVC expands and contracts a great deal with changes in temperature, which requires you to install special expansion fittings on long PVC runs. (The *NEC*® has specific requirements for this expansion and contraction.) PVC is used for PV output circuits that run underground or on the outside of a building.
- **Liquid-tight flexible nonmetallic (LFNC):** LFNC is appropriate when you need to use a flexible conduit in an outdoor location. It's easy to work with and used to connect PVC conduit to a disconnect or inverter. It's more

expensive than PVC conduit, and the fittings used aren't cheap either, which means you'll generally run PVC as close as possible to the desired endpoint and use LFNC to make the final connection.

# Delving into Disconnects

The conductors used in PV systems carry the current from the array down to the inverter and then to the loads. For safety and maintenance reasons, you need to install a way for those conductors to disconnect themselves from all sources of power. The exact location and specifications for disconnects depend on the specifics of the system installed, but all PV systems, regardless of their size, must possess the ability to disconnect the conductors.

Figure 10-3 shows two types of disconnects used in PV systems:

- Figure 10-3a is a disconnect integrated into an inverter. These can often disconnect both the PV output circuit (DC) and the inverter output circuit (AC) at the same time. They're an excellent way to meet the *NEC®* disconnect requirements for inverters, but they don't meet the requirements of utilities that insist upon visible, lockable disconnects.
- Figure 10-3b shows two disconnects positioned outside of an inverter. Each disconnect is dedicated to a single circuit (PV output and inverter output, respectively). Installing two disconnects requires more space around the inverter and increases your installation time, but if the utility calls for a visible, lockable disconnect on the inverter output circuit, these disconnects satisfy both the utility and the *NEC®* requirements.

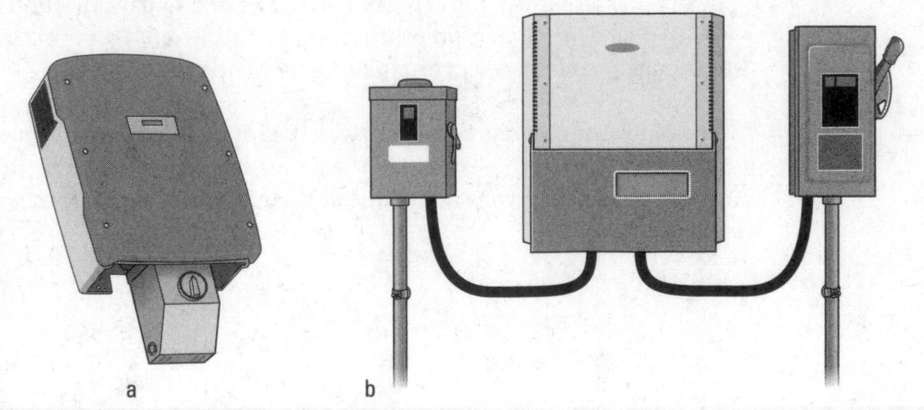


**FIGURE 10-3:** A disconnect integrated into an inverter (a) and separate disconnects (b).

According to the *NEC®*, DC and AC disconnects must be provided to stop power flowing in the conductors and allow for safe access to all components of the PV system. Generally, this guideline means supplying disconnects at the inverter location(s) as well as at points of interconnection with the utility (this last part applies only to utility-interactive systems).

REMEMBER

One common (and unpopular) required disconnect is a visible, lockable disconnect at the utility meter location. The utility typically requests this disconnect for its personnel. It must be physically placed next to the meter (so the utility can access it 24 hours a day) and electrically placed between the inverter and the utility point of interconnection. The idea is that if a utility power outage occurs and a utility worker wants to, she can disconnect the PV array from the grid and lock the disconnect for safety. When you ask the utility to send a net-metering agreement, look through the packet you receive for disconnect requirements. If you have any doubt or want to install the disconnect(s) slightly differently than specified by the utility, be sure to contact the utility for clarification.

REMEMBER

The *NEC®* doesn't require any disconnects at the PV array, but some local jurisdictions do. Always check with the electrical inspector to see whether your client's jurisdiction has any additional requirements for disconnects.

***Note:*** Disconnects can also serve as the form of overcurrent protection, like a circuit breaker that makes the utility interconnection for a utility-interactive inverter. See the next section for the full scoop on overcurrent protection.

# Perusing Overcurrent Protection Devices

When conductors are installed in PV systems, they need protection from the possibility of too much current passing through them. Enter overcurrent protection devices (OCPDs). When too much current begins to pass through a conductor, an OCPD opens the circuit, preventing additional current from passing through until someone manually resets or replaces the device.

The following sections cover the two types of OCPDs used in PV systems — circuit breakers and fuses — so you can get a feeling for exactly how they protect conductors and where you should install them. (I show you how to size these devices in Chapter 13.)

## Circuit breakers

Circuit breakers are the devices you most likely have installed in the main distribution panel at your home. The circuit breakers used in PV systems are very similar. They may have a different look based on the manufacturer, but for the most part they operate in the same way: When too much current passes through a circuit breaker, the breaker trips, opening the circuit and stopping the current from flowing. Most breakers are thermally activated, so they trip when they reach a certain temperature. The nice feature about circuit breakers is that when they trip, you can simply reset them instead of replacing them.

REMEMBER

Circuit breakers are manufactured for specific enclosures. Consequently, you can't install just any circuit breaker inside a load center; instead, you have to buy the breaker that was manufactured for the exact panel you're using. Circuit breakers are also manufactured for specific current and voltage levels as well as the current type (AC or DC). Always make sure the circuit breakers are listed for the circuit you want to install them in.

REMEMBER

Circuit breakers are typically used as the OCPD on the AC side of PV systems (in other words, the inverter output circuit) as well as on the DC side of low-voltage, battery-based arrays (meaning the PV source circuit, inverter input circuit, and inverter output circuit).

WARNING

Don't use circuit breakers on the DC side of high-voltage PV installations because they don't carry the proper ratings for high-voltage applications (which is anything greater than 150 VDC).

## Fuses

When you're dealing with a high-voltage DC circuit or when the OCPD needs to be placed outside of a load center, you need to install a fuse. Fuses come in all shapes and sizes, and they can fit any PV system requirement.

Fuses work pretty similarly to circuit breakers, which makes them a good substitute when you can't use circuit breakers. Inside every fuse is a filament that remains intact as long as it doesn't get too hot. If too much current passes through the fuse, the fuse overheats and pops, and the current is interrupted. ***Note:*** A popped fuse must be replaced; it can't be reset like a circuit breaker.

Like circuit breakers, fuses have specific listings associated with them. The main one you need to be aware of is the DC rating. Many fuses carry both AC and DC current ratings but at very different voltage levels. You need to make sure you order the fuses with the proper DC voltage and current ratings when you use them on the DC side of a system (typically the PV source circuit).

The most common locations for fuses are in a combiner box for the PV source circuit and inside an inverter for the connection between the PV output conductors and the inverter. Fuses are also good for when an inverter's output circuit can't be connected to the utility through a circuit breaker. In this last example, a fused disconnect provides both overcurrent protection and a method for disconnecting the inverter in one device (I cover disconnects earlier in this chapter).

# Focusing on Ground Fault Protection

To help reduce the risk of fires when a PV array has experienced a fault, ground fault protection (GFP) devices have become a requirement for all but a few PV installations. When you deal with grid-direct systems, you need only consider the proper installation techniques because GFP comes standard in grid-direct inverters (see Chapter 9). When you're working with battery-based systems, however, expect to have an additional component to install. (Fortunately you only have one option in terms of current ratings. Check the *NEC*® for its specific requirements regarding interrupting faults.)

GFP is added to PV systems at the inverter inside a DC wiring box to reduce the risk of fire if an array's conductors become compromised. If either the positive or the negative conductor is damaged, that conductor can come into contact with a metal component in the PV system such as a module, the racking, or the inverter. This contact between the conductor (which contains the current from the PV array) and metal pieces allows current to flow and presents a fire (and shock) hazard.

The GFP merely reduces the fire risk and alerts the system owner of the problem; it does nothing to reduce the risk of shock. In fact, after the GFP has been activated, the shock hazard is even higher because there's now a voltage potential between anything metal and ground. So if you were to grab the portion of the array that was faulted, the current would flow through your body, causing a shock throughout your body at best and death at worst.

A GFP works by purposely connecting one of the current-carrying conductors across the GFP circuit. This circuit is kept closed through a small-amperage fuse (generally just 1 A). The circuit is looking for any current that's flowing from one of the current-carrying conductors to ground. Under normal circumstances, there's no current to ground, but as soon as a wire comes into contact with a grounded portion of the array (anything metallic), current then flows to ground. This improper current flow causes the fuse inside the GFP to pop and a visible sign to appear, like a warning light or a circuit breaker in the off position, indicating the presence of the ground fault. So when the GFP is open, a reference to ground is no longer present for the conductors. This situation is known as *floating*, and it's a dangerous one for whoever's troubleshooting the array.

If the GFP has been activated, you must track down and fix the ground fault before the array can continue operation. This process can take some time and skill because tracking down a ground fault can be difficult and complicated. (One resource for you to research the methodology and safety precautions is *SolarPro*, `solarprofessional.com`, a magazine that has covered this topic in great detail.) After the ground fault has been fixed, you can replace the fuse with a properly rated replacement and restart the system.

# Looking at the Basics of Labels

Proper labeling is growing increasingly important as PV systems become more popular. Inspectors are now requiring PV system installers to do a better job of meeting Article 690 of the *NEC*®, which lists a number of labeling requirements for PV systems. I outline the exact language needed on system-specific labels in Chapter 18, but in this section, I outline the general requirements and the generic labels needed for every job.

PV system labels give safety personnel, such as firefighters, quick information they can use in an emergency. Consequently, you should always design your system labels so they present the necessary information simply and accurately.

Here are some of the major labeling requirements you must meet in any PV system you design (you can purchase premade versions of the generic labels, but you'll need to have the system-specific ones made at a local sign shop):

- The DC disconnect needs labeling with specific information about the different voltage and current levels (see Figure 10-4a).
- The point of interconnection with the utility needs labeling indicating voltage and current levels (see Figure 10-4b).
- The building exterior needs a plaque or label identifying the secondary power source (the PV array) and all disconnects for the system (see Figure 10-4c).

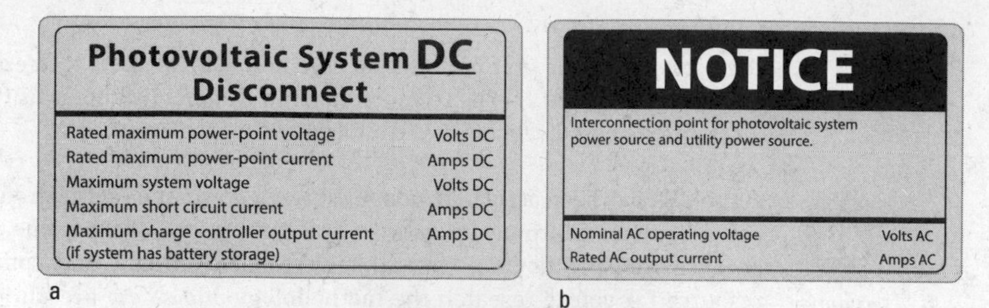


WARNING

POWER TO THIS SERVICE IS ALSO SUPPLIED FROM PHOTOVOLTAIC EQUIPMENT LOCATED ON THE WEST ROOF

DISCONNECTING MEANS LOCATED INSIDE THIS PANEL AND ON ROOF AS SHOWN

ROOF TOP SOLAR ARRAY

ROOF TOP SOLAR DISCONNECT

MAIN SERVICE PANEL

c

**FIGURE 10-4:** Sample labels for PV systems.

# 3 Sizing a PV System

## IN THIS PART . . .

Here's your chance to walk through the process of properly sizing both grid-direct and battery-based PV systems. The different methodologies in this part are presented based on the type of system you're going to install and your goals for the system.

Chapter 11 details the process for sizing and specifying a grid-direct PV system, the most popular type of PV system currently installed. If you're installing a battery-based system, either utility-interactive or stand-alone, then Chapter 12 is your go-to resource for sizing and specification assistance. And because all PV systems require safety components, Chapter 13 shows you what you have to consider when sizing and specifying the different safety components.

IN THIS CHAPTER

» Understanding the role budget and space play in limiting an array's size

» Evaluating energy production and consumption

» Calculating power values and temperature-adjusted voltages

» Checking the maximum current input as the final piece of the puzzle

# Chapter 11
# Sizing a Grid-Direct System

So you've used the information in Chapter 2 to determine that a grid-direct system is the best fit for your client, and you've familiarized yourself with the elements of such a system in Part 2. Now the fun really starts — it's time to *size the system!* By that, I mean it's time to match the array to the customer's needs and to the inverter and associated safety equipment.

When you set off on the task of sizing a grid-direct PV system you need to account for many variables. As you find out in this chapter, the overall process results in a PV system that's based on the client's budget, the available area, the annual energy production and consumption at the site, and your choice of materials for the job (notably, the array and the inverter).

REMEMBER

In comparison to stand-alone PV systems (which I show you how to size in Chapter 12), grid-direct sizing is much less complicated. You also don't have to make nearly as many assumptions and leaps of faith. Because the utility grid is always (well, almost always) present, and because a grid-direct inverter operates in parallel with the utility, you don't have to size the array to meet electrical consumption at any given time. If the loads within the building require more power than the array can provide, the utility can supply the difference. And even though the PV system goes down when the utility goes down, the system automatically restarts when the grid comes back to life. A grid-direct system, therefore, should be viewed as a way to supplement (rather than replace) the utility grid.

*Note:* Sizing a grid-direct PV system requires you to know a few temperatures and be able to make calculations with those temperatures. The calculations get pretty messy when you try throwing Fahrenheit into the mix, which is why all the temperature-related calculations presented in this chapter use Celsius.

## First Things First: Evaluating the Budget and the Available Array Area

Whenever you're looking to install a grid-direct PV system, the overall system size will generally be limited by one or more factors, but the first two you need to consider are your client's budget and the amount of space available for the PV array. People with unlimited budgets and unlimited space for a PV system are out there, but those clients are few and far between.

REMEMBER

Before you establish anything else, find out what your client's budget is. Of course, everyone would love to offset 100 percent of their electrical energy consumption with solar power, but most people simply can't do that. So before you spend too much time sizing the best-possible system for your client's desires, find out what his budgetary realities are first and work from there. By establishing your client's budget in the beginning of the process, you can prevent yourself from wasting a lot of time and energy designing a system that never has a chance of being installed.

The next limitation to consider is the area available for mounting the array. For the majority of grid-direct PV systems, this area is the roof of the house or business. (Other options include the ground outside of a building or the top of a pole.) To determine the amount of space available for the system, you need to perform a site survey; Chapter 5 goes over the site-survey process and the major considerations, such as shading issues and various roof obstacles, you need to make when evaluating where to place the array.

After you conduct a site survey, you can calculate the available square footage by measuring the length and width of the array area and multiplying those two values together. You can then take that area and multiply it by the module's *power density*, or the number of watts per square foot for the module you're thinking about using based on the client's budget and what's available from your suppliers

(Chapter 6 has details on power density and different types of modules). The resulting number gives you an idea of the total power in watts (W) that can be placed in the available area.

WARNING

I recommend you don't attempt to use every last square inch of a roof surface when installing an array. Squeezing a few more modules up there whenever possible is tempting, but don't do it! Cramming the roof will not only make the installation more difficult but it'll also make it tricky (and possibly downright dangerous) to try and maneuver on the roof when the system needs maintenance. Also, wind loading at the edge of the roof is much greater (requiring more mechanical support), and some jurisdictions have requirements based on access for firefighters and other safety personnel. (The same goes for array sites other than roofs — when it comes to placing PV modules in a given spot, less is definitely more.)

Be sure to consider "dead" spaces (the areas around the array that can't be used due to shading or the need to maintain sufficient paths for access) as well as the small spaces between the modules when determining the overall area available for the PV array. Suddenly that decent-sized roof may be very limited to account for all of these issues. To account for dead spaces, simply subtract the area needed for the access paths from the overall area and use one of the shading-analysis tools in Chapter 5 to determine the area that's unavailable due to shading. By taking these areas out of the picture, you can establish a realistic idea of the total power rating of the potential array.

REMEMBER

The size of the PV array may change over the course of your design process. For example, you may estimate that you can fit 24 modules on a roof, but as you progress, you may realize that the string configuration requirements (covered later in this chapter) won't allow that many modules. Or perhaps you have to consider a different module type entirely. Consider your initial estimate of the available space a starting point and use it only for reference.

# Estimating the Site's Annual Energy Production

When you know roughly how much space you have for the array, you can begin to estimate the annual energy production for that site. This estimate is helpful to have when designing grid-direct PV systems because you need to evaluate the type of agreement the utility will be willing to enter into. (I review some common agreements and how they affect your decision-making process in the later "Looking at contract options with the utility" section.)

You can estimate the annual energy production in a variety of ways. For larger commercial systems or in situations where advanced techniques are required, you can use modeling programs to evaluate multiple scenarios and change parameters to estimate the energy production (PV*Sol and PVsyst are two popular programs). For residential and small commercial systems, one of the best ways to estimate energy production is to use the PV Watts tool (available at `www.nrel.gov/rredc/pvwatts`).

PV Watts is a free Web-based tool provided by the National Renewable Energy Laboratory. Many people in the solar industry use PV Watts on a regular basis. To use it correctly, you need to know

- The amount of power the PV array you're thinking of using can potentially produce (in kilowatts [kW])
- The tilt, orientation, and shading effects of the array (see Chapter 5)
- Some basic information about the equipment you're thinking about using (the PV Watts program makes some good assumptions about the performance of the equipment)

The PV Watts tool takes the information you enter and creates a *total solar resource factor* (TSRF), the percentage of the solar resource available for that specific site in relation to a perfectly oriented and shade-free array, as I describe in Chapter 5. It then provides a simple report that estimates monthly and annual energy production values measured in kilowatt-hours (kWh). It also calculates the dollar value for that energy, which is nice because this is a value everyone can relate too (unlike peak sun hours or system efficiencies).

TIP

I suggest you use the first version of the PV Watts calculator until you become familiar with the program. Then you can move on to Version 2 if you like. (I rarely use Version 2 because Version 1 gets me all the information I generally need.) Either way, PV Watts is relatively self-explanatory; it even has helpful links for commonly asked questions and ways to manipulate the program based on your needs.

REMEMBER

The annual energy production you calculate at this point is only an estimate. Don't treat it (or let your client treat it) as the promised production value. The energy produced from the array will be affected by installation methods to an extent, but ultimately, the available solar resource dictates how much energy is produced. So if your client experiences an abnormally cloudy or sunny season, the energy-production number you gave him will be off.

# Sizing the Array to Meet Your Client's Energy Consumption

I promise that if you're in the PV industry for more than two weeks, you'll be asked this question before you finish telling someone what you do for a living: "How big of a PV array do I need to power my average-size house?" My favorite response: "Gosh, I'm not sure. What color is your house?"

Neither question makes much sense, does it? Sure, I can figure out what the average American household's energy consumption was two years ago, but that number rarely means much to anyone individually. The energy consumption of an individual, family, or business has little to do with the size of a house or commercial building and more to do with people's lifestyles. Does your client have all-electric heating? How does he heat his water? Does he have an electric car hidden in the garage, just charging away? Is his business home to a welding shop or a number of refrigerators?

After you establish the maximum array size and estimate the site's annual energy production, you need to look at some utility bills (preferably all 12 bills from the previous year) to get a sense of how much energy the house or business consumes and what the impact of a rooftop or yard-based PV system will be. You also want to guarantee that the energy the system produces provides the maximum financial benefit for your client based on the agreement with the utility. I explain what you need to know in this section.

## Determining annual energy consumption

Grid-direct PV systems have the advantage of built-in energy record keeping from the utility provider. To determine the annual energy consumption for the household or business in question, simply collect the last 12 months' worth of bills from the utility. This snapshot will give you a great idea of the amount of energy consumed annually (so long as you look at the total kilowatt-hours consumed). Of course, if you have access to more than one year's worth of utility bills, go ahead and take a look at them all. The extra information can only add to your knowledge of your customer's energy habits.

TIP

One thing to ask is whether there are any recent changes to the electrical consumption. For example, did your client install a solar hot-water system a few months ago? Or did he just put an electric water heater in the master bedroom suite? Such changes affect future electrical consumption, which is why you should always base your estimates on the most recent information.

*Note:* If you can't obtain the energy records for the client's current home or past home because the current electrical consumption is dramatically different, you may have to estimate the annual energy consumption by using the same process I describe in Chapter 12 for sizing battery-based systems.

## Looking at contract options with the utility

Your client has a few options for entering into a contract with the utility, but the most common approach is net metering. In a *net-metering agreement*, the utility agrees to "pay" your client the exact amount it charges him for energy — given that your client doesn't produce more energy than he uses in a given time period (typically a year). The exact restrictions are included in the interconnection agreement provided by the utility. Usually, if a PV user produces more energy than he consumes in a year under a net-metering agreement, the utility can say thanks very much for the extra energy and move on. The PV user (in this case, your client) doesn't get any extra accolades or cash for producing more energy than he consumes.

REMEMBER

Many utilities allow customers to essentially bank up their kilowatt-hours when they produce more than they consume — kind of like when your extra calling-plan minutes roll over. At some point, though, the utility will start over; this is known as the *true-up period.* Some utilities make this true-up period occur monthly, which means your client doesn't get the full benefit of the energy produced in any month that he doesn't use that same amount of energy. As the installer and designer, be sure to confirm with the utility how its true-up period works before getting too far into the design process.

Another component of net metering to consider is the *time-of-use metering option* (TOU). With TOU, the utility charges your client different rates based on the time of day he uses energy. Typically, TOU rates are highest during the middle of the day when overall consumption peaks. This means that for a PV system installed under a net-metering agreement that uses TOU, the times of the day when the PV array is producing the greatest amount of energy correspond to the times when the utility rates are at their highest. If your client can maximize PV production and minimize consumption to correspond to peak energy rates, say he's at work all day with as many appliances powered down as possible, the overall effect can be beneficial financially.

The other contracting system that may be available is the *feed-in tariff system* (FIT). With a FIT contract, your client doesn't really care about annual energy consumption because there's no direct relationship between his PV array size and his energy consumption due to the fact that he receives more money per kilowatt-hour generated from the PV array than he's charged from the utility. So if your client signs a FIT contract with the utility, you want to design the PV system so it maximizes the amount of energy the array produces regardless of your client's energy appetite.

*Note:* Regardless of the utility contract, I encourage you to advise your client to conserve energy as much as possible even though he doesn't really have to worry about comparing his energy consumption to the size of the array. Energy conservation should be the mantra of anyone installing a PV system. Besides, conserving energy helps your client financially — the less energy he consumes, the less he has to buy.

## Using consumption and contract options to select an array's needed power value

At this point, you've collected data about how much energy your client consumes and established a starting point for the total amount of energy the proposed PV array will produce annually. Now you can begin comparing the two numbers and working with your client to establish the best option based on the utility agreement he has to enter into.

If your client consumes less energy annually than the PV array will likely produce, he's in great shape. He can either reduce the overall wattage of the array or be prepared to overproduce annually and not receive full financial credit for every kilowatt-hour produced by his system.

A more likely scenario is that the energy produced by the PV array is less than the energy consumption of the people in the building. In this case, you can work with your client to help him reduce his overall energy consumption (common techniques for this include changing to compact fluorescent light bulbs and installing better insulation). You can then compare the estimated production to the assumed reduced energy consumption and establish the percentage of electrical energy that will be provided by solar power and the associated dollar savings. For example, if your client's PV system can produce 2,000 kWh but the client consumes 5,000 kWh annually, the PV array will offset 40 percent of his total energy consumption.

REMEMBER

Keep the true-up period that the utility uses in mind (see the preceding section for the scoop on this). If the utility looks at the values monthly, you need to evaluate the production and consumption ratio on a monthly, rather than annual, basis. Make sure your client is aware of how the interconnection agreement works so he isn't surprised if he doesn't get full credit from the utility because he produced twice as much energy as he consumed.

Applying the right consumption numbers and contract options may seem like a lot of upfront work, but after you go through the process a few times, you can establish some good working numbers based on the equipment you like to use, local utility requirements, and the site-specific information you gather. Having these numbers in mind helps you finalize the proposed array's design, including its size.

## Getting Ready to Match an Inverter to an Array

After you nail down the right PV array wattage for a client's grid-direct system, you're ready to establish the relationship between the inverter and the PV array. To do so, plan to look at all the electrical characteristics — power, voltage, and current.

I suggest looking at power, voltage, and current in that order so you can narrow down your inverter choices with each step. Another benefit to following this sequence? The final check, current levels, typically becomes a simple verification.

Every inverter manufacturer has online PV sizing calculators to help you with matching an inverter to a PV array. Even though I use them on a regular basis, I do so as a check of my calculations, not the source of my calculations. As a PV system designer and installer, it's vital that you perform your own calculations instead of relying on the online calculators. Here's why:

- **The potential for error:** The data plugged into the online calculators is entered by humans who're fully capable of making mistakes. Trust me; I've seen some of these mistakes firsthand.
- **The inevitable disclaimer:** Every inverter manufacturer requires you to accept a disclaimer before you can access its tool. In essence, the disclaimers all say the same thing: The information here may or may not be accurate, and if you use the tool, we (the manufacturer) aren't responsible for any problems.
- **The *National Electrical Code® (NEC®)*:** One section of the *NEC®* says that "you shall do the calculations." At some point, the local electrical inspector may ask you to show him these calculations. If you can't, you run the risk of having your system fail inspection. (*Shall* in *NEC®* terms is pretty strong language. It's equivalent to the sideways look you used to get from mom and grandma — something not worth messing with.)

REMEMBER

Occasionally, you may want (or need) to use multiple inverters in a system. In grid-direct systems, the most common reason for having multiple inverters is that the desired PV array wattage is greater than a single inverter can handle. Many of today's inverters are rated at less than 10 kW, but fewer options exist between 10 kW and 30 kW, so if the PV array wattage falls into this range, you may need to design for multiple inverters. Another popular reason for having more than one inverter in a grid-direct system is that sometimes the PV modules are installed in different locations or sometimes they're all on a roof but pointing in different directions.

# Matching Power Values for an Array and an Inverter

As I note in Chapter 9, all inverters are rated by their maximum continuous power output, which is measured in watts or kilowatts. (More often than not, this number is incorporated into the inverter's model number, giving you a quick idea of the inverter's rating.) This value is the AC power output. Inverters actually limit their power output, so you can use this value to figure out the maximum power input coming into the inverter from the PV array.

When you look at PV module ratings, you find that the power output ratings are based on standard test conditions (STC) where the cell temperature is 25 degrees Celsius and the intensity of the sun is equal to 1,000 W/m$^2$. PV modules rarely operate in these conditions due to the motion of the sun across the sky and increased PV operating temperatures (see Chapter 6). So, the power output from the array (in other words, the power input coming into the inverter) is typically considerably less than the rated values.

TIP

On top of that, both the inverter and the array (along with all the other equipment in a grid-direct system) have efficiency losses that must be taken into account. Fortunately, you can quickly estimate all of these system inefficiencies by adopting the industry standard that any PV system will operate at approximately 80 percent overall system efficiency, which means that about 80 percent of the array's rated value will be "processed" by the inverter and pushed on to the load center. Even though a PV system as a whole can be more efficient or less efficient at times, approximately 80 percent is a good industry standard to use when relating the PV array output power to the inverter's output power.

So how do you get started? Well, because you've already established the approximate wattage of the PV array you want to install (if you haven't, refer to the earlier "Sizing the Array to Meet Your Client's Energy Consumption" section), you can now take that array wattage and relate it to the inverter's rated output power. Because inverters are limited in the amount of power they can process, it doesn't make a lot of sense to try and make an inverter work any harder than it can. You therefore need to put enough, but not too much, PV power into the inverter in order to have the inverter efficiently produce AC power on the output side.

The industry standard for turning PV-rated DC power into AC output power is 80 percent. You can use this value to help determine the size of the inverter based on the array you're working with. For example, if you go through the process I describe in the beginning of this chapter and decide that your client is a good candidate for a 5 kW array, you can use that array size to narrow down your inverter choices. Because the overall system efficiency will be around 80 percent, you can multiply the 5 kW array size by the 80-percent efficiency rating to calculate the minimum inverter rating.

5 kW array × 80% efficiency = 4 kW minimum inverter rating

This calculation tells you that for a 5 kW array, you should have at least a 4 kW inverter. This inverter-sizing methodology is within the specifications of most inverters. Of course, you should confirm with the inverter's manufacturer that it supports this sizing method.

To find the maximum rated input power from the PV array, refer to the inverter's specification (spec) sheet. You can put less power on an inverter than what the calculated maximum array size suggests, but you should put on more only in very rare circumstances (and only after careful consideration of the effects). If you were to put on more power, the inverter would limit the output to its rated value, so your client wouldn't get any benefit from the extra modules anyway.

Believe it or not, I encourage you to put fewer modules on inverters than what they can handle. If you think about the last example, if you went ahead and put 5,000 W on that 4,000 W inverter, you'd be limiting the overall system efficiency to 80 percent. On days when the efficiencies are higher than normal (say, a cool day with high irradiance conditions), the inverter would have to limit its output, resulting in a lost opportunity. Also, I don't like to see inverters pushed so hard for long-term reliability. An inverter's internal temperature rises as the number of watts out of the inverter rises. As with all electronics, the hotter an inverter runs, the shorter its life will be.

When I design a PV system, one of my first goals is to multiply the PV array's power value by 87 percent rather than the 80-percent maximum value. Doing so gives me a larger inverter wattage that's closer to the PV array's wattage. By keeping the PV array size closer to the inverter's rating, you keep the inverter cooler, increase its life, and allow the array to operate at maximum efficiency values at all times.

So in the earlier example, the minimum-size inverter for a 5,000 W array would be:

$$5{,}000\text{ W} \times 87\% = 5{,}000\text{ W} \times 0.87 = 4{,}350\text{ W}$$

However, a 4,350 W inverter isn't an option (because inverters come in nice round numbers like 4 kW or 4.5 kW), which means you need to look for an inverter that's close to this value. You may be able to find a 4,500 W inverter, or you may need to jump all the way up to a 5,000 W inverter. (Note that this efficiency level allows the system to operate at approximately 87 percent, a value that won't limit the array's power output. The real world may of course interfere, but the 87-percent efficiency level is still a solid starting point.)

# Coming Up with the Right Voltage Values for Your Array and Inverter

After you know the necessary power values for the PV array and the inverter, you can look at operating voltages in real-world conditions to further narrow your search for an inverter.

- On the inverter's DC side (which is connected to the array), there's a minimum number of volts required to allow the inverter to produce the maximum amount of power from the array and a maximum voltage that can be applied without damaging the inverter.
- On the inverter's AC side (which is connected to the loads), a minimum and maximum voltage value is dictated by testing requirements and anti-islanding values (see Chapter 9 for details on anti-islanding).

REMEMBER

The minimum to maximum voltage values on both the AC and DC sides define the inverter's two voltage windows (a *voltage window* is a range of allowed voltages). If the voltage values fall outside of these windows, the inverter won't work; in the case of too much DC voltage, the inverter can actually be damaged. Keeping the voltages inside these windows is an important design consideration. After all, it's your job to keep the PV array operating and maximizing your client's investment.

To keep the voltage values within the inverter's voltage window on the DC side, you must define the adjusted voltage values of the PV array and the DC voltage window based on temperature. In Chapter 6, I explain how PV modules are affected by temperature, so head there if you need to see those relationships. (You also need to remember the relationships between open circuit voltage, $V_{oc}$, and maximum power voltage, $V_{mp}$, for this portion of the design; I cover these in Chapter 6 as well.) For the inverter's AC voltage, you must choose an inverter that matches the available AC voltages. I explain what you need to know in the sections that follow.

## Establishing the inverter's AC voltage

On the AC side, you need to determine the nominal AC voltage that the utility power is operating at (and which you'll be interconnecting to) and make sure you *specify* (choose) an inverter that operates at that same voltage. A qualified person (such as an electrician), should verify this voltage, preferably during your initial site survey (covered in Chapter 5).

TIP

The nominal AC voltage for residential systems is usually 120/240 VAC. For commercial systems it's either 120/208 VAC or 277/480 VAC. (*Note:* You should still check with the individual utility, just to be safe.)

REMEMBER

Many inverters allow you to *field select* (change the voltage settings on-site) the nominal AC voltage, but not all do. For those that operate at a specific nominal voltage, you need to make sure you specify the correct inverter for the voltage present. For example, if you're installing an inverter at a residential location where the voltage is 120/240 VAC, you can't install an inverter that's meant for a 120/208 VAC system. You can't mix up the inverters and expect the inverter to ever work.

Completing this step of matching the inverter to the AC voltage present helps reduce the pool of eligible inverters even further. Now you can evaluate your PV array in relation to very specific models of inverters and the DC requirements for those units.

## Defining the inverter's DC voltage window

In Chapter 9, I introduce you to the concept of an inverter's voltage window. This window consists of a maximum input voltage that you must stay under if you want to avoid damaging the inverter and a minimum voltage you must stay above to keep the inverter operating at the array's maximum power point. As a PV system designer, you need to identify these values from a spec sheet and make sure the array can operate within this window throughout the year. If you allow the array to move outside this window, you risk damaging the inverter or shutting down the system for the day.

As you can see in Figure 11-1, the DC voltage window can be viewed in relation to the PV array in an IV curve (refer to Chapter 6 for an explanation of these curves). The goal is to keep the $V_{oc}$ portion of the IV curve below the maximum inverter voltage value and the $V_{mp}$ portion within the operating voltage range. Each of the voltages gets adjusted based on temperature, so this window can actually become very narrow very quickly.

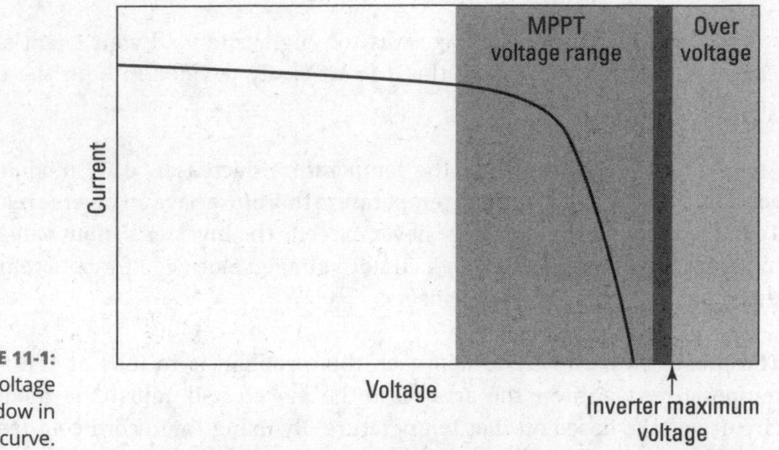


**FIGURE 11-1:** The DC voltage window in an IV curve.

REMEMBER

The minimum and maximum DC voltage values dictate the number of modules that can be placed in any string that's connected to an inverter. Typically, the calculations I show you in the next two sections result in multiple string lengths — different numbers of modules you can place in series for proper inverter operation.

WARNING

Most inverters have just a single maximum power point tracker installed. Unless you're using an inverter that has multiple built-in maximum power point trackers, you must keep the strings of PV modules equal in size. Therefore, if you calculate that an inverter can accept multiple strings of 11 modules or strings of 12 modules, all the strings must be either 11 or 12 modules in length. In other words, the string lengths must remain equal, or else the inverter can't operate correctly.

## Calculating the modules' maximum DC voltage contribution

When working with the DC side of an inverter, I like to consider the inverter's maximum DC voltage input first because providing too much voltage to the inverter can cause damage. You see, all inverters have capacitors inside them that act as shock absorbers. They accept the array's power and are able to smooth out any bumps along the way, which means the capacitors are one of the very first things connected to the PV array. If too much voltage is connected to these capacitors, they'll eventually become compromised and fail, meaning the inverter can't operate.

WARNING

You should also know that all inverters record the maximum voltage values ever seen by the inverter. So if you design or install your client's system incorrectly and apply too much voltage to the inverter, that information will be recorded. If your client sends the inverter in for a warranty repair, the manufacturer will check the stored data, see that the voltage was too high, and void your client's warranty. This is an expensive mistake that can be easily avoided in both the design and installation processes.

As I explain in Chapter 6, as the temperature decreases, a PV module's voltage actually increases (voltage and temperature therefore have an inverse relationship). To make sure an array's voltage never exceeds the inverter's input value, you need to adjust the modules' STC open circuit voltage value ($V_{oc}$) for cold temperatures; this value is reported on spec sheets.

The most common way to approach this problem is to look at the record cold temperature wherever the array is to be placed and adjust the modules' open circuit voltage based on that temperature. By using the record cold temperature, you're accounting for the fact that it takes very little irradiance to produce voltage from a module (*irradiance* is the intensity of the solar radiation striking the earth). And because current won't be flowing immediately (because the irradiance value is very small when the sun breaks over the horizon), the inverter will be connected to this initial high voltage.

In reality, this is a conservative approach because, as I note in Chapter 6, a PV module reaches approximately 90 percent of its full voltage at 200 W/m$^2$. By the time the module is producing full voltage, the array has sufficient irradiance to produce current, and the voltage will automatically drop to maximum power voltage values. The exact values are difficult to calculate accurately, so in all the calculations I show you in the following sections, I use the record cold temperatures and say that the voltage will jump to full open circuit voltage as soon as the sun breaks over the horizon each morning.

With that in mind, you're ready to calculate a PV module's adjusted open circuit voltage based on temperature. First things first: Keep the big picture in mind. You must determine the maximum voltage produced by the module based on the record cold temperatures and then use this adjusted voltage to determine the maximum number of modules you can place in a series string without exceeding the inverter manufacturer's requirements for the maximum DC voltage input. I walk you through the process in the sections that follow.

## Talking about temperature coefficients

In order to accurately calculate a PV module's adjusted open circuit voltage, you need to know how the module's manufacturer measures how its modules' voltage values will react at temperatures less than and greater than the STC of 25 degrees Celsius. This number is known as a *temperature coefficient*, and it can be reported for both $V_{oc}$ and $V_{mp}$. (***Note:*** The two different voltage values have two different temperature coefficients, but many PV module manufacturers report only the temperature coefficient for $V_{oc}$. When I get to adjusting the $V_{mp}$ in the later "Crunching the numbers" section, I show you how to estimate this rarely provided coefficient fairly accurately.)

REMEMBER

A typical $V_{oc}$ temperature coefficient value for crystalline modules is −0.35%/°C. This value indicates that for every degree change in the PV module's temperature, the voltage changes by slightly more than a third of a percent. The negative number in the coefficient is very important; it shows the inverse relationship between temperature and voltage. (Keep in mind that −0.35%/°C is a close approximation for all crystalline PV modules, but it's by no means an absolute value. As for thin film modules, the temperature coefficient values vary widely among technologies.)

According to STC, the base temperature for all PV modules is 25 degrees Celsius, so you have to look at the change in temperature as it relates to 25 degrees Celsius. For example, if the module is on a rooftop and the temperature at dawn is 15 degrees Celsius, the module's voltage will be 10 degrees Celsius less than STC: 15°C − 25°C = −10°C (the negative sign indicates a temperature less than STC). So if you were asked what the percentage change in voltage is due to temperature, you could run the numbers like so:

$$-10°C \times -0.35\%/°C = +3.5\%, \text{ or a rise of } 3.5\%$$

Another way you may see temperature coefficients reported is as a certain number of volts per degree Celsius. The exact number is dependent on the $V_{oc}$ for a particular module and should only be used for that module. To see what I mean, suppose an array was in a 15 degrees Celsius environment. If I told you that the temperature coefficient for the module was −0.158 V/°C, you could take that information and tell me how many volts the module would be reading off of the STC of 25 degrees Celsius:

$$15°C - 25°C = -10°C$$

$$-10°C \times -0.158\ V/°C = +1.58\ V, \text{ or a rise of } 1.58\ V$$

TIP

You can use either value reported as long as you use them correctly in the equations. However, the second temperature coefficient mentioned, the number of volts per degree, is generally easier to visualize because you can imagine the numbers of a meter changing.

## Working the steps

To determine the change in voltage due to temperature, you need to apply the temperature-adjustment equation. Here's how to do just that:

1. **Collect the temperature coefficient for the PV module.**

   You can find this information on the manufacturer-provided spec sheet that comes with the module.

2. **Collect the record cold temperature for your client's location in degrees Celsius.**

   The Web site `weather.com` is a great resource for this data.

3. **Calculate how many degrees Celsius less than STC the site will be on that record cold day.**
4. **Multiply the temperature coefficient for the module by the number of degrees calculated in Step 3.**

   The result of this equation is the change in voltage the module will produce.

5. **Add the number of volts calculated in Step 4 to the $V_{oc}$ for the module at STC.**

   What you're left with is the adjusted maximum module voltage based on the area's record cold temperature.

To help you grow more at ease with the temperature-adjustment equation, try the following example. The PV module in question is a typical crystalline module located in a place that has a record cold temperature of −5° Celsius. The module specifications you need to collect are as follows:

- $V_{oc}$ = 45.0 V
- Temperature coefficient = –0.158 V/°C

At this point, you have all the information you need to determine the maximum voltage for this module at the area's record cold temperature. Follow the previously outlined steps to find that

1. The given temperature coefficient is –0.158 V/°C.
2. The lowest recorded site temperature is –5°C.
3. The number of degrees this temperature is from STC is –5°C – 25°C = –30°C.
4. The voltage change is therefore –30°C × –0.158v/°C = +4.7 V.
5. The adjusted module voltage is thus 45.0 V + 4.7 V = 49.7 V.

TIP

But what if the module manufacturer gives you the temperature coefficient in the percent per degree Celsius? The easiest thing to do is convert that percentage into a number of volts per degree Celsius by multiplying the percent per degree Celsius by the open circuit voltage. For example, if the module you're using is rated at 45 $V_{oc}$ and the temperature coefficient is given as −0.35%/°C:

$$45\ V_{oc} \times -0.35\%/°C = 45\ V_{oc} \times -0.0035/°C = -0.158\ V_{oc}/°C$$

REMEMBER

The steps for working the temperature-adjustment equation can be used with any PV module so long as you find the temperature-adjustment data specific to the module manufacturer — the PV technology (crystalline versus thin film) doesn't matter.

WARNING

The different PV module technologies all have their own temperature-adjustment factors, which means you can't swap the values of a crystalline module for those of a thin film module just because the manufacturer doesn't supply the information. If you do, you'll wind up with false information that may cause you to design and install a PV system that'll damage the inverter on the first cold morning.

## Using NEC® info in a pinch

If you're using a crystalline PV module, either single or multicrystalline, and the module manufacturer doesn't supply a temperature coefficient, the *NEC®* states that you can use the values presented in Table 690.7 (see Figure 11-2). This table allows you to look up a multiplier based on a temperature range and use that to determine the temperature-adjusted voltage.

If, for example, you're using a crystalline PV module with a $V_{oc}$ value of 45 V and the manufacturer doesn't supply a temperature coefficient, you can turn to Table 690.7 in the *NEC®*. If the record cold temperature is −5 degrees Celsius, just find −5 degrees Celsius in the table and look straight across to find the appropriate multiplier, which is 1.12. To determine the temperature-adjusted maximum voltage, simply multiply the module $V_{oc}$ value at STC by the multiplier. Here's what the equation looks like:

$$45\ V \times 1.12 = 50.4\ V$$

REMEMBER

Obviously, using the *NEC®* table is a lot easier than performing the temperature-adjustment equation yourself. So why would you ever want to take the time to run the calculations when you can easily find what you need in the *NEC®* table? Well, there are a few reasons:

- First off, Article 690.7 of the *NEC®* has a requirement that when the information is provided from the manufacturer, "you shall calculate the new voltage."

- Second, you may want to use thin film technology in the PV system you're designing, in which case the table doesn't even apply. If you're using any PV technology other than crystalline, you have to use the calculations I show in the previous sections to determine the maximum voltage.
- Finally, the *NEC*® will always be more conservative in its method. Therefore, relying solely on the table may result in a maximum voltage that doesn't work well in the system you want to install.

Table 690.7 Voltage Correction Factors for Crystalline and Multicrystalline Silicon Modules

Correction Factors for Ambient Temperatures Below 25° C (77° F).
(Multiply the rated open circuit voltage by the appropriate correction factor shown below.)

| Ambient Temperature (°C) | Factor | Ambient Temperature (°F) |
|---|---|---|
| 24 to 20 | 1.02 | 76 to 68 |
| 19 to 15 | 1.04 | 67 to 59 |
| 14 to 10 | 1.06 | 58 to 50 |
| 9 to 5 | 1.08 | 49 to 41 |
| 4 to 0 | 1.10 | 40 to 32 |
| −1 to −5 | 1.12 | 31 to 23 |
| −6 to −10 | 1.14 | 22 to 14 |
| −11 to −15 | 1.16 | 13 to 5 |
| −16 to −20 | 1.18 | 4 to −4 |
| −21 to −25 | 1.20 | −5 to −13 |
| −26 to −30 | 1.21 | −14 to −22 |
| −31 to −35 | 1.23 | −23 to −31 |
| −36 to −40 | 1.25 | −36 to −40 |

**FIGURE 11-2:** *NEC*® Table 690.7.

*Reprinted with permission from NFPA 70®, National Electrical Code®, Copyright © 2007, National Fire Protection Association, Quincy, MA 02169. This reprinted material is not the complete and official position of the NFPA on the referenced subject, which is represented only by the standard in its entirety.*

## Applying the adjusted open circuit voltage to come up with the right number of modules in a string

After you calculate the module's temperature-adjusted voltage, you need to see how that temperature-adjusted voltage will affect the number of modules you can place in any one string that will connect to your grid-direct inverter.

REMEMBER

As I explain earlier in this chapter, every inverter has a maximum allowable voltage it can receive from the array. For grid-direct inverters, this value typically ranges from 500 VDC to 600 VDC, although it can be even lower than that (the maximum allowable voltage really depends on the components used by the inverter manufacturer). To determine the maximum number of modules you can place in a series string, you need to look up the maximum input voltage from the spec sheet of an inverter you're thinking about using and divide that value by the module's maximum $V_{oc}$ value (which you calculated based on the record cold temperature in the area).

Say you want to use an inverter that has a 600 VDC maximum input value and you select the temperature-adjusted module $V_{oc}$ from the earlier "Working the steps" section. You can calculate the maximum number of modules like so:

$$600 \text{ VDC} \div 49.7 \text{ V} = 12.07$$

This equation tells you that you can place 12.07 modules in a string without exceeding the 600 V rating of the inverter. But because you can't buy fractions of a module, you have to round this number down to the nearest whole number, which is 12 modules.

*Note:* I used the maximum voltage I calculated in this example, but you can also run the numbers by using the maximum voltage you find with the help of the *NEC*® table in the earlier "Using *NEC*® info in a pinch" section. To save you flipping some pages, that voltage was 50.4 V. If 50.4 V was in fact the correct adjusted voltage, the maximum number of modules you could place in a series string would be

$$600 \text{ VDC} \div 50.4 \text{ V} = 11.9$$

In this case, you have to round down to 11 modules in a string. If you were to round up to 12, the 600 V value would be exceeded, and you'd run the risk of killing the inverter. This is an example of why taking the time to perform all the calculations can be beneficial: The array using 11 modules in a string may not be the best scenario for your client's installation, and having the ability to add an extra module could make a big difference.

At this point, all you've done is define the maximum number of modules you can place in a string. You haven't necessarily determined the right number of modules for the system because the number of modules in a string doesn't need to exactly match the physical layout of any one row or column of modules. Therefore, you must repeat this process for voltage loss due to high temperatures (see the next section) and then use that info, combined with the initial criteria you established — the client's budget, the available area, and so on (see the earlier sections in this chapter) — to define the whole array.

## Figuring out the modules' minimum DC voltage contribution

On the other end of the voltage window is the minimum power point tracking voltage. If the voltage from the array ever drops below this value — usually due to high heat conditions that occur on the sunniest days (as in the days system owners expect their arrays to produce the maximum amount of energy possible) — the inverter won't be able to continue operating and will shut down. Although this scenario won't damage the inverter, it may damage your reputation. If your client sees his inverter shut down on a bright, sunny day because the voltage isn't high enough, expect to have a not-so-pleasant conversation with him (and possibly lose recommended business from your client if you can't salvage his opinion of your work).

A PV module's voltage decreases as the temperature increases (see Chapter 6 for more on this), which means that in the heat of the day — while the array is operating — the modules' voltage will be reduced. In this situation, you need to look at the maximum power voltage, $V_{mp}$, and make adjustments from there because the array will be operating and the voltage will be reduced from the $V_{mp}$ value. I explain what you need to do in the next sections.

### Keeping long-term performance considerations in mind

PV modules are great and can perform for many years, but they aren't perfect. The typical power output warranty for a module says that the manufacturer guarantees that the module will produce at least 80 percent of its original power output in 25 years. This verbiage means that over the course of 25 years, the module's power output will be reduced by less than 1 percent per year — at the expense of both voltage and current over the course of the module's life. Being aware of this voltage loss is critical because voltage is so important when matching PV arrays to inverters.

REMEMBER

So what does long-term voltage loss mean to you as you go about sizing systems? Simple. You don't want to design a PV array for connection to an inverter that barely fits inside the voltage window today. For example, if an inverter needs 250 VDC to operate, you don't want to design your PV array to operate at 255 VDC at the hottest time of the year starting in the first year. If you do, you're not allowing for much voltage degradation before the array can't keep up with the inverter's requirements.

TIP

In the next few sections, I show you how to calculate the minimum number of modules you need to place in a string to allow the inverter to operate when the module temperature is elevated. When you go through these steps and arrive at the minimum number of modules needed, add one module to that value and consider that your minimum (unless of course adding one module makes you have too many modules in a string; see the previous section for more on that). By adding just one module, you build in protection against falling out of the inverter's voltage window due to natural degradation, thereby preserving the long-term performance of the system.

## Picking the temperature to use in your math

Just like you need to calculate the adjusted $V_{oc}$ for the record cold temperatures, you need to calculate the adjusted $V_{mp}$ for hot temperatures. But which hot temperature should you use? After all, the *NEC*® doesn't have any requirements when it comes to keeping an array above an inverter's minimum voltage because you can't cause any damage with a low-voltage scenario like you can when too much voltage is applied. You also can't rely on a table to tell you the multiplier value because no such table exists.

Here's what to do: Start by defining the ambient summertime temperature, which is the high temperature at the array location during the summer. Then increase the estimated module voltage above the ambient temperature based on the method used to hold the PV array.

You have to know how to use the temperature-adjustment equation I present in the earlier "Working the steps" section. The method used to calculate an array's voltage in hot weather is exactly the same as for cold weather; you merely use different numbers. Here are your ambient temperature options:

- **The average high summertime temperature for the array's location:** If you use the average high summertime temperature for the array's location, you'll design around a realistic value that'll be reached often. Then again, there'll be many days of the year that the average high temperature is exceeded. If you use the average high temperature and don't allow for those extremely hot days, the array may just turn off on those record hot days.
- **The record high temperature:** By designing with the record high temperature in mind, you ensure that the PV array will operate in all conditions, which isn't a bad situation in my opinion. On the other hand, using the record high temperature can lead to overly conservative arrays because the calculated voltage loss is on the extreme end and doesn't really happen on a regular basis.

REMEMBER

- **The American Society of Heating, Refrigerating, and Air-Conditioning Engineers' (ASHRAE) 2-percent design temperature:** This is the value that I like to use. ASHRAE's *2-percent design temperature* says that for the given site, the temperature rises above the reported value only 2 percent of the time. In other words, 98 percent of the time, the temperature is less than or equal to the number given. Using this value allows you to design for nearly every situation without requiring excessive calculations.

TIP

In 2009, Solar ABCs, a group that advocates for the solar industry on a wide variety of topics, released a document titled "Expedited Permit Process for PV Systems." In this document, Solar ABCs publishes the average 2-percent temperature data for June through August for a number of cities across the United States. This resource (found at `www.solarabcs.org/permitting/Expermitprocess.pdf`) gives you a quick reference for the 2-percent data.

## Accounting for mounting in your math

After you know which hot temperature you intend to use for your calculations, you need to consider how you plan to mount the PV array because the mounting method affects the module's temperature. To account for the mounting method, add a certain number of degrees to the ambient temperature based on the array's proximity to a mounting surface. The values shown in this section are estimates based on measurements made at multiple locations over a number of years. They can be used in the design process and represent rises in a PV module's temperature at the hottest time of the year.

REMEMBER

To estimate a PV module's temperature at the hottest time of the year, you should add one of the following temperatures to the ambient temperature:

- 35 degrees Celsius for arrays with fewer than 6 inches between the backs of the modules and the mounting surface
- 30 degrees Celsius for arrays with more than 6 inches between the back of the array and the mounting surface
- 25 degrees Celsius for arrays that are mounted on top of a pole

By adding these values to the ambient temperature, you can arrive at a best estimate for your array's *operating temperature* — the temperature you expect to see at the module level if you were to measure the temperature in the middle of the summer.

## Crunching the numbers

When you know the temperature you're going to estimate for the modules when they're operating in the summer, you can begin the process of adjusting the $V_{mp}$ value. To do so, you need to apply a specific temperature coefficient to the $V_{mp}$ as the array grows hotter.

TIP

PV module manufacturers often report the temperature coefficient for $V_{oc}$, but they rarely provide it for $V_{mp}$. Also, they almost all universally report the temperature coefficient for power. Although voltage isn't the only factor affecting the temperature coefficient for power, it's what affects it the most. Therefore, if the module manufacturer doesn't supply a temperature coefficient for voltage, I suggest taking the value reported for power and substituting it for the voltage coefficient. Doing so results in slightly more conservative values when it comes to the adjusted voltage, but your numbers will be reasonably close to the exact answer (and in most scenarios, the end result will be exactly the same).

REMEMBER

For crystalline modules, the coefficient for power (and unless specified otherwise, for voltage) is very close to −0.5%/°C, which is a slightly greater change when compared to the coefficient that applies to $V_{oc}$.

Here's how to apply all the factors when calculating the adjusted $V_{mp}$ values for your modules:

1. **Collect the temperature coefficient for your PV module in volts per degrees Celsius.**
2. **Collect the 2-percent design temperature for your client's location in degrees Celsius.**
3. **Determine how many degrees Celsius to add to the ambient temperature based on how you plan to mount the array.**
4. **Add the ambient and mounting method temperatures to estimate the operating temperature of the array in the summertime.**
5. **Calculate how many degrees Celsius greater than STC the modules will be on 98 percent of the days.**
6. **Multiply the temperature coefficient for the module by the number of degrees calculated in Step 5.**

   The result of this equation is the change in voltage that the module will produce.

7. **Add the number of volts calculated in Step 6 to the $V_{mp}$ for the module at STC.**

   Because the Step 6 number will always be negative, you must subtract the calculated value instead of adding it in order to find the adjusted maximum module voltage based on that temperature.

Suppose the $V_{mp}$ for a particular module is 37.2 V and the manufacturer reports that the temperature coefficient for power is −0.5%/°C. Because the manufacturer doesn't provide the temperature coefficient for voltage, you must apply the power coefficient to the voltage. To calculate the number of volts per degree Celsius, multiply the coefficient by the $V_{mp}$.

$$37.2\text{ V} \times -0.5\%/°\text{C} = 37.2\text{ V} \times -0.005/°\text{C} = -0.186\text{ V}/°\text{C}$$

The array location has an ASHRAE 2-percent design temperature of 33 degrees Celsius, and the array will be mounted parallel to a roof with only 4 inches of space between the roof and the backs of the modules. Apply this information to the preceding steps to get the following:

1. The reported temperature coefficient is −0.5%/°C.
2. The ambient design temperature is 33°C.
3. The altered ambient temperature (to accommodate the mounting method, which calls for fewer than 6 inches between the backs of the modules and the mounting surface) is 68°C (33°C + 35°C = 68°C), the estimated module temperature when the array is operating.
4. The number of degrees off of STC is therefore 68°C − 25°C = 43°C.
5. The voltage loss due to temperature is thus 43°C × −0.186 V/° = −8.0 V.
6. The new adjusted voltage for the module at the design temperature is therefore 37.2 V + −8.0 V = 37.2 V − 8.0 V = 29.2 V.

As you can see, the loss of voltage will be 29.2 V in the summertime — that's pretty significant. If you continue to look at how the voltage will be reduced as the module ages, you'll soon recognize the importance of keeping an eye on this calculation in the design process.

## Totaling the minimum number of modules needed in a string

After you have the adjusted $V_{mp}$ value, you can calculate the minimum number of modules needed in any string in order to operate the inverter. Continuing with the earlier example, if the inverter you're thinking about using has a minimum input of 250 VDC, the minimum number of modules necessary will be that minimum voltage divided by the temperature-adjusted module voltage.

250 VDC ÷ 29.2 V = 8.56, or 9 modules minimum

Although this calculation defines the absolute minimum number of modules needed to keep the inverter running, a good practice is to make your minimum number of modules be the calculated number plus one. This practice gives you plenty of room for degradation and allows for the array to operate significantly hotter than anticipated and still keep the inverter running. So I'd want to use at least 10 modules in the strings for the earlier example.

# Bringing It All Together: Combining Your Power and Voltage Information

This section is your chance to put all the information you've acquired (specifically, matching the array's power to the inverter and defining the string length based on the maximum and minimum number of modules allowed in each string) and put it to use for your client's site.

Earlier in the sizing process, you determined the number of modules you can fit on the roof and possible inverter sizes (see the earlier "Matching Power Values for an Array and an Inverter" section). Now, with the string lengths defined, you can see whether that array size is a real possibility. For example, in the earlier "First Things First: Evaluating the Budget and the Available Array Area" section, you determined that the roof (or ground, the physical site isn't crucial, just the space limitations) could hold 24 of the modules and that the inverter you based your calculations on can handle the power from 24 modules. Can you really fit all 24? The minimum string length was 9 modules (or 10, if you follow my advice), and the maximum was either 11 or 12, depending on how you calculated the temperature adjustment (see the earlier "Working the steps" and "Using *NEC*® info in a pinch" sections).

If you took the time to calculate the adjusted $V_{oc}$ by hand instead of using Table 690.7 from the *NEC*®, then yes, you can fit all 24 modules on the roof by placing two strings of 12 modules in parallel. If, however, you only used Table 690.7 and calculated a maximum of 11 modules in series, then no, you could place only 22 modules on the roof because you'd be limited to strings of 11 modules (because the space is limited to 24 modules or less, two strings of 11 is the closest you can get).

Unless you're using an inverter that can handle different maximum power point inputs, the strings must be the same length and face the same direction. Otherwise, they'll have different maximum power points, and the inverter won't be able to harvest all the potential power.

What if the roof could hold only 18 modules? In that case, the array and inverter relationship in this example will work, but you can probably do better. I suggest finding a different inverter that has a lower DC input voltage than 250 VDC, just to make sure the system doesn't shut off in high heat conditions or in a number of years as the modules degrade.

## One Last Check: The Inverter's Maximum Current Input

The maximum amount of power input and the DC voltage window are generally the only considerations you need to make when sizing a PV array to the DC side of the inverter. One final check to make, though, is the maximum current input from the array to the inverter.

All inverter manufacturers list the maximum current input allowable on their inverters, but this value isn't always the easiest specification to find because the manufacturers tend to list it in the backs of their installation manuals. Be sure to find out the maximum current input value so you can make sure the array size you've calculated doesn't exceed it.

To verify that the array you've sized doesn't exceed the inverter's maximum current input, divide the value reported by the inverter manufacturer by the short circuit current rating of the array at STC.

IN THIS CHAPTER

- » Performing a load analysis
- » Determining the battery bank's capacity
- » Calculating required power output for a PV array and a charge controller
- » Making sure the inverter fits
- » Adding a generator to the system

# Chapter 12
# Sizing a Battery-Based System

In addition to grid-direct PV systems, which I show you how to size in Chapter 11, the other major PV system type you'll likely have the opportunity to design and install is a battery-based system. You can install one of these systems on the grid *(utility-interactive)* or off the grid *(stand-alone)*. The basics of the design and installation process for both are similar, including the fact that sizing any battery-based system is equal parts art and science.

The science comes into play when you analyze specific site information and calculate the power values of the battery bank and the array. The art part is apparent when you work with clients to define realistic load profiles so you can size the entire system (people generally don't like to admit to how much television they watch). The art of sizing battery-based systems is also apparent in the assumptions and estimations you have to make given that the variables affecting system sizing are always changing.

In this chapter, I show you how to approach sizing battery-based systems for utility-interactive and stand-alone applications. I walk you through the calculations to make when sizing the elements of either system and give you guidance on some of the assumptions and estimations you'll need to make. (Chapter 2 has details on the configuration of a battery-based system and guidelines on selecting the right type of battery-based system for your client.)

Although battery-based systems are currently installed far less frequently than grid-direct systems, customers often ask about them. If you can be informed on the subject, you may be able to make yourself stand out from other PV system designers and installers and add a valuable service to your business.

As with grid-direct systems, the customer's budget and available space for a PV array are the initial considerations you should make. After all, you don't want to spend a ton of time designing a system the client can't afford or put on her property due to space constraints. Establish these limitations early on following the process I describe in Chapter 11.

# Get Loaded: Looking at Loads in a Battery-Based System

A *load* is any piece of electrical equipment people want to use in their homes and offices. When sizing a battery-based system, you need to establish exactly what loads your client wants to run and how long she plans to run those loads. This information serves as the basis for all of your other calculations throughout the design process.

For utility-interactive, battery-based systems, the battery bank provides power for *backup loads* (loads that the client wants to have on regardless of the utility availability). In this scenario, you have two load centers: the main distribution panel (MDP) and the backup subpanel. Any of the loads connected to the backup subpanel will always be available, whereas the loads connected to the MDP will only be powered when the grid is present. For stand-alone, battery-based systems, the battery bank is designed to power all the electrical loads the client wants to run.

The sections that follow help you and your client take a critical look at the loads the battery bank will serve in both utility-interactive and stand-alone systems. They also help you figure out the energy needs of the loads that will rely (at least in part) on the battery bank.

## Evaluating the loads that the battery bank must serve

After determining your client's budget and available space, your next task when sizing any battery-based system is to evaluate the loads the batteries will be serving. When I say *loads,* I mean all the loads — everything from the barely there energy drain (think small cellphone chargers) all the way up to the electrical hogs (think air-conditioning units). When using batteries to power loads, you have to generate and store every watt-hour (Wh) used, which means you need to find out whether a more efficient alternative exists. For example, compact fluorescent light bulbs (CFLs) produce the same amount of light as incandescent lights, but they use a lot less power. If you can convince your client to replace her incandescent bulbs with CFLs, her battery bank will be able to deliver power for more loads due to the reduced power requirements. Ultimately, the less energy your client consumes, the less expensive the system will be to install and maintain.

Following are some points to consider about the common loads powered entirely or partially by battery banks:

- **For well pumps:** In many situations, a well pump is the sole source of water for a home. Well pumps can be large electrical loads with the potential to cause problems for the inverters and batteries in a battery-based system. With advancements in inverter technologies, however, inverters are much better about running these large pumps.

  TIP

  Look at the well pump's power draw and try to determine the energy consumption (see the next section for details). If the pump is being replaced or hasn't been installed, try to work with the pump supplier to get the most energy-efficient version available.

- **For refrigeration and lighting:** Refrigeration and lighting will be present in most homes or offices, so you need to be able to account for them accurately. In the next section, I show you how to account for the energy consumed by these appliances and how to make sure you design the battery system to handle these loads.

- **For phantom loads:** The small loads that are on 24/7 are the so-called *phantom loads.* Many televisions and entertainment centers draw power even when they're "off," and chargers for small electronic devices and digital clocks incorporated in microwaves and stoves can cause major problems. If these small loads are always present, then the inverter can never turn off and must always supply power. Therefore, the inverter requires a small amount of power to produce a low level of power, causing it to operate at its worst efficiency level.

  TIP

  The solution to phantom loads? When no major loads are running, allow the inverter to go to sleep by unplugging the phantom loads. By removing these small loads, the inverter can go to sleep and wake up (and operate more efficiently) when the larger loads are turned on.

REMEMBER

One type of load that should never be placed on a battery bank is anything that uses electricity to generate heat (these are called *resistive loads*). That rules out water heating, space heating, and electric stoves (incandescent lights also fall into this category). In a stand-alone, battery-based system, petroleum-based fuels generally support resistive loads, with propane being one of the most common in off-grid applications. For people connected to the grid with a utility-interactive, battery-based system, these loads can still be present; you just can't back them up with the batteries.

In a utility-interactive, battery-based system, loads served by the batteries should generally be kept to a minimum; the batteries should only supply power to loads that are truly necessary (which means you need to have a frank conversation with your client to help her evaluate what's really necessary in her daily life). This is because the utility grid is the primary power source, and the batteries are merely the backup power source.

WARNING

When talking about utility-interactive, battery-based systems, people often refer to loads the battery bank needs to back up when the grid fails as *critical loads.* Strike this phrase from your vocabulary when talking about these systems. The *National Electrical Code®* (*NEC®*) defines *critical loads* as loads used for life support, public safety, and other truly vital functions. If you list the system as backing up critical loads when it really isn't, a persnickety system inspector or plan checker may make your life harder.

REMEMBER

When deciding what loads to power in utility-interactive, battery-based systems, the loads you should always consider include water-supply loads (well pumps and pressure pumps); lighting; refrigeration; and maybe one or two outlets to plug small appliances into, like a radio and a cellphone charger. (You must be careful, though; if you make it too easy to pull power from the battery bank, the system owner will do exactly that and then be disappointed in the lack of time that the batteries back up those "few" loads.)

For a stand-alone battery-based system, a load is just a plain old load, not a backup load or a critical load. People who live in off-grid homes typically have major lifestyle differences from people who live in homes that are on the grid. They have to evaluate the necessity of anything that requires electricity to run because they don't have unlimited access to electricity. This doesn't mean that your client can't lead a "normal" life (whatever that means anymore); she just has to become selective in her electrical consumption.

Some clients with stand-alone systems may want to run some loads straight from the DC electricity stored in the battery bank. Generally, the loads you can run from DC are limited to lighting and refrigeration. These DC loads are nice in the sense that they can pull power directly from the battery bank without the help of an inverter, which increases the overall efficiency of the system because the loads use the same type of electricity produced by the PV array and stored in the batteries. However, DC loads are found in specialty locations and must be matched to the voltage available from the batteries.

TIP

Keep DC loads separate on your list of loads. When you have to account for system efficiencies, you'll use the DC loads in your calculations.

## Calculating the energy required during an outage for utility-interactive systems

For a utility-interactive, battery-based system, you need to know how much energy your client will need to use over a very short period of time. After all, most people experience power outages that are measured in hours (or a few days at the most). On top of the short duration, typically only a few loads need to be backed up during an outage. If, however, your client insists on powering the entire house/office or a number of major loads, you need to incorporate a generator into the system design. I cover generators and how to include them later in this chapter.

REMEMBER

When designing a utility-interactive, battery-based system, look at the loads the client wants to run during an outage and estimate how much energy they'll consume during the specific amount of time required. You can do this by multiplying the total run watts for a load by the number of hours it needs to run during a utility outage to find the total energy consumption during the outage. Generally, 24 hours is enough, except for those clients who want to have multiple days of backup.

## Determining the average daily energy consumption for stand-alone systems

After you've identified all the loads (both AC and DC; turn to Chapter 2 for the full scoop on loads), you need to evaluate how much energy each load consumes in order to begin the process of sizing all the required components. Going through the load analysis may seem like a real pain, but if you don't take the time to estimate each load's energy consumption, the installed system will be either grossly under- or oversized for your client's needs. Both situations result in a waste of time and money.

REMEMBER

To determine the amount of energy consumed by each AC load in kilowatt-hours (kWh), you need to know the number of watts the load draws, the amount of time it runs each day, and the number of days it's used each week. Certain loads may only run a few times a week, whereas others may run daily. By averaging out the loads over a week's time, you can establish a consistent pattern of energy consumption. Use the following equation:

$$\text{Energy (in watt-hours)} = (\text{Watts} \times \text{Hours per day} \times \text{Days per week}) \div 7 \text{ days per week}$$

REMEMBER

When estimating weekly energy consumption, include all the watts drawn. For example, if you're looking at lighting, don't just calculate the energy based off one light — look at all the lights that will be on at the same time.

Here's another example: Pretend that the washing machine isn't run each day in your client's home. If you calculated the washing machine's energy consumption based on the days it runs, you'd end up with a value that's higher than normal. By instead averaging out the washing machine's energy consumption over the course of a week, you wind up with a daily energy consumption that's slightly higher than the reality for the days the washer doesn't run and slightly lower than the reality for the days it does run. This results in a good approximation for the week.

Imagine a washing machine that draws 175 W for 45 minutes. If it's run four days a week, you can determine the average daily energy value by multiplying the power draw by the number of hours to get

$$175 \text{ W} \times 0.75 \text{ hrs} = 131 \text{ Wh, or } 0.131 \text{ kWh}$$

So the washing machine uses 0.131 kWh each time it's run, and it's on four days per week. Therefore, the average daily energy consumption is 131 Wh × 4 days ÷ 7 days = 75 Wh per day.

TIP

With all the loads that clients want their battery-based systems to support, tracking the individual energy-consumption amounts can get confusing. I recommend using a spreadsheet to help keep all the calculations straight. Having a spreadsheet also makes it easier to make major changes should the client decide any are necessary. Keep your load-analysis spreadsheet simple, like the one in Figure 12-1. In this chart, you see a few loads that are to be powered by a battery-based PV system. The first column denotes the itemized loads; the second column indicates how many of each item there are. The third column indicates the power draw (in watts) of an individual load; the next two columns represent the number of hours per day and the number of days per week each load is estimated to run, respectively. The second-to-last column denotes the number of watts each load draws multiplied by the number of items; the final column calculates the average daily energy consumption of each load by multiplying the total run watts by the number of hours per day multiplied by the days per week and then divided by seven.

To get a good feel for the table, look at the television row. This client has a nice new big-screen TV that draws 200 W when it's on, and she has admitted that the TV is on for 4 hours on the days that her family watches it. So each day anyone in the client's family watches TV, the set consumes 200 W × 4 hours = 800 Wh. But there's one day every week that the family doesn't watch TV, so to determine the average daily energy consumption for the week, you multiply the 800 Wh by the 6 days per week the client's family actually watches it and divide that number by 7 days per week. Doing so puts the average daily energy consumption for the TV at 686 Wh, or 0.686kWh.

| AC Load Description | Quantity | Run Watts | Hours/Day | Days/Week | Total Watts | Total Watts/Week | Total Whrs/Day |
|---|---|---|---|---|---|---|---|
| 19 cubic feet fridge/freezer | 1 | 120 | 11.00 | 7.00 | 120 | 9240 | 1320 |
| Television sets | 1 | 200 | 4.00 | 6.00 | 200 | 4800 | 686 |
| DVD/VCR | 1 | 21 | 2.00 | 4.00 | 21 | 168 | 24 |
| Stereos | 1 | 50 | 3.00 | 7.00 | 50 | 1050 | 150 |
| Well pump 1½ hp, 240 VAC | 1 | 1300 | 2.00 | 7.00 | 1300 | 18200 | 2600 |
| Compact fluorescent lights | 10 | 20 | 4.00 | 7.00 | 200 | 5600 | 800 |
| Solar DHW system | 1 | 40 | 5.00 | 7.00 | 40 | 1400 | 200 |
| Total | | | | | 1931 | | 5780 |

Total Watt hours/day divided by 1000 = kWh/day 5780/1000 = 5.780 kWh/day

Total kWh/day x 365 = kWh/year 5.78 x 365 = 2109.07 kWh/year

**FIGURE 12-1:** An example of a load-analysis spreadsheet.

If your client wants to use any DC appliances, such as lighting or refrigeration, you need to account for the energy consumption for those loads as well. You calculate this consumption the same way you do for AC loads, but keep the numbers separate in your tables; you have to account for the efficiency losses associated with converting the DC in the batteries to AC for the AC loads. After you take the efficiency losses into account, you can add the two values together to find the total energy consumption of all the loads.

REMEMBER

The inverter itself is a load that's always present unless advanced energy-management techniques are used when programming the inverter. This load isn't large, but if you don't consider it as being left on 24 hours a day, the inverter's idle draw can have a significant effect on the rest of your system sizing and the overall performance of the system. You can find the exact amount of energy an inverter draws when it's sitting idle by looking at the manufacturer-provided specification (spec) sheet. Simply grab the spec sheets of the inverters you're most likely to use, and estimate each inverter's power consumption from its respective spec sheet.

# Sizing the Battery Bank

After you know what your client's electrical lifestyle is on an average day, you need to translate that into the amount of energy stored in her battery bank (also known as the battery bank's *capacity*). For any battery-based system — whether utility-interactive or stand-alone — when you size the battery bank, you take the view that no other source of power exists (at least for a certain amount of time) and that the battery bank is the primary source of energy (the PV array, a generator, or the utility merely replenishes the battery bank when it discharges). Consequently, you need to size the battery bank to run the electrical loads your client wants, when she wants — which means you need to establish some criteria that you expect the battery bank to follow. All of the following dictate the battery bank capacity you're looking for:

- The efficiency of the inverter
- The number of days you expect the battery bank to last without recharging
- The batteries' operating temperature and voltage
- How much of the battery bank your client is willing to use
- The voltage at which you want the battery to operate

The next sections provide greater details on these variables. They also explain how to put them all together so you can accurately determine the battery capacity needed and create the battery bank.

REMEMBER

When you buy batteries to make up the entire battery bank, you have a few options. The most common battery type for battery-based PV systems is a 6 V nominal battery. (This battery has three individual cells in it that are all wired internally to deliver 6 V across the terminals.) You then take these batteries and wire them in a series-parallel arrangement to achieve the voltage and capacity characteristics you're after. Other options include 12 V nominal batteries as well as individual 2 V cells in their own plastic cases; these cells look like batteries, but because there's only one cell, technically they're cells and not batteries. (Batteries also come in 4 V and 8 V nominal arrangements, although these are less common.)

## Inverter efficiency

There'll always be some losses associated with turning DC into AC, which is why no inverter can deliver 100 percent of the energy from a battery bank to the loads. However, if the inverter can be more efficient at inverting, the battery bank can be smaller. Consider the AC loads attached to the proposed inverter and the inverter's size (in terms of power output) in order to maximize efficiency levels.

What I mean by this is don't put in a 4 kW inverter if all the client will ever draw is 1 kW. Instead, try to match the loads and the inverter. (I explain inverter sizing in more detail later in this chapter.)

Inverter manufacturers list the efficiencies of their units on all of their spec sheets. What you need to remember is that the number listed by the manufacturers is the peak efficiency value. As such, it'll almost always be an impressive value that's somewhere near 95 percent. Although 95-percent efficiency may be possible, it's not achievable on a frequent basis. Most battery-based systems are regularly closer to 90-percent efficiency. Like all variables, this percentage will vary, but 90 percent is a fair value that represents a typical operating efficiency for an inverter.

## The days of autonomy

The number of days your client wants her battery bank to sustain her electrical lifestyle is known as the *days of autonomy.* In other words, it's the number of days the client expects her battery bank to provide her with her average daily energy requirements without needing to be recharged by the PV array and the charge controller, generator, or utility. This number is completely up to the system owner but you (as the system designer) should offer suggestions that will keep your client satisfied. The local climate usually plays a major role in this decision, as does the available budget for the project. As you can imagine, the more days of autonomy, the more batteries you need and the higher the system cost climbs.

Many stand-alone residential applications use two or three days of autonomy as the starting point, whereas most utility-interactive systems use just a single day. For commercial applications, the grid is typically present, so one day of autonomy should suffice (although your commercial client may want more than that based on her business and the amount of lost revenue associated with a power outage). You can consider adding more days of autonomy, but then you have to play a balancing game with the size of the battery bank and the size of the PV array (and your client's checkbook). I explain how to size the PV array to the battery bank later in this chapter.

## The temperature used for battery operation

As I explain in Chapter 7, the temperature that batteries operate at affects their capacity. The colder a battery is, the less capacity it can deliver. Why? Because the efficiency of the chemical reaction occurring inside the battery increases and decreases at different temperatures. Battery manufactures publish the exact effects that temperature has on their batteries, so you should be able to find that data for the battery you're considering in order to apply the correct *temperature derate factor* (the percentage of the capacity you can expect from a battery based on the temperature). I show you how to apply the temperature derate factor later in this chapter.

TIP

Because most systems use lead-acid batteries and the technology is pretty consistent among the different manufacturers, I use a single temperature derate factor: 90 percent. This percentage corresponds to a battery temperature of approximately 60 degrees Fahrenheit and indicates that at that temperature, the battery will only be able to deliver 90 percent of its *rated value* (the battery's capacity at 77 degrees Fahrenheit).

## The depth of discharge

*Depth of discharge* (DOD) is the amount of energy drawn from the battery bank; it's generally given in terms of a percentage, as I note in Chapter 7. The higher the DOD value, the more energy has left the battery bank. As with days of autonomy (which I explain earlier in this chapter), DOD can (and should) be dictated in the system-design process because it affects the overall size of the battery bank. When you look at a typical chart provided by battery manufacturers that shows the number of cycles versus DOD, it becomes apparent that the smaller the DOD is, the greater the number of cycles (a *cycle* is the period from when the batteries' capacity is drawn down to when it's recharged). Although this fact probably isn't surprising, it doesn't mean you should try to baby the batteries and design a system around a small DOD. What you really need to do is evaluate where on the curve the maximum amount of energy will be delivered over the battery bank's life.

To determine the ideal DOD to use with a battery bank, look at the whole picture in graph form; a graph shows a battery bank's number of cycles against the percentage of its discharge. Figure 12-2 shows an example.

If the battery bank in Figure 12-2 is rated at 400 amp-hours (Ah), you can use that information to estimate the energy delivered over the course of the battery bank's life. From the graph in Figure 12-2, you can see that this battery bank will last for approximately 3,200 cycles if the DOD is only 50 percent. The number of cycles is reduced to approximately 2,100 when the DOD is 80 percent. So which DOD delivers more energy over the life of the battery bank? Run the numbers to figure it out:

400 Ah × 50% DOD per cycle × 3,200 cycles = 640,000 Ah

400 Ah × 80% DOD per cycle × 2,100 cycles = 672,000 Ah

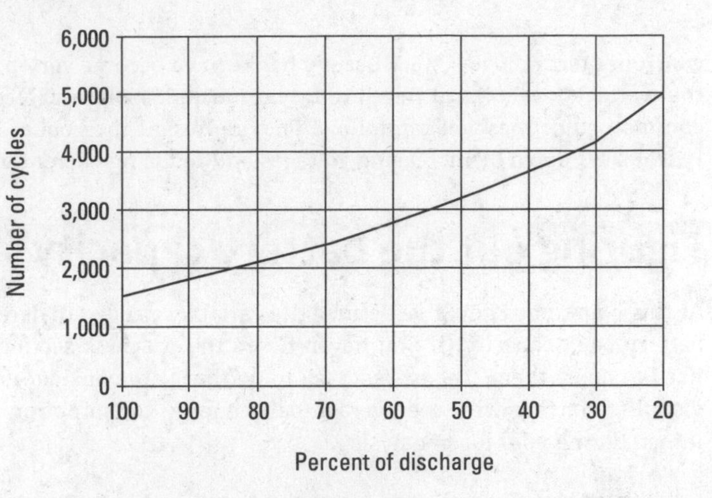


**FIGURE 12-2:** A battery bank's number of cycles versus its depth of discharge.

Even though the idea of reducing the DOD looks good at first glance because it increases the overall life of the battery bank, it results in fewer amp-hours delivered. Because the battery bank's job is to store and deliver energy, you may want to tell your client to consider discharging the batteries more often to maximize her investment (and reduce the system's initial cost). When evaluating the DOD, most battery bank designs use a value that's somewhere between 50 percent and 80 percent, but there's really no exact "right" answer. You have to evaluate the options and make a suggestion based on the information in the charts from the manufacturer of the batteries you use in the bank.

Be careful to never exceed an 80-percent DOD in your design. Repeatedly reducing a battery bank's capacity more than 80 percent harms the batteries and causes premature failure of the bank.

## Nominal voltages

For any battery-based system you install, you need to look at battery bank nominal voltages of 12, 24, or 48 VDC. (*Nominal voltage* is a reference voltage; see Chapter 3 for an introduction.) These voltages correspond to the inverter input requirements for the majority of commercially available inverters. It also corresponds to the nominal voltages of lead-acid battery cells, which are 2 V nominal.

Recently, battery-based systems have moved to higher voltages, which means that even modestly sized battery systems are commonly wired for 48 VDC. If you're wiring a battery bank for a very small PV system (less than 200 W or so) or a system that's using DC loads directly (such as recreational vehicles), you may consider wiring the batteries for only 12 V. Systems using inverters that produce relatively small AC power levels (less than 2,000 W) may be able to justify using a 24 V battery bank, but with the advancements made in inverter and charge controller technologies, 48 V battery banks have become very popular. (Note that the wattage levels listed here are by no means absolute values. Rather, they're common guidelines you can follow. They represent the goal of keeping *conductor* [wire] sizes down by increasing voltages and reducing current values.)

## Figuring out the battery capacity you need

At this point, you should've defined the variables needed to determine the overall battery bank capacity (if you haven't, see the previous sections). Now you just need to apply them. The easiest way to do that is to consider each variable individually, starting with the average daily energy consumption value you determined during your load analysis.

REMEMBER

If you follow the steps in this section, you can estimate the required capacity for a battery bank in both a utility-interactive and stand-alone system.

1. **Determine the average daily AC watt-hours (or kilowatt-hours) consumption level.**

   I explain how to do this in the earlier "Determining the average daily energy consumption for stand-alone systems" section. For the purposes of providing an example, refer to Figure 12-1 to find that the average daily energy consumption of my sample client is 5,780 Wh, or 5.78 kWh.

2. **Divide the watt-hours value from Step 1 by the estimated inverter efficiency.**

   This step increases the required capacity due to the fact that an inverter loses some of its stored capacity during the process of turning DC into AC (10 percent loss is common). Continuing with the example, you find that 5.78 kWh ÷ 0.9 = 6.42 kWh (90 percent is a fair inverter efficiency to estimate).

3. **Add any energy consumption from DC loads to the watt-hours value in Step 2.**

   This value represents the total daily energy consumption for all the loads connected to the battery bank. If the client has three 20 W DC lights that she runs for two hours each day, the total DC energy consumption is 3 lights × 20 W × 2 hours = 120 Wh, or 0.12 kWh. The total energy consumption is therefore 6.42 kWh + 0.12 kWh = 6.54 kWh.

4. **Multiply the energy value from Step 3 by the desired days of autonomy.**

   Doing so tells you the amount of energy the battery bank needs to store (two or three days is a pretty typical value). My example client has a stand-alone, battery-based system and wants three days of autonomy, so that makes the new energy value 6.54 kWh × 3 days = 19.62 kWh.

5. **Divide the value calculated in Step 4 by the temperature compensation value provided by the battery manufacturer.**

   Ninety percent of manufacturers estimate the adjusted capacity at 60 degrees Fahrenheit. Apply the manufacturer's value here for the estimated temperature of the battery bank you're considering. So if the example battery bank will be stored at 60 degrees Fahrenheit, perform this calculation: 19.62 kWh ÷ 0.9 = 21.8 kWh.

6. **Divide the value from Step 5 by the allowable depth of discharge.**

   The greater the DOD, the smaller the battery bank can be because you'll be using more of the capacity (approximately 50 to 80 percent). This client and I settled on a DOD of 75 percent, so the math looks like this: 21.8 kWh ÷ 0.75 = 29.1 kWh.

7. **Divide the value from Step 6 by your desired nominal voltage for the battery bank.**

   Batteries are rated in amp-hours, not watt-hours. By using the nominal battery bank voltage, you can determine the required amp-hours for the battery bank (use a 12 V, 24 V, or 48 V value here). The system in my running example will be installed at 48 V to keep the current values at a minimum and reduce the conductor sizes. Here's the math: 29.1 kWh ÷ 48 V = 0.606 kAh, or 606 Ah.

## Strung along: Wiring the battery bank

As soon as you know what the capacity of the battery bank should be and the nominal voltage (see the "Nominal voltages" section earlier in this chapter), you're ready to evaluate the different battery options and decide which one is best for the battery bank you're constructing.

REMEMBER

When wiring batteries into a battery bank, you need to consider the voltage of each individual battery as well as each battery's capacity because you're creating a string of batteries by placing multiple batteries in series (see Chapter 3 for details on both series and parallel configurations). As you make the series connections, the voltage will increase while the capacity (measured in amp-hours) remains constant. To increase the capacity of the bank, you must place more strings in parallel with the first in order to keep the voltage constant while increasing the capacity. I'm only talking about the design aspects here; for the installation and safety concepts, jump to Chapter 15.

If you install a single string of batteries in series and in a few years your client has an issue with any one battery (or cell), the system will likely need to be shut down until that one battery is replaced. That may seem extreme, but when you have a single string, the electrons have to flow through each battery to complete the circuit. If one battery is dead or shorted, the current can't get past that battery. This fact is why many PV pros regard placing two strings in parallel as a more desirable solution. With two strings in parallel, you can keep all the parallel connections equal in length (and *resistance*, which is simply resisting the flow of current; see Chapter 3) and the battery bank as a whole can perform well. If any one battery or cell peters out, you can just remove one string from the bank, allowing the system to continue limping along until the situation is corrected.

WARNING

As much as possible, avoid placing more than two strings in parallel. As soon as you begin placing three or more strings in parallel, it becomes increasingly difficult to keep the parallel conductors' resistance equal. The individual strings have the tendency to charge and discharge at different rates, causing imbalances in the bank and reducing the overall life of the batteries.

To determine the specifications for the batteries, I like to first look at the required capacity of the battery bank. Because you want to wire the battery bank with either one or two strings of batteries, you need to find a battery with a capacity amount equal to or half of the capacity you calculated in the preceding section. When you have an idea of the battery capacity needed, you can evaluate different battery spec sheets to decide on the battery for your system. Find the batteries with the correct capacity first and then look at how many you need based on your nominal voltages.

TIP

When you're evaluating battery specifications, use the C/20 capacity value reported by the battery manufacturer (I cover C rates in Chapter 7). The C/20 rate is a good value to use unless you know that your discharge rate will be something dramatically different.

REMEMBER

To determine the number of batteries required in a string, divide the nominal battery bank voltage by the individual batteries' nominal voltages.

From the example in the last section, I calculated that the battery bank would need to have a capacity of 606 Ah at 48 V. And because I want to have two strings of batteries in my bank, I need to look for a battery that has a C/20 rate of 303 Ah (which is difficult to find, so I may need to settle on a battery with a C/20 rate of 300 Ah or buy into a bigger 350 Ah battery). Batteries with this level of capacity are commonly found in 6 V nominal options. So if you're going to wire a bank for 48 V and each battery is 6 V, you know the battery strings should be eight batteries long. Here's the math:

$$48\text{ V} \div 6\text{ V} = 8\text{ batteries per string}$$

# Sizing the PV Array

When it comes to sizing the PV array in a battery-based system, a number of considerations are required. This is one area where the type of system — utility-interactive or stand-alone — makes a big difference in your approach.

- For utility-interactive systems, the PV array generally operates more like it does in grid-direct systems; the utility is present most of the time, and the inverter sends excess power into the grid to run the meter backward. The major difference between utility-interactive, battery-based systems and grid-direct systems is that in the former system, a battery bank is ready to power any loads on the backup load panel. Consequently, the PV array's primary responsibility is to produce as much power as possible at all times to offset loads and send energy back into the grid.
- For stand-alone systems, the actual performance of the PV array in relation to the battery bank is more important because the array is the primary source of power and the primary battery charger. The owner of this type of system may be willing to adjust the tilt of her array seasonally to optimize for the sun's position in the sky with the goal of maximizing performance and minimizing generator run time.

## Sizing the array in a utility-interactive system

If a battery-based PV system will be used simply to back up a few loads for a home or business that's connected to the utility grid (in other words, the system is utility-interactive), the process of sizing the array works similarly to the process of sizing an array for a grid-direct system (see Chapter 11). Of course, any system that's connected to the utility requires interconnection agreements (see Chapter 11 for more on these).

REMEMBER

In a utility-interactive, battery-based system, the batteries spend the majority of their lives full and waiting to go to work during a utility outage. Also, the battery bank is typically sized to last the duration of the outage, which means the PV array isn't used as a daily battery charger. Instead, it's merely used to reduce the power required by the utility (and, if the system owner is lucky, send power back into the grid). Given this fact, you should size the PV array in a utility-interactive, battery-based system based on the client's available budget, the area available for the array, and the annual energy consumption.

## Sizing the array in a stand-alone system

When the battery-based system you're sizing is of the stand-alone variety, the PV array needs to produce an amount of energy equal to your client's average daily energy consumption (as calculated in the earlier "Determining the average daily energy consumption for stand-alone systems" section); if it doesn't, the battery bank will never be able to recharge fully. In addition, the array should be able to help recharge the battery bank after there has been little to no charging by any source (such as the PV array or a generator) and the battery bank has dipped into the reserve supplied by your client's desired days of autonomy. (I cover days of autonomy earlier in this chapter.)

In reality, the amount of energy consumed isn't a constant value; it changes throughout the year. Typically, people use more energy during the winter, which happens to correspond to the time of year with the lowest solar resource (if your client has large cooling loads, such as air conditioners, this may be different). This situation presents a problem for you as a PV system designer. If you design the PV array around the scenario of high consumption and low solar resource, you'll end up with a PV array that's very large. Come summertime, when the energy consumption is reduced and the solar resource is increased, the PV array will be oversized and have the batteries charged very early in the day, which is bad because the PV array will be underutilized those times of the year, and the initial system cost will be outrageous.

To determine the appropriate array size in watts, you need to gather some information about the site and make some assumptions regarding the operation of the system. These values will help you estimate the array size needed based off of the total energy consumption you calculated in the very beginning of the process (as I explain earlier in this chapter).

### Examining efficiency values

REMEMBER

For the array to produce enough energy for the loads, you need to look at the losses within the system. This means considering two main efficiency values: the batteries' charging and discharging efficiency and the efficiency of the PV array to deliver the energy. These values vary, but you can use the following estimations, which are based on typical equipment and technologies:

- **Battery efficiency:** A common value for battery efficiency is 85 percent, which represents the fact that you can never get 100 percent of the energy used to charge the battery when discharging the battery. This efficiency value represents losses internal to the battery as well as the ability for the charge controller to charge the battery.

» **PV array efficiency:** This value is affected by, among other things, the temperature of the array, how dirty the modules are, voltage losses in the wiring, and the age of the array. You can look at all the individual losses and estimate their effect on the total array, but I prefer to estimate the average value at 75 percent. This percentage is on the conservative side in terms of estimating the losses, but for a stand-alone situation, this approach allows you to supply all the energy required more often.

## Considering the total available solar resource

Another consideration you must make is the *total solar resource factor* (TSRF), which is a combination of shading effects and the effects of the array's tilt and azimuth. You determine what this value is when you conduct the site survey. Flip to Chapter 5 for a review of the TSRF.

## Settling on a number of peak sun hours

Most off-grid clients who want a stand-alone system quickly come to the realization that they'll need to use a generator part of the time. Your goal when designing a stand-alone, battery-based system for them is to minimize that run time. Consequently, you need to choose a certain number of peak sun hours for your design. The addition of an external charging source, like a generator, allows you more flexibility with the solar resource data.

TIP

When designing an array for an off-grid system that has a generator, I typically use the average number of peak sun hours for the site (see Chapter 4). If, however, your client doesn't want a generator, you need to design the system based on the lowest amount of solar resource, which is typically the value found in the middle of winter.

## Running the numbers

After you define the variables related to array sizing, you're ready to estimate the PV array size in watts. Use the following steps:

1. **Gather the total energy value calculated in the load analysis.**

   This is the same total energy value you used when sizing the battery bank (see the earlier "Figuring out the battery capacity you need" section). In that example, the total energy consumption (of both AC and DC loads) was 6.54 kWh.

2. **Multiply the estimated battery and PV array efficiencies.**

   Based on what I tell you in the earlier "Examining efficiency values" section, take 85% × 75% to get 64%. Even if you choose not to use these numbers, what you wind up with after completing this step is the total efficiency of the PV array in charging the batteries.

3. **Multiply the efficiency value from Step 2 by the TSRF that you determined during the site survey.**

   The client in this example has a TSRF of 90 percent (in other words, she loses 10 percent of the potential resource due to shading and the array tilt and orientation). Here's the calculation: $0.64 \times 0.9 = 0.57$.

4. **Divide the total energy value found in Step 1 by the total efficiency value found in Step 3.**

   Doing so gives you the total daily amount of energy the array needs to produce. In this case, that's 6.54 kWh ÷ 0.57 = 11.5 kWh.

5. **Divide the energy value from Step 4 by the peak sun hours value you decided to use.**

   The result of this equation is the array size in watts. So if the average peak sun hours is 4.2 for the client's site, then the array needs to produce 11.5 kWh in 4.2 hours, or 11.5 kWh ÷ 4.2 = 2.73 kW = 2,730 W.

REMEMBER

The array size you calculate by using the preceding steps is the size needed to produce the amount of energy consumed on an average day with an average amount of solar resource available. The actual performance of the system once installed will vary over the different seasons.

When you know the appropriate array size in watts, you can calculate the number of modules needed. Divide your previously calculated array wattage by the standard test conditions (STC) rating of the modules you want to use. For example, if the module you want to use is rated at 195 W, you can divide 2,730 W by 195 W to find that you need 14 modules. (*Note:* More often than not, this calculation doesn't result in a whole number. You need to round up to the next whole number and oversize the array a bit. Don't worry; you're better off designing a system that will produce more energy than not enough.)

# Sizing the Charge Controller

After you size the battery bank and PV array, your next step is to size the charge controller. For both types of battery-based systems, the most common controller choice is the maximum power point tracking (MPPT) controller, although you can also use a pulse-width modulation (PWM) controller; I cover both types in Chapter 8. The sections that follow walk you through the steps of sizing a charge controller for any battery-based system.

REMEMBER

As with inverters in grid-direct systems, the charge controller connected to the PV array in a battery-based system needs to be large enough to handle all the power provided by the array. Consequently, you may end up with multiple charge controllers in your system. Each charge controller is connected to a dedicated PV array, but they all connect to the same battery bank. (I walk you through the process of determining how many charge controllers to *specify* [select] in the following sections.)

## Voltage specifications

All charge controllers have a voltage window that you must stay within. Specifically, they have a maximum input voltage that they can accept and a minimum voltage value that they need to stay above. Your job is to look at the temperature-adjusted voltages from the PV modules to correctly account for the charge controller's window. You evaluate the voltage window for charge controllers the same way you do for inverters (see Chapter 11).

REMEMBER

When you know the temperature-adjusted voltages for the PV modules, you can use the number and type of modules you decided on (see the earlier "Running the numbers" section if you haven't yet) to figure out the number of modules you can place in series for each string. Just like with inverters, all the strings connected to the same charge controller must be the same length. But unlike grid-direct inverters, the charge controller in a battery-based system usually operates at a much lower voltage. You therefore have to use fewer modules per string. For example, it's common to see strings of just three or four modules connected to a charge controller; in a grid-direct system, you say see a string of eight to ten modules running a grid-direct inverter. After you know how many modules are in each string, you can determine how many strings to place in parallel based on the power requirements for the desired PV array and finalize the charge controller specification by looking at the amperage requirements (see the next section).

After you determine the required voltage window based on the PV array you're using, you can narrow down your choice of charge controller a bit by looking at the relationship between a PV array's voltage and a battery bank's voltage. This is another area where the difference between MPPT and PWM controllers makes itself known.

» Most, but not all, MPPT controllers have the ability to take a higher voltage on the input side (the PV array) and reduce that voltage to a smaller amount on the output side (the battery bank). By using MPPT controllers that have the ability to step down the array voltage, you give yourself more design options and open up the possibility of using a greater number of PV modules. (On a charge controller without this feature, the voltage window is extremely narrow, so you're forced to use a PV module (or string of modules) that has the same nominal voltage as the battery bank.) Also, by wiring the array at a "high" voltage, you can reduce the size of the wires running between the array and the controller. (To deliver the same amount of power, a higher-voltage array needs to push less current through the conductors, and the amount of current flowing through directly affects conductor sizing.)

You also get the benefit of using modules that have been manufactured with voltages that don't correspond to traditional battery-charging voltages, which opens up your choice of modules even more. Many PV manufacturers now make their modules with the grid-direct market in mind, which means their modules can't be connected to a PWM controller and effectively charge a battery. If you try to use a grid-direct style module with a PWM controller, you'll end up with either a PV module that doesn't have enough voltage to push the current into the battery or too much voltage that may damage the controller.

- A PWM controller, unlike an MPPT controller, needs the PV array nominal voltage equal to the battery bank's nominal voltage. When looking at the voltage specifications for a PWM controller, you really don't need to go through the trouble of adjusting the module voltages for temperature and then comparing that to the controller's voltage window. Because the nominal input must equal the nominal output, this work has essentially been done for you. The downside is that this convenience means you can't use just any old module you want for your system. You must use modules that have been manufactured specifically as 12 V or 24 V nominal and wire them in series as required by the battery bank.

REMEMBER

Essentially, the features of MPPT charge controllers make it easy to justify the added expense of one when looking at PV systems with large arrays meant to provide the energy for a home or small commercial building. The main benefit of an MPPT controller, compared to a PWM controller, is the ability to use the full power output of the PV array regardless of the batteries' charging voltage. In other words, you can get closer to the full benefit of the array's power, which equates to more efficient battery charging.

## Power or amperage specifications

In addition to voltage specifications, the other half of charge controller sizing is the power or current specifications. You need to consider these limitations when sizing a controller to make sure you get the full benefit of the controller. Depending on the charge controller technology used, you have to look at either the power or the current values from the PV array. *Note:* After completing the math described in the next sections, you may end up with a final PV array size that's different from what you calculated earlier. The number of modules per string and the number of strings may dictate that you adjust the array size.

### Some simple math: PWM charge controllers

REMEMBER

For PWM controllers (or for MPPT controllers that use a PV array nominal voltage equal to the battery bank's nominal voltage), determining the current specifications is easy as long as you know the specifics for the PV array. For example, if you're using a PV module that has a nominal voltage of 24 V and the battery is wired for 48 V, the modules will be wired with two in each string. If you have five strings in parallel, you can calculate that the maximum power current, $I_{mp}$, for that array will be the individual modules' $I_{mp}$ value times 5. With a PWM controller, the current rating of the controller must be at least that much current. For example, if the 24 V nominal PV module being used has a $I_{mp}$ value of 6 A and 5 strings are in parallel, the total current output from the array will be 5 strings × 6 A = 30 A. So the PWM controller should have a minimum operating current rating of 30 A.

For safety, you should always verify that the charge controller can handle the short circuit current from that same configuration by multiplying the module's $I_{sc}$ value by the number of strings in parallel and comparing the result to the controller's maximum short circuit current input value.

### A little tougher: MPPT charge controllers

When you're working with an MPPT controller and you want to step down the array's voltage to a lower value, figuring out the current specifications becomes slightly more difficult. In this situation, you're mainly concerned with the current value leaving the controller. You need to make sure you don't apply too much power to the controller and try to push more current out of the controller than its rating. If you do, then the controller will *current limit* itself (send out a certain amount of current even though there's more available from the array) and the "extra" power will be turned into heat. This situation won't damage the controller, but it isn't ideal.

REMEMBER

To guarantee that you don't exceed the charge controller's abilities, relate the controller's limitations based on its output current and the battery bank voltage. If you multiply the controller's maximum output current value by the battery bank's nominal voltage, you get a certain number of watts, which is a power value that you can then relate directly to the PV array's size. How so? Well, you want to make sure that the PV array doesn't exceed this wattage value, or else you run the risk of applying too much power to the controller and creating too much heat.

Consider this example. A common MPPT charge controller has a current output value of 60 A. If you want to use an MPPT controller to charge a 24 V nominal battery bank from a PV array, you need to ensure that the array doesn't exceed 24 V × 60 A = 1,440 W. By keeping the array below this power value, the controller can take all the available voltage and current from the array and effectively push that into the battery bank.

If in this calculation you find that the controller's power value is less than the PV array size you calculated, you can either look for a controller with a larger current output value or buy multiple controllers and wire the PV array into subarrays so that each subarray's power doesn't exceed the controller's limits.

## A check before you move on: Comparing the array size to the battery capacity

After you finish sizing the charge controller but before you go too much further, you should check that the PV array's power output and the battery bank's capacity are sized within reason. What you're concerned with here is the PV array's charging ability as compared to the battery bank's capacity.

You want the PV array's charging ability to be somewhere between a C/10 rate and a C/20 rate (the number listed represents the hours needed to charge the battery bank). If the PV array is too large (above the C/10 rate), it'll charge the battery bank too fast, and the batteries won't be able to charge efficiently because they won't be able to take all the current sent by the array. If the PV array is too small (below the C/20 rate), the charge rate will be small, and the PV array will never be able to fully recharge the battery bank. To determine where you stand, use the battery bank's capacity and figure out what the C/10 and C/20 values are. Your PV array's charging current (the output of the charge controller) should be somewhere between those two numbers. (*Note:* The slower C/20 rate is typically more realistic due to budget and space restraints.)

Here's an example to help you see what I mean: If you have a 600 Ah battery bank, you want the charging current from the PV array to land somewhere between 60 A and 30 A (600 Ah ÷ 10 hours = 60 A and 600 Ah ÷ 20 hours = 30 A). If the battery bank is at 48 V nominal, then the PV array would be between 1,440 W and 2,880 W in size.

# Sizing the Inverter

The final step in sizing a battery-based system is sizing the inverter. When specifying an inverter for any battery-based application, you need to consider the voltage for the loads, the maximum power draw, charging capabilities (from an

AC source), and the ability for the inverter to supply power when certain loads *surge* (draw a large amount of power for a very short duration).

REMEMBER

When I refer to *inverters*, I'm really referring to *inverter/chargers* (this is conventional terminology in the solar industry). Always verify that the inverter you're working with is really an inverter/charger. Sure, you could design the system with a regular inverter and have the client buy a separate battery charger for use in conjunction with an AC power source (see the later section on generators), but this isn't a great solution. Besides, with all the reliable inverter/charger choices out there these days, why add the hassle of specifying and installing an additional device?

REMEMBER

The battery-based systems you install can use either a single inverter or multiple ones. When installing more than one inverter, you have to make sure the inverters communicate with each other and work together (the different inverter manufacturers have specific methods for this communication). Bear in mind that if you do add multiple inverters, their power output values will be additive. So if you have two 3 kW inverters installed in a system, they can work together to provide 6 kW of power to loads.

## Viewing voltage output

Your load analysis (I explain how to conduct one earlier in this chapter) helps you know what voltages your client's loads require. For most residential applications, this voltage is 120/240 VAC, which is the voltage required by AC household loads. Small commercial buildings typically have 120/208 VAC voltage requirements. ***Note:*** The need for 240 VAC is limited to a select few loads, with the primary one being water pumping.

REMEMBER

Most inverters are available with 120 VAC output only; these have the ability to be *stacked* (connected together) to supply 120/240 VAC to the load centers (and some can be stacked to provide 120/208 VAC for commercial applications). A small number of inverters are available with a standard output of 120/240 VAC; these don't need to be stacked to provide the right voltage. So for residential and small commercial applications, an inverter's output voltage isn't a restriction.

## Calculating the power draw

Inverters don't care about energy consumption. They're simply concerned with delivering voltage and current (power) to loads. How long those loads run doesn't concern them, which is why you need to take care to determine an inverter's power output requirements by using the data that you gathered during your load analysis regarding the power draw of individual loads.

REMEMBER

Inverters for battery-based PV systems are always rated at their continuous power output value (just like grid-direct inverters), which means you need to make sure that the inverter(s) you select can provide the amount of power required by your client's loads. You need to estimate which of the loads will be running at the same time and then add up those power values to determine the minimum power rating for the inverter.

Refer to the load analysis shown in Figure 12-1, and you'll see that if every load were turned on in that home, the total power draw would be 1,931 W — that's almost 2 kW. As the system designer, if you even *think* there's a chance that all those loads would run at the same time, you'd buy an inverter/charger rated at a minimum AC output rating of 2 kW.

TIP

If you add up the loads and they require more power than a single inverter can supply, you can consider adding more inverters and allowing them to work together to power the loads. Most inverters have this ability, to a limit. To be safe, just verify that the inverters you want to use can be stacked.

TIP

Consider future loads that may require power so that the inverter installed today can handle growth in the next few years. Consider the users; the potential for growth in a system designed for a retired couple is a lot different from that of a system designed for a young family.

## Staying in charge

REMEMBER

When connected to an AC power source (such as a generator or the utility), an inverter stops turning DC into AC and becomes a battery charger, taking the AC source and recharging the batteries for you. Inverters have a limited charging capability, though, so you should consider that value in your design. Ideally an inverter can charge the batteries at a C/10 rate.

All inverters list their maximum charging capabilities in amps so you can directly compare the charger portion of the inverter/charger to the battery bank when you divide the batteries' capacity (in amp-hours) by 10 (the number of hours required to recharge the battery).

## Looking at surge ratings

Any load with a motor (such as refrigerators, washing machines, and well pumps) causes a brief power surge when it starts operating. If the inverter can't deliver enough power to the loads during that brief surge period, the entire system may crash, and all the loads may go out. Fortunately, today's inverters can surge three to four times their rated output to start motor loads.

REMEMBER

In order to account for inevitable surges, you need to estimate the amount of power the inverter will be providing just prior to the surge and then estimate what the power draw will be when the surge happens. By adding these values together, you can verify whether the inverter can handle regularly occurring surges. For example, if the inverter is delivering 2 kW of power to all the loads and the refrigerator kicks on, requiring a surge of 1 kW, the inverter has to be able to deliver 3 kW during that surge time.

TIP

The amount of surge is specific to the appliance and is typically noted on the listing label as the maximum current draw. If you can't get to the listing label on the load, you can either estimate the surge by multiplying the load's power draw by three or you can use a clamp meter and record the current draw as the load starts (see Chapter 3 for guidelines on using a clamp meter).

### Evaluating inverter and array power output

For utility-interactive, battery-based systems, you need to consider all the items presented in the preceding sections plus the relationship between the array's power output and the inverter's power output. These systems want to send as much power back to the utility as possible, so you need to make sure the inverter can handle the power output of the array under all scenarios.

REMEMBER

Presumably, the batteries will be full and the inverter will want to take all the array's power and send it somewhere. First, it'll satisfy any loads within the house or office, and then it'll push the current back into the grid. Either way, the inverter will be taking the array's power and processing it. Chapter 11 explains how to size an inverter's power output based on an array's output.

## Incorporating a Generator

Most people installing a battery-based PV system want to incorporate a generator into the system. Why?

» For utility-interactive systems, a generator provides peace of mind. In other words, it guarantees that the building will have power regardless of the length of a power outage.

WARNING

Many clients who have the grid present and either have an existing generator or plan to incorporate a generator want to back up their entire home or office through the MDP. Be careful, though; if the generator isn't hooked up correctly, the utility-interactive inverter may see the generator as the utility and try to send power back into the generator. Although this scenario is good for a utility, it's dangerous for a generator. Work with the inverter manufacturer to make sure that the inverter can't send power back to the generator.

» For off-grid homes, many people consider a generator to be a necessity because producing 100 percent of energy needs solely from a PV source is difficult. Generators are used in off-grid systems primarily to charge the batteries when the PV array can't keep up.

*Note:* Generators are also perfectly suited for equalizing flooded batteries as a part of the required maintenance for any battery-based system that uses flooded batteries.

The sections that follow outline the fundamental features of generators as well as the basic requirements for sizing a generator for off-grid systems so you can recommend the right generator for your client's needs.

## Generator features

When the time comes to determine the best generator for your client's system, consult with a reputable generator dealer and consider the following features:

» **Engine speed:** Lower speeds generally equate to higher overall life. Look for a generator that operates at 1,800 rotations per minute (RPM).

» **Fuel source:** Try to match the generator's fuel source to one that your client will have handy. For instance, many off-grid homes use propane for cooking and water heating. Propane is a good fuel source for a generator as well.

» **Remote start:** The inverters used in battery-based PV systems can incorporate remote and even automatic generator charging. Make sure the generator has a remote-start feature as well. If the generator doesn't, your client will have to walk to the generator to turn it on every time (which isn't a lot of fun in the middle of winter).

» **Output voltage:** You need to match the generator's voltage to that of the inverter. A generator output voltage of 240 VAC is most common, but some generators can be configured to 120 VAC. (I explain how to size an inverter earlier in this chapter.)

Generators are a source of frustration for many people because they require regular maintenance and are prone to Murphy's Law — they break down only when you absolutely need them. So, when working with a generator distributor, verify that it offers full warranties for generators used in any stand-alone, battery-based PV system. Some generator manufacturers have a blanket statement that their warranties don't apply for stand-alone applications. Make your distributor aware that this generator will still be used as a backup, only it'll be backing up a PV system and not the grid.

## Generator sizing

The generator portion of the sizing calculations for battery-based systems is often the part of the design that doesn't receive enough attention — largely because a generator may already be in place.

You may install a stand-alone, battery-based PV system where a generator is already in place and the owner doesn't want to switch to a different one. Another scenario I've seen is where the generator used by the construction crew to build the home becomes a permanent resident and is incorporated into the PV system. Using generators that aren't fully designed into the PV system is far from ideal, but it's a reality for many systems.

When the generator *can* be properly designed into the system, there are a few key parameters to keep in mind:

- The amount of current available for running loads and charging batteries is a major consideration. After the generator is turned on, the inverter locks onto that power source and passes generator power through to the loads in the house or office and uses whatever's left over to charge the batteries. Therefore, the generator's power output needs to equal at least the amount required by any simultaneously running loads plus the maximum amount of power the inverter can use to charge the batteries. So if the home draws 2 kW and the charger needs 3 kW to properly charge the battery bank, the generator should be sized at 5 kW at an absolute minimum.

  Encourage your clients to run their major electrical loads — washing machines, vacuums, and the like — when the generator is operating in order to drown out the noise from the generator.

- Generators are rated by their power output, a value that's typically at 240 VAC. If you're running a single inverter at 120 VAC, you'll probably only get half of the inverter's rating, which means that a 5 kW rated generator can only deliver 2.5 kW when operating at 120 VAC. If you were to run only 120 V off of a 240 V generator, that'd cause damage to the generator eventually because the generator's output wouldn't be properly balanced. A select few generators can be rewired to get their fully rated output at 120 VAC, so there may be a way around that issue — but I have an easier method for sizing a generator.

TIP

Base the generator's power output off of the inverter's power output. As an inverter manufacturer taught me years ago, a generator's power output is based off of the unit being pushed downhill with the wind at its back. (This is a kind way of saying that the rating system is overly optimistic.) If you size the generator's output by three to four times the inverter's output, you should be able to meet the needs of your client's loads and battery-charging requirements. So if you size an inverter at, say, 2 kW, the generator should be a minimum of 6 kW to 8 kW.

IN THIS CHAPTER

» Calculating the correct conductor size

» Making sure you have the right size (and type) of conduit

» Protecting the conductors with properly sized overcurrent protection devices

# Chapter 13

# Sizing Conductors, Conduit, and Safety Components

After you identify all the components your client's PV system requires, you need to determine the correct sizes for all the *conductors* (wires), conduit, and mandatory safety equipment that must be installed (I walk you through what all of these are in Chapter 10). Some of these components may need to be special ordered, so sizing them correctly during the initial design process is critical. After all, one missing piece can bring an entire job to a grinding halt, frustrating everyone involved.

In many cases, the methods used in this chapter are the same as the ones used by electricians for sizing standard electrical equipment. However, there are a few places where you need to look at the process a little differently. Don't worry, though, because I share what they are in this chapter. I give you methods for making sure the conductors used in the PV system you're designing are large enough for proper system performance and sized to meet the requirements of the *National Electrical Code*® (*NEC*®). I also explain how to size conduit and overcurrent protection devices. (For details on the installation of these components, see Chapter 17.)

## Conductor Sizing 101

When I start sizing the safety components of a PV system, I like to size the conductors used throughout the system based on *National Electrical Code*® (*NEC*®) requirements to ensure that they're large enough to safely allow current to flow through them. To do this correctly, you must conduct some *circuit identification*. In other words, you need to pinpoint where in the system the conductors go because the conductors from one part of the system to another don't have the same requirements. For example, the conductors from the PV array to the inverter are subject to different requirements than the conductors from the inverter to the AC main distribution panel (MDP). ***Note:*** The actual conductors used can be the same on the DC and AC sides of the system as long as they're installed properly (I cover different conductor types and where they go in Chapter 10).

Following are the proper circuit designations and where they're located in relation to the installed equipment (see Chapter 10 for more details about different types of circuits):

- **The PV source circuit** comes from the array and ends in a junction box or combiner box.
- **The PV output circuit** comes from the junction box or combiner box and goes to a DC disconnect.
- **The inverter input circuit** goes from the DC disconnect to the DC input of the inverter.
- **The inverter output circuit** comes from the AC side of the inverter and goes to the MDP.

In the next sections, I show you how to determine the proper conductor size for the PV systems you design. I also present the safety factors you should use during the conductor sizing process.

REMEMBER

The method I show for calculating conductor sizes is a simplified version from what other PV pros may choose to use. When you use my method, you'll occasionally end up with a conductor that's a size larger than you absolutely need in order to satisfy *NEC*® requirements. I like the idea of simplicity, though (I'm sure you do, too), and unless you're installing a very large system, the difference in cost (larger conductors are more expensive than smaller conductors) is outweighed by the time you save.

## Defining the PV circuits' maximum and continuous current

Starting at the PV array, you can look at the individual circuits and size the conductors as they work their way to the inverter; these circuits are known collectively as the *PV circuits.* To size the conductors in these circuits, you have to define the maximum amount of current they'll ever carry.

To define the maximum current that'll ever be produced by the PV circuits (and therefore the maximum current that'll be carried by the conductors), you need to look at Article 690.8(A)(1) of the *NEC*®, titled "Circuit Sizing and Current." The maximum current value is as follows:

> The number of modules (or strings) wired in parallel × The short circuit current, $I_{sc}$, rating of the module (or string) × 1.25

*Note:* The 1.25 factor is in place to recognize that PV modules produce current based on the available irradiance (*irradiance* is the intensity of the solar radiation striking earth; flip to Chapter 4 for more information). The *NEC*® also recognizes that a PV array can (and will) produce more current than its modules are rated for if given the opportunity.

If you keep reading to Article 690.8(B)(1) of the *NEC*®, you'll see that a second safety factor of 1.25 is applied. It's there because the *NEC*® dictates that the conductors and overcurrent protection devices (OCPDs) used in electrical systems "shall not continuously carry more than 80 percent of the maximum current." The *NEC*® is calling for you to oversize the conductors so that when current is run continuously (more than three hours at a time), the conductors won't overheat, fail, and cause fires.

To make sure the conductors and OCPDs don't carry more than the predetermined limit of 80 percent of the maximum current, you need to multiply the maximum current by 1.25 (1.25 is the inverse of 80 percent). Doing so tells you the continuous current requirement.

If you want to take a shortcut, you can multiply the two 1.25 safety factors together first to find out the overall safety factor to apply to PV circuits, which is 1.56 ($1.25 \times 1.25 = 1.56$).

At this point, all you've done is define the *ampacity* (amount of current flow) required by the conductors on the PV side of the system. As I explain in the later "Putting together the details to determine conductor sizing" section, you have to adjust this ampacity value based on the locations of the conductors.

For example, if you have an array that consists of ten modules in series with each module short circuit current rated at 5.5 A, the maximum current from that series string of ten modules will be 5.5 A × 1.25 = 6.9 A. To define the ampacity required for the conductor connected to this string, multiply the maximum current by 1.25 again to get 6.9 A × 1.25 = 8.6 A. These results mean that the individual string is rated at 5.5 A but that the conductors and any OCPD connected to that string must be rated for at least 8.6 A.

## Calculating non-PV circuits' maximum current

Unlike the PV circuits covered in the preceding section, inverter input circuits on battery-based systems and inverter output circuits (anything from the inverter's AC output to the load panels) aren't subject to varying current levels based on irradiance. The maximum current levels for battery-based inverter input circuits or any inverter output circuits are actually based on the equipment to which they're connected. (*Note:* The inverter input for a grid-direct inverter is on the PV side and falls into the previous section's requirements.)

The maximum current of an inverter is defined by its manufacturer and is dictated by the maximum power the inverter can produce and the voltage it can churn out. Look for this value on an inverter's specification (spec) sheet listed as maximum output current, as well as in the inverter's manual.

To calculate the ampacity required by the conductors and OCPD, you only have to apply one 1.25 factor so you don't continuously run more than 80 percent of the maximum current across these components. Here's the formula:

Maximum output current × 1.25

For example, a 4 kW inverter operates at 240 VAC and has a maximum current output of 16.7 A. According to the preceding formula, the conductors and OCPDs connected to this inverter's output circuit would need a minimum rating of 16.7 A × 1.25 = 20.9 A.

For battery-based inverters, the inverter input conductors run between the batteries and the inverter (or inverters if the system you're designing calls for more than one; I outline circumstances that call for multiple inverters in Chapter 11). When applying the *NEC®* factors for these circuits, you need to consider the maximum amount of current that could flow across the wires. According to Article 690.8 of the *NEC®*, the maximum current for these conductors is calculated by the following formula:

Maximum continuous power output ÷ Minimum operating voltage

Performing this calculation tells you the largest amount of current that can continuously run through the conductors.

If the specified inverter has a continuous power output rating of 3 kW and it can operate all the way down at 42 VDC, the conductors between the battery bank and the inverter need to be able to handle at least 3,000 W ÷ 42 VDC = 71.4 A.

## Considering conditions of use with some handy tables

At this point, you've calculated the maximum amount of current each conductor will need to carry based on the *NEC*® requirements for each circuit (if you haven't, refer to the earlier sections in this chapter for help). Now you need to take the *conditions of use* into consideration; in other words, you have to adjust the maximum current levels you found based on the locations where the conductors will be used (on the roof, in the attic, through the basement . . . you get the idea) as well as how many will be run together.

REMEMBER

As conductors heat up, their ability to pass current is reduced, which is why you need to consider the conditions that'll increase the temperature of the conductors and reduce their ampacity values. The two conditions to account for are the number of current-carrying conductors running inside a single conduit and the ambient temperature to which the conductors are exposed.

Prepare to turn to the following four tables in the *NEC*® when determining the proper application of the conditions of use:

- Table 310.15(B)(2)(a), titled "Adjustment Factors for More Than Three Current-Carrying Conductors in a Raceway or Cable"
- Table 310.15(B)(2)(c), titled "Ambient Temperature Adjustment for Conduits Exposed to Sunlight On or Above Rooftops"
- Table 310.16, titled "Allowable Ampacities of Insulated Conductors Rated 0 Through 2000 Volts, 60°C Through 90°C (140°F Through 194°F), Not More Than Three Current-Carrying Conductors in Raceway, Cable, or Earth (Directly Buried), Based on Ambient Temperature of 30°C (86°F)"
- Table 310.17, titled "Allowable Ampacities of Single-Insulated Conductors Rated 0 Through 2000 Volts in Free Air, Based on Ambient Air Temperature of 30°C (86°F)"

The following sections provide basic instruction on using each of these tables.

## The tables in Article 310.15: Temperature considerations for conductors

Conductors used in PV systems must be evaluated under the conditions typical of where they'll be used, and the tables in Article 310.15 of the *NEC®* can guide you on what *derate factor* (the values used to reduce conductor ampacity values from the starting points listed in the tables) to apply based on a conductor's location.

Use the (B)(2)(a) table if three or more conductors are being run together and the (B)(2)(c) table if the conductors are exposed to the sun on or above a roof. (I explain how to apply these factors in the later "Putting together the details to determine conductor sizing" section.)

In 2008, Table 310.15(B)(2)(c) was added to account for increased temperatures on rooftops. This addition also addresses the American Society of Heating, Refrigerating, and Air-Conditioning Engineers (ASHRAE) temperature I refer to in Chapter 11 — specifically, the expected summertime temperature. By using the temperatures listed in the ASHRAE tables I direct you to in Chapter 11, in conjunction with the *NEC®* table when appropriate, you can establish a proper design temperature to use in your calculations.

## Tables 310.16 and 310.17: Conductor ampacity values

The values listed in Tables 310.16 and 310.17 (which are represented in Figures 13-1 and 13-2; both figures show only a portion of the tables) represent the amount of current each type and size of conductor is rated for. The columns marked "Size AWG or kcmil" represent the physical size of the conductor. *AWG* stands for *American Wire Gauge*; the AWG convention is that large numbers represent small wires and small numbers represent big wires. After 0000 (pronounced *four aught*), the big wires are represented by their cross-sectional area instead of the AWG convention. The other columns list different types of conductors by abbreviation.

The ampacity values are based on an ambient air temperature of 30 degrees Celsius (86 degrees Fahrenheit) and no more than three current-carrying conductors (the ground wire isn't a current-carrying conductor) run together. You adjust these values when applying the appropriate conditions of use.

Note that the tables listing the conductor ampacities also include different columns labeled with different temperatures. These are the temperature limitations of the conductors due to the insulation protecting the conductive material. Table 310.16 in Figure 13-1 differentiates between copper and aluminum conductors. Aluminum has less ampacity than copper, so to carry amounts of current that are equal to their copper counterparts, aluminum conductors must be physically larger. In all the examples I give in this book (and in most PV installations), copper conductors are used exclusively.

**Table 310.16 Allowable Ampacities of Insulated Conductors Rated 0 Through 2000 Volts, 60°C Through 90°C (140°F Through 194°F), Not More Than Three Current-Carrying Conductors in Raceway, Cable, or Earth (Directly Buried), Based on Ambient Temperature of 30°C (86°F)**

| | Temperature Rating of Conductor [See Table 310.13(A)] | | | | | | |
|---|---|---|---|---|---|---|---|
| | 60°C (140°F) | 75°C (167°F) | 90°C (194°F) | 60°C (140°F) | 75°C (167°F) | 90°C (194°F) | |
| Site AWG or kcmil | Types TW, UF | Types RHW, THHW, THW, THWN, XHHW, USE, 2W | Types TBS, SA, SIS, FEP FEPB, MI, RHH, RHW-2, THHN, THHW, THW-2, THWN-2, USE-2, XHH, XHHW, XHHW-2, ZW-2 | Types TW, UF | Types RHW, THHW, THW, THWN, XHHW, USE | Types TBS, SA, SIS, THHN, THHW, THW-2, THWN-2, RHH, RHW-2, USE-2, XHH, XHHW, XHHW-2, ZW-2 | |
| | COPPER | | | ALUMINUM OR COPPER-CLAD ALUMINUM | | | Site AWG or kcmil |
| 18 | — | — | 14 | — | — | — | — |
| 16 | — | — | 18 | — | — | — | — |
| 14+ | 20 | 20 | 25 | — | — | — | — |
| 12+ | 25 | 25 | 30 | 20 | 20 | 25 | 12+ |
| 10+ | 30 | 35 | 40 | 25 | 30 | 35 | 10+ |
| 8 | 40 | 50 | 55 | 30 | 40 | 45 | 8 |
| 6 | 55 | 65 | 75 | 40 | 50 | 60 | 6 |
| 4 | 70 | 85 | 95 | 55 | 65 | 75 | 4 |
| 3 | 85 | 100 | 110 | 65 | 75 | 85 | 3 |
| 2 | 95 | 115 | 130 | 75 | 90 | 100 | 2 |
| 1 | 110 | 130 | 150 | 85 | 100 | 115 | 1 |

**FIGURE 13-1:** A portion of *NEC®* Table 310.16.

*Reprinted with permission from NFPA 70®, National Electrical Code®, Copyright © 2007, National Fire Protection Association, Quincy, MA 02169. This reprinted material is not the complete and official position of the NFPA on the referenced subject, which is represented only by the standard in its entirety.*

**Ampacities of Insulated Copper Conductors***

| | TYPE OF INSULATION | TW, UF | RHW, THHW, THW, THWN, XHHW, USE, ZW | TBS, SA, SIS, FEP, FEPB, MI, RHH, RHW-2, THHN, THHW, THW-2, THWN-2, USE-2, XHH, XHHW-2, ZW-2 |
|---|---|---|---|---|
| | AWG | 60°C Rated | 75°C Rated | 90°C Rated |
| CONDUCTOR IN FREE AIR | 18 | — | — | 18 |
| | 16 | — | — | 24 |
| | 14 | 25 | 30 | 35 |
| | 12 | 30 | 35 | 40 |
| | 10 | 40 | 50 | 55 |
| | 8 | 60 | 70 | 80 |
| | 6 | 80 | 95 | 105 |
| | 4 | 105 | 125 | 140 |
| | 3 | 120 | 145 | 165 |
| | 2 | 140 | 170 | 190 |
| | 1 | 165 | 195 | 220 |
| | 0(1/0) | 195 | 230 | 260 |
| | 00 (2/0) | 225 | 265 | 300 |

* Based on ambient temperature of 30° C (86° F) and not more than three current-carrying conductors when in a raceway, cable, or earth (directly burned).

**FIGURE 13-2:** A portion of *NEC®* Table 310.17.

*Reprinted with permission from NFPA 70®, National Electrical Code®, Copyright © 2007, National Fire Protection Association, Quincy, MA 02169. This reprinted material is not the complete and official position of the NFPA on the referenced subject, which is represented only by the standard in its entirety.*

Conductor sizes are also limited by the rating of all the *terminals*, connection points where a conductor is physically connected to a termination such as a disconnect, fuse holder, or circuit breaker, that the wires are connected to. These terminals are where the conductors from the PV system connect inside the various combiner boxes, disconnects, inverters, and other pieces of electrical equipment. When you make calculations related to terminals, be sure to use Table 310.17.

REMEMBER

The equipment used in PV systems that are less than 600 V (which are all the residential and commercial systems out there) is either 60 degrees Celsius (140 degrees Fahrenheit) or 75 degrees Celsius (167 degrees Fahrenheit). This fact means that when you use conductors rated at 90 degrees Celsius (194 degrees Fahrenheit), you need to evaluate the rating of those conductors at the terminal's maximum value. Occasionally, this check will require you to size the conductor even larger to accommodate the temperature limitations of the terminal. The majority of the terminals used in equipment associated with PV systems are 75 degrees Celsius (167 degrees Fahrenheit), but don't assume this is the case. Instead, verify the proper ratings so you don't get caught installing the wrong equipment. To avoid any possible issues in the field, I suggest you only install conductors that are rated for 90 degrees Celsius (194 degrees Fahrenheit). Doing so gives you the maximum flexibility in sizing and appropriately installing the conductors.

REMEMBER

When installing equipment with different ratings, you must use the lowest rating. For example, if the conductors are connected to a disconnect on one end with terminals rated at 60 degrees Celsius (140 degrees Fahrenheit), and a combiner box at the other end with 75 degrees Celsius (167 degrees Fahrenheit) terminals, you must use the 60 degrees Celsius terminals for the basis of your conductors' size.

## Putting together the details to determine conductor sizing

REMEMBER

To calculate the required conductor sizes, follow these steps:

1. **Determine the maximum circuit current.**

   See the earlier "Defining the PV circuits' maximum and continuous current" section for the how-to.

2. **Calculate the continuous current requirement.**

   For help determining the continuous current requirement, refer to the related section earlier in this chapter. (Note that the continuous current requirement defines the OCPD value you need to use in the later "Sizing Overcurrent Protection Devices and Disconnects" section.)

3. **If the conductors will be installed in conduit and there will be more than three current-carrying conductors in the conduit, divide the Step 2 value by the appropriate derate value from Table 310.15(B)(2)(a); if the conductors will be exposed to sunlight on the roof, add the appropriate temperature adjustment to the ASHRAE ambient temperature value from Table 310.15(B)(2)(c), based on the height of the conductors off of the roof surface.**
4. **Divide the value in Step 3 by the appropriate temperature-correction factor for the expected temperature the conductors will be exposed to from the 90 degrees Celsius (194 degrees Fahrenheit) column in Table 310.16 to find the corrected conductor ampacity required.**

   Choosing the temperature-correction factor from the 90 degrees Celsius (194 degrees Fahrenheit) column assumes you're installing only conductors with 90 degrees Celsius (194 degrees Fahrenheit) ratings, as I suggest in the preceding section. If you want to use conductors rated at 75 degrees Celsius (167 degrees Fahrenheit), be sure to use that column of Table 310.16 instead.
5. **Use the value from Step 3 to find the smallest 90 degrees Celsius/194 degrees Fahrenheit (or 75 degrees Celsius/167 degrees Fahrenheit) conductor that exceeds the ampacity requirement in Table 310.16.**
6. **Verify that the conductor selected has a large enough ampacity based on the lowest terminal ratings that will be used.**

   To do this, look at the ampacity value listed in the appropriate column in Table 310.17 (which consists of terminal temperature listings) and check that the ampacity value listed for the chosen conductor is greater than or equal to the continuous current value in Step 2.

I know how tricky these calculations can be to make, so I want to share an example with you. For the purposes of this example:

- The array has eight strings in parallel, and each string has an $I_{sc}$ of 5.9 A.
- All the strings are placed in parallel inside a combiner box.
- The THWN-2 conductors are in conduit and exposed to temperatures of 45 degrees Celsius, but they aren't exposed to sunlight on a roof.
- The terminals inside the combiner box, as well as the disconnect on the other end of the conductors, are all rated at 75 degrees Celsius.

With this information, you can apply the previously described steps to determine the minimum conductor size necessary to satisfy the *NEC*®:

1. **Find the maximum current.**

   8 strings in parallel × 5.9 A per string × 1.25 = 59 A

2. **Determine the continuous current**

   59 A × 1.25 = 73.75 (round up to 74 A)

3. **Derate for more than three current-carrying conductors.**

   In this example, all the PV source circuits are placed in parallel and a single PV output circuit leaves the combiner box. Therefore, no derate factor is applicable.

4. **Apply the conditions of use.**

   74 A ÷ 0.87 = 85 A

   You find the appropriate temperature derate factor from the bottom of Table 310.16. The 90 degrees Celsius (194 degrees Fahrenheit) column, which you use for THWN-2 conductors, has a temperature derate of 0.87 at 45 degrees Celsius.

   ***Note:*** If the conductors were exposed to sunlight on a roof, you'd also used a factor from Table 310.15(B)(2)(c).

5. **Check Table 310.16, using the value from Step 3.**

   4 AWG in the 90 degrees Celsius (194 degrees Fahrenheit) column at the top of Table 310.16 is the smallest conductor that exceeds 85 A. A THWN-2 conductor therefore has an ampacity value of 95 A.

6. **Confirm that the conductor ampacity is large enough.**

   According to Table 310.17, 4 AWG conductors rated at 75 degrees Celsius (167 degrees Fahrenheit) — the terminal ratings — have an ampacity of 85 A. This value exceeds the continuous current value of 74 A from Step 2. You now know that a 4 AWG conductor installed in this location meets the *NEC*® requirements for ampacity and can safely carry current from this array.

You can follow this process for all the circuits used within a PV system. When calculating for circuits other than those on the PV side (the inverter output circuit, for example), you start at Step 1 with the maximum current value as given by the manufacturer as I describe in the earlier "Calculating non-PV circuits' maximum current" section. You then proceed through the rest of the steps as normal.

## Accounting for voltage drop after you size your conductors

The *NEC®* requires that the conductors you use in any PV system must be large enough to carry current from the source to the load safely, but no *NEC®* requirement dictates that you must deliver the power as efficiently as possible. As a PV system designer and installer, though, you want to make sure that the power supplied by the PV system is never reduced because the conductors are too small.

No matter how large you size the conductors in your system (see the previous section for guidelines), some amount of voltage drop will occur. As current flows in the conductors, resistance in the conductors fights that current flow. As you discover in Chapter 3, voltage has a direct relationship to current (which is measured in amps) and resistance via Ohm's Law, which can be stated in several ways:

- Volts = Amps × Resistance ($E = I \times R$)
- Amps = Volts ÷ Resistance ($I = E \div R$)
- Resistance = Volts ÷ Amps ($R = E \div I$)

This interplay between current flow and resistance means you need to evaluate the conductors used in your system and make sure they aren't so small that excessive resistance causes too much loss.

Because the *NEC®* doesn't dictate the maximum amount of voltage drop, you get to decide which amount suits you best. Within the solar industry, the level of voltage drop is referred to in terms of the DC and AC sides of the system.

- The accepted amount of DC voltage drop is 2 percent from the PV array down to the inverter input. This isn't a hard value, though. If you want to, you can evaluate the financial effects of running larger conductors to reduce voltage drop and find that a 3-percent drop makes sense.
- A tolerable amount of voltage drop on the AC side is between 1 percent and 1.5 percent for utility-interactive systems (either grid-direct or battery-based). If the PV system is of the stand-alone, battery-based variety, between 2 percent and 3 percent voltage drop is okay.

I show you how to calculate the voltage drop values for DC and AC circuits in the next sections. In both examples, I use Ohm's Law and some given properties for copper conductors (but you can solve for any of the pieces of information as long as you can define all the other parts of the equation). ***Note:*** In the real world, you may need to take slight differences in the nature of AC circuits into account when calculating the AC voltage drop. For all practical purposes, though, using Ohm's Law as I show you is accurate enough.

If you calculate the voltage drop and come up with a gauge that's different than the calculations in the preceding section, use the larger of the two conductors. This way you always satisfy both requirements.

## DC voltage

The first step in calculating the voltage drop from the PV array down to the inverter is to break down the individual parts of the PV circuit(s).

To estimate the total voltage drop, you need to consider the longest PV source circuit length (from one of the strings to the combiner) and the PV output circuit (from the combiner to the inverter) and make two separate calculations that you then add together.

*Note:* By considering the longest PV source circuit length, you'll be slightly more conservative for the shorter lengths, which results in slightly better performance in those circuits.

Before you can run your calculations, you need to collect the following information about the PV system:

- **The maximum power output rating of the array from the modules' spec sheets:** You use the $V_{mp}$ value because you're most concerned about the voltage drop when the array is producing its greatest amount of power.
- **The maximum current value for each circuit from the modules' spec sheets:** You use the $I_{mp}$ value for the same reason you use the $V_{mp}$ value.
- **The total circuit length:** This number tells you how far the electrons have to travel to complete their circuit. You generally have to estimate this number by measuring the distances for the proposed equipment locations. Convert this distance into kilofeet (a goofy measurement, I know) because the table you have to reference supplies conductor resistances in terms of resistance per thousand feet of conductor. To convert to kilofeet, divide the estimated distance by 1,000.
- **The resistance of the conductor:** Refer to the "Conductor Properties" table in the *NEC*® (a portion of which is in Figure 13-3) to see the different resistance characteristics of conductors. When you have an idea of the conductor size to use (because you did the *NEC*® calculations in the previous section or because you're installing modules with preinstalled conductors), go to this table and use the resistance per kilofeet to swap into the equation and solve for voltage drop.

**Conductor Resistances***

| AWG | SOLID COPPER | STRANDED COPPER |
|---|---|---|
| 18 | 7.77 | 7.95 |
| 16 | 4.89 | 4.99 |
| 14 | 3.07 | 3.14 |
| 12 | 1.93 | 1.98 |
| 10 | 1.21 | 1.24 |
| 8 | 0.764 | 0.778 |
| 6 | — | 0.491 |
| 4 | — | 0.308 |
| 3 | — | 0.245 |
| 2 | — | 0.194 |
| 1 | — | 0.154 |
| 0 (1/0) | — | 0.122 |
| 00 (2/0) | — | 0.0967 |

* in Ω/kft at 75°C (167°F)

**FIGURE 13-3:** A portion of the *NEC*® "Conductor Properties" table (Chapter 9, Table 8).



Calculating the DC voltage drop requires you to apply the information you collected regarding the module specs and the conductor properties (length and resistance) to a form of Ohm's Law a little differently than Herr Ohm intended (although if given the chance, I think he'd approve). The difference is that a length value is associated with the equation:

$$V_{drop} = I_{mp} \times R_c \times L$$

$V_{drop}$ = The maximum number of volts you're willing to lose based on the array's $V_{mp}$ (it's the variable you solve for)

$I_{mp}$ = The maximum power current rating of the circuit

$R_c$ = The resistance of the conductor (Ω/kft)

$L$ = The total circuit length in thousands of feet (kft)

Suppose you have an array that consists of three strings of ten modules and all the strings are arranged on a roof in rows of ten. These three strings are all wired to a roof-mounted combiner box that runs down from the roof to the inverter mounted outside next to the utility meter and MDP. This setup means you have three PV source circuits running from the strings to the combiner box and a single circuit running the rest of the way out.

Each PV module is 36.2 V and 4.9 A, and the strings are wired with 12 AWG conductors. With all of this information, you can measure the total circuit length for the circuit and calculate the total voltage drop. In order to determine the voltage drop in the PV source circuit, you need to collect the following module and conductor information:

- **$V_{mp}$** for the modules is 36.2 V, so the string $V_{mp}$ = 362 V.
- **$I_{mp}$** for the module is 4.9 A, which means the string $I_{mp}$ is also 4.9 A.
- **The total circuit length** is 75 feet, or 75 ÷ 1,000 = 0.075 kilofeet.
- **The resistance, $R_c$, in the 12 AWG wire used** is 1.98 Ω/kilofeet (from the *NEC*® table).

Here's how to calculate the number of volts that will be lost in this circuit (in other words, the voltage drop, or $V_{drop}$), using the preceding information:

$$V_{drop} = I_{mp} \times R_c \times L$$

$$V_{drop} = 4.9\text{ A} \times 1.98\ \Omega/\text{kilofeet} \times 0.075\text{ kilofeet}$$

$$V_{drop} = 0.73\text{ V}$$

This number indicates how many volts will be lost due to the resistance in the conductor. Your next step is to relate this voltage drop to a percentage of the entire circuit. To determine the voltage drop in a percentage, divide the calculated loss by the original $V_{mp}$ value for the array:

$$V_{drop}\,\% = V_{drop} / V_{mp}\text{ of the array}$$

So when you apply the values calculated in the example, you get the following:

$$0.73\text{ V} \div (36.2\text{ V} \times 10) = 0.73\text{ V} \div 362\text{ V} = 0.002 = 0.2\%$$

This result indicates that a 0.2-percent voltage drop exists in the PV source circuits. To determine the voltage drop in the PV output circuit, repeat this process. If when you add these two voltage drop percentages together, the total voltage drop is less than your desired amount of 2 percent, the conductor sizes check out. If the total voltage drop is greater than your desired amount, you need to upsize one of the conductors to minimize the loss.

## AC voltage

When you go to look at the voltage drop on the AC side, you use the exact same equation that I feature in the preceding section with different numbers for the voltage and current values.

- The nominal AC voltage replaces the maximum power output rating of the array.
- The maximum inverter current output value tells you the current input.
- You can reference the conductor resistance from the table in Figure 13-3.
- The circuit length is still the total circuit length (twice the physical distance between the inverter and the utility interconnection).

Say you have an inverter that's operating at 240 VAC and a maximum current output of 24 A. The inverter output circuit is 10 AWG, and the inverter is 75 feet from the MDP. To find the AC voltage drop in this example system, you need to collect the following information:

- **The AC operating voltage** is 24 VAC.
- **The maximum current output** is 24 A.
- **The inverter output circuit** is 10 AWG, with a resistance of 1.24 Ω/kilofeet.
- **The total circuit length** is 150 feet or 0.15 kilofeet.

You can apply these values to the voltage drop equation in the preceding section:

$$V_{drop} = I_{max} \times R_c \times L$$

$$V_{drop} = 24\ \text{A} \times 1.24 \times 0.15$$

$$V_{drop} = 4.5\ \text{V}$$

To determine what this is in terms of a percentage value:

$$V_{drop}\ \% = V_{drop} \div V_{nominal}$$

$$V_{drop}\ \% = 4.5\text{V} \div 240\ \text{V} = 0.19 = 1.9\%$$

This percentage exceeds the recommended maximum voltage drop of 1.5 percent for utility-interactive systems, so in this case, I recommend looking at the next-largest size of conductor to minimize the effects of excessive voltage drop on this circuit.

# Sizing Conduit

Almost all conductors require proper protection. The exceptions are conductors used in PV source circuits and a type of conductor called ground wiring.) For PV systems, most conductors are run inside a *conduit*, which is either a metal or PVC tube that protects conductors from their surroundings.

REMEMBER

The exact type of conduit required depends on where you're placing the conductors. For PV output circuits run inside a building before a readily accessible disconnect, the conductors must be contained in a metallic conduit body. If the same conductors are kept outside the building, PVC would be an acceptable conduit material. You can also run PVC conduit underground to connect PV arrays to inverters. (Chapter 10 details conduit types and uses.)

After you establish the appropriate conduit type and the size and number of conductors you need to run, you can use the tables in Annex C of the *NEC*® to look up what size conduit is required. When you find the appropriate conduit table, you can then go to the appropriate section of that table — the one that lists your specific conductor type (for example, THWN-2). The tables then list all the available conductor sizes in a column. Find your conductor size and then move across the table to find the conduit that can hold the minimum number of conductors you want to use.

TIP

If you find a conduit size that can hold the exact number (or maybe one extra) of conductors you plan to run, go ahead and bump up to the next conduit size. Doing so will cost your client a little bit more money, but this move saves you time in the field, which in turn saves the client money on your labor costs, making it a worthwhile extra equipment expense. Look at it as an investment in your sanity. After all, just because a table says you can pull eight conductors through a conduit doesn't mean you actually *want* to do that.

# Sizing Overcurrent Protection Devices and Disconnects

When wiring the components of a PV system, it's your responsibility to protect the conductors from the possibility of too much current flowing through them. Enter overcurrent protection devices (OCPDs). These devices come in the form of either circuit breakers or fuses. The exact requirements and locations for OCPDs vary based on the current's source (see Chapter 10 for details). In the sections that follow, I provide guidelines for basic OCPD sizing, as well as specifics for sizing OCPDs on PV and inverter circuits.

*Note:* You size disconnects exactly the way you size OCPDs, but for simplicity's sake, I just refer to OCPDs in the following sections.

## Beginning with a few basics

Believe it or not, in the process of sizing the PV system's conductors, you also defined the required size of the OCPDs for the conductors. (It's based off the continuous current requirement in Step 2 of the process described in the earlier "Putting together the details to determine conductor sizing" section.)

TIP

If you can find an OCPD that's listed for 100 percent of its rating, use the maximum current rating of your circuit as the size of your OCPD (Step 1 of the conductor sizing process). OCPDs with 100 percent ratings aren't common, though. Check the spec sheet provided by the manufacturer to verify whether a particular OCPD can really handle as much current as an array produces.

When you calculate an OCPD size, most likely it'll be a value that isn't commonly available. Fortunately, the *NEC*® allows you to round up to the next available OCPD size. These commonly available sizes can be found in Article 240.6 of the *NEC*®. In the example I use earlier to show you how to calculate the conductor size, the continuous current value was 74 A. OCPDs sized at 74 A aren't regularly available, so the OCPD size you'd need to use for the circuit in question would be rated at 80 A (the next available size).

WARNING

If the circuit uses small conductors, be aware that the *NEC*® limits the size of OCPD you can connect these wires to, regardless of the conductors' ampacity. As reported in Article 240.4(D) of the *NEC*®, the limitations are as follows:

- 14 AWG connected to 15 A or less
- 12 AWG connected to 20 A or less
- 10 AWG connected to 30 A or less

To put these limitations in perspective, if you run the calculations and determine that a 35 A OCPD is required for your installation, then you can't use a 10 AWG conductor, even though in some situations that conductor has an ampacity value of 40 A.

TIP

Manufacturers of PV modules and inverters are required to specify the maximum OCPD their equipment can be connected to, known as the *maximum series fuse rating* for PV modules. It's listed with every module you ever buy. So when you're buying OCPDs for the PV strings, know that the manufacturer has defined the largest amperage rating its OCPD can have.

## Placing protection on PV circuits

Although you may be used to placing OCPD on each and every circuit you wire in a regular electrical system, placing OCPD in every PV source and output circuit isn't an absolute requirement. When your system meets Article 690.9 of the *NEC*®, you can actually place two strings of PV modules in parallel without having to put

OCPDs on either string. As soon as you add a third string, though, you should place OCPDs on the strings.

To meet *NEC®* requirements eliminating OCPDs on PV circuits, you first have to size your conductors to meet Article 690.8. Next, you can't have any potential *backed currents,* current that can come back from sources such as batteries and inverters. (This is true when using grid-direct inverters, but if you're using battery-based inverters, you won't meet the requirements, so you have to use OCPD on all PV circuits.) Finally, the short circuit current ($I_{sc}$) from all sources can't exceed the ampacity of the conductors. (This last requirement is where the limit of two strings in parallel comes into play.)

The idea is, if you have three strings in parallel and a *fault* occurs on one of the strings (meaning that string becomes a load and starts accepting current), the $I_{sc}$ values from the other two strings (the external sources) will exceed the conductors' ampacity and the modules' rating.

REMEMBER

Follow this general rule for grid-direct systems: When you have one or two strings, skip the OCPD; when you have more than two strings, put an OCPD on each string. (Note that battery-based systems require an OCPD on every string.)

TIP

Rooftop combiner boxes are a common and acceptable location for series string OCPDs. These devices are affected by temperature though, so check the manufacturer's specifications for proper OCPD sizing when installing an OCPD in high-temperature locations (like rooftops).

## Protecting inverter circuits

As for the AC output of your inverter, you always need an OCPD for this circuit. A variety of options exist for installing this protection, with the most common being a circuit breaker located in the MDP. This circuit breaker serves as both a way to disconnect the inverter and a method of protecting the conductors. (Some installations call for a fused AC disconnect to be installed rather than a circuit breaker; I cover such installations in Chapter 17.)

To determine the amperage rating of the OCPD connected to the inverter, you need to refer to the inverter manufacturer's specifications. Many inverter manufacturers list this value on their spec sheets. If you encounter one that doesn't, you may need to dig into the installation manual; if nothing else, that should indicate the maximum OCPD sizing you can use in conjunction with the inverter.

# 4 Installing a PV System

## IN THIS PART . . .

This part is dedicated to taking your system design and putting it into service. It covers all the factors you need to consider to stay safe during and after the installation.

Chapter 14 explains how to obtain the permits you need to begin the installation process, and Chapter 15 covers all sorts of installation-related safety information. For insight into installing the structural and electrical components of a system, turn to Chapters 16 and 17. Last but not least, Chapter 18 fills you in on how to turn the system on, handle the local building department's inspection, and maintain the system. (Your client is responsible for the bulk of the maintenance, but you need to know how to do it too so you can pass the information along and conduct an annual maintenance visit.)

IN THIS CHAPTER

» Dealing with permitting toward the start of a project

» Drawing the right stuff

# Chapter 14 The Permitting Process

Before you can even begin a PV project, you must obtain the proper permits from the local building department. Most jurisdictions are able and willing to work with individuals and companies who want to install PV systems. And even for those locations where PV systems aren't as common, the local building departments have multiple resources available to them (although you may need to help guide it and supply any information the local folks need). This chapter reveals the details you should have ready for the permit office to ensure that this portion of your job goes smoothly.

Keep in mind that you need to document your installations as well as you possibly can. In this chapter, I also tell you all about how to make drawings of your systems to help in permitting, design, construction, and inspection. (I don't expect you to become an expert in making these drawings, but you should find a program you like or work with somebody who can help you create clean, professional drawings.)

## Obtaining Permits before You Install a PV System

The permitting process for PV systems calls for you to collect specification (spec) sheets for the products you plan to use, document your design, and present all of this information to the local building department (city or county, depending on the location of the site). In some parts of the United States, building departments are very familiar with PV systems; they may even have a standard set of procedures in place for handling permits. In other locations, you may as well be speaking Greek when you come in and ask for a permit for the photovoltaic system you want to install. No matter the situation, I suggest you show up at the building department with multiple supporting documents and take some time to talk with the person issuing the permits (and even one of the inspectors if you have the opportunity — addressing any unusual electrical topic sooner rather than later will help you install the entire system up to the inspector's expectations).

Of course, no such thing as a "typical" permitting process really exists. The requirements from one location to the next (such as the timeline associated with obtaining the permits) can vary dramatically, making matters rather difficult for PV system designers and installers who work in multiple jurisdictions. To help make the process move as quickly as possible, call the building department ahead of time and verify the documentation you need to bring to satisfy its requirements.

TIP

The cost of permits is another huge variable from location to location. One community may charge a small fee for a permit, but the fees may be hundreds of dollars more just one town over. My advice to you? Get to know the fee structure the local building department uses for determining permit fees. If you quote a client a price for his PV system without including the costs of permit fees, it may be an unpleasant surprise for someone to pay later on. Your best bet is to contact the building department well ahead of time, obtain an estimated permit cost, and include that number in the total system cost you quote to your customer. (Turn to Chapter 5 for more information on discussing costs with customers.)

Although the actual processes involved in obtaining permits vary, the type of information required for both residential and commercial applications is pretty consistent. Most jurisdictions mandate permits for both the electrical and mechanical elements of PV systems because they want to make sure the systems you install will be safe for many years to come. Your job is to obtain the necessary licenses and certifications and present well-organized, understandable packets of information for use in the permitting process. The following sections help you figure out what information to gather and give you some insight into the differences between permits for residential applications and permits for commercial applications.

## In the beginning: Having the right licenses and certifications

The first step in the permitting process is to make sure you have all the licenses and/or certifications required by the jurisdiction you're working in. Here's the major difference between a license and a certification:

» *Licenses* are issued by authorities who maintain minimum qualification and education requirements. Typically these authorities are a state office that requires license holders to meet a set of minimum qualifications and pass a written exam before being issued a license.

» *Certifications* can be just as difficult to obtain as licenses, but they don't have the same legal status licenses do. They're issued by independent national organizations.

In order to obtain a permit (commonly referred to as *pulling a permit*), building departments need to make sure you're properly licensed. These requirements vary greatly from state to state, so check in with the state's licensing board for the exact requirements in the state(s) in which you're working.

REMEMBER

In the PV industry, one particular certification is considered the standard: the North American Board of Certified Energy Practitioners (NABCEP)–Certified PV Installer. To qualify for this exam, you must meet the NABCEP's criteria for on-the-job experience as a lead PV installer and meet a minimum education requirement. Passing the NABCEP's installer exam doesn't give you the legal right to work in any state, but it does show consumers and rebate programs that you're educated and experienced in working with PV systems — a perception that can make you stand out from your competition. (For more information about the NABCEP, check out `www.nabcep.org`.)

TIP

Be sure to check with the rebate programs in your client's area to verify the eligibility requirements before you get too far down the line. They may require that the installer hold a certification such as NABCEP to be eligible for rebate money.

*Note:* Time for full disclosure here. I'm a certified PV installer, and I teach a number of courses under the NABCEP program. So, I'm a firm believer in the organization's mission, and I support its efforts. I appreciate the level of professionalism the NABCEP has brought to the whole industry and how it has helped move PV installations and the PV-installing trade into the mainstream.

## Home grown: Permitting for residential systems

For residential systems, the permitting process is relatively painless (aside from the cost, of course, but if you've done your due diligence by including the permit cost in your quote to the client, that's not your responsibility). Of course, if you don't have the right documentation, the permitting process can be a complete pain in the you-know-what. Rest easy, though, because I explain what you need to obtain mechanical and electrical permits for residential PV systems in the sections that follow.

## Permits for mechanical components

The other half of the permitting process is generally for the mechanical and structural aspects of the PV system. Before issuing this permit, the building department wants to make sure that the PV array is properly supported and that the structure it's attached to can withstand the new loading. (Chapter 16 goes through some of thc common locations for, as well as issues involving, mounting an array.)

For homes that have been built since the mid-1970s, building codes have been fairly consistent. In nearly all situations, the building is able to take on the addition of a rooftop PV system. However, this fact doesn't mean you don't have to calculate the building loading or show the effects of the array on the roof; I'm just saying that you generally won't have problems. When possible, bring supporting documentation (pictures work great) regarding the existing roofing system so the building department can better determine the appropriateness of the existing structure.

For homes built before the mid-1970s, or whenever you come across a roof that you suspect isn't properly supported, you need to find a structural engineer in the local area who's willing to help you document the existing roof and guide you in any requirements necessary for adding additional support. The local building department can tell you what the exact requirements are in that area.

For PV arrays mounted at locations other than the roof, many jurisdictions want to see your plan for installing the array and the racking manufacturer's installation instructions (which typically include drawings that show the array and support footings installed). These jurisdictions are typically much more comfortable when the manufacturers can show the resistance to wind loading because, to them, you're installing a huge sail just waiting to take off in a windstorm. Some building departments may even require a structural and/or geotechnical engineer to review the installation method and have you submit a wet-stamped drawing set (I fill you in on wet stamping in the next section).

REMEMBER

Regardless of the PV array location, your drawing package should include an overall plan view of the site with the equipment locations called out. Think of a *plan view* as a bird's-eye view that gives anyone looking at it a quick reference to the location of the site, the existing equipment, and the proposed equipment.

TIP

One relatively easy way to create this drawing is to use satellite views that you can obtain for free online and use a basic computer drawing program to indicate all the equipment locations. Two commonly used satellite programs are Google Earth and Bing Maps (I describe these programs in Chapter 5). A standard drawing program is Paint, available on any Windows-based computer. (Flip to the later "Not Just Pretty Pictures: Creating Drawing Sets" section for more details on drawings.)

## Permits for electrical components

The electrical components of a PV installation require permits as well as an inspection by someone from the building department. Before issuing electrical permits, the local building department wants to make sure the proposed equipment can be integrated into the home's electrical components. First, though, you must go through the electrical sizing and design process I outline in Part 3. You also need to make sure the system you plan to install meets the requirements set by the *National Electrical Code*® (*NEC*®). (I cover the major highlights of the *NEC*® and installing electrical components in PV systems in Chapter 17 if you need to jump there.)

REMEMBER

When going to get your electrical permits, bring in copies of the spec sheets for all the electrical components you plan to install (the spec sheets help confirm the voltage and current values you use in the design process) as well as a simple one-line drawing of your system design. A *one-line drawing* uses boxes to represent the different components and lines to represent the conductors and conduit used to connect all the components. It's called a one-line drawing because you connect the individual components with a single line, even though that single line represents multiple conductors.

REMEMBER

Your electrical one-line drawing can be a simple electrical drawing. At a minimum, your one-line drawing should include

- The site address and the installer's (your) contact information
- The make and model of all specified equipment
- The design temperatures (cold and hot) you used for voltage correction
- The wire and conduit sizes for all circuits
- The general location of all the equipment
- Specifications for the point of interconnection with the utility (if you're installing a grid-direct system or a utility-interactive, battery-based system)
- A note indicating the lack of the utility if you're installing a stand-alone, battery-based system
- Required information for labels that will be installed

Figure 14-1 shows an example of a one-line drawing used to obtain residential permitting. (See the later section titled "Not Just Pretty Pictures: Creating Drawing Sets" for more information on drawings.)

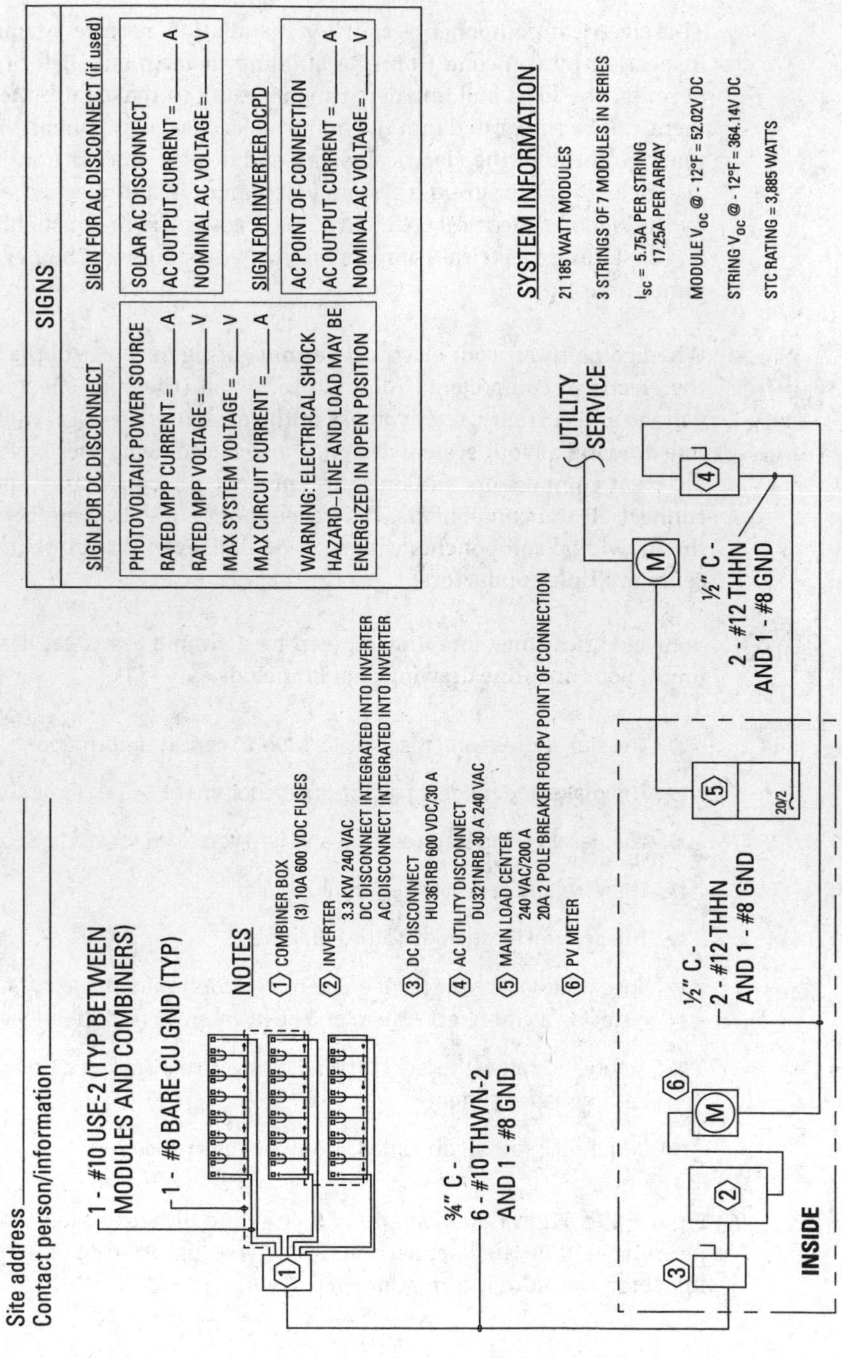

Site address ____________
Contact person/information ____________

1 - #10 USE-2 (TYP BETWEEN MODULES AND COMBINERS)

1 - #6 BARE CU GND (TYP)

3/4" C -
6 - #10 THWN-2
AND 1 - #8 GND

1/2" C -
2 - #12 THHN
AND 1 - #8 GND

1/2" C -
2 - #12 THHN
AND 1 - #8 GND

UTILITY SERVICE

INSIDE

OUTSIDE

20/2

NOTES

1. COMBINER BOX
   (3) 10A 600 VDC FUSES
2. INVERTER
   3.3 KW 240 VAC
   DC DISCONNECT INTEGRATED INTO INVERTER
   AC DISCONNECT INTEGRATED INTO INVERTER
3. DC DISCONNECT
   HU361RB 600 VDC/30 A
4. AC UTILITY DISCONNECT
   DU321NRB 30 A 240 VAC
5. MAIN LOAD CENTER
   240 VAC/200 A
   20A 2 POLE BREAKER FOR PV POINT OF CONNECTION
6. PV METER

SIGNS

SIGN FOR DC DISCONNECT

PHOTOVOLTAIC POWER SOURCE
RATED MPP CURRENT = ______ A
RATED MPP VOLTAGE = ______ V
MAX SYSTEM VOLTAGE = ______ V
MAX CIRCUIT CURRENT = ______ A

WARNING: ELECTRICAL SHOCK
HAZARD–LINE AND LOAD MAY BE
ENERGIZED IN OPEN POSITION

SIGN FOR AC DISCONNECT (if used)

SOLAR AC DISCONNECT
AC OUTPUT CURRENT = ______ A
NOMINAL AC VOLTAGE = ______ V

SIGN FOR INVERTER OCPD

AC POINT OF CONNECTION
AC OUTPUT CURRENT = ______ A
NOMINAL AC VOLTAGE = ______ V

SYSTEM INFORMATION

21 185 WATT MODULES

3 STRINGS OF 7 MODULES IN SERIES

$I_{SC}$ = 5.75A PER STRING
17.25A PER ARRAY

MODULE $V_{OC}$ @ - 12°F = 52.02V DC

STRING $V_{OC}$ @ - 12°F = 364.14V DC

STC RATING = 3,885 WATTS

**FIGURE 14-1:** A one-line drawing used to obtain electrical permits.

TIP

Many rebate and tax credit programs require that the installation application include drawings and documentation similar to the permitting office. However, they're interested in the system's performance as well as its safety. For this reason, you may need to include information on your one-line drawing such as conductor lengths, voltage drop calculations, and system-performance estimations.

Most of the time you won't be required to hire a professional electrical engineer to review and wet stamp your drawings. A *wet stamp* is when an engineer applies her stamp to a drawing or document, indicating that she has reviewed all portions of the material and approves of the design. It generally holds a lot of weight in a permitting office because it indicates that someone (other than the building department staff) has reviewed the information and is willing to take the responsibility (and liability) for its accuracy. However, you may need to bring in outside expertise to verify the proper way to install electrical components in special situations. The most common scenario is for the home that was built with less-than-conventional methods (or maybe before a convention existed). In this case, you should consider hiring an engineer to help calculate the steps needed to install the system's electrical parts.

TIP

If you're working in a jurisdiction that has never seen a PV system, or if you're dealing with an oddball electrical issue, ask to speak with one of the inspectors while you're getting your electrical permits so you can bring up the issue at hand or discuss some of the more interesting parts of the installation, like grounding (covered in Chapter 17). I've found that grounding is an area where asking for forgiveness after the fact rather than permission upfront leads to trouble.

## Big business: Permitting for commercial systems

Unlike with residential systems, many jurisdictions include a mandatory plan review for PV systems installed on commercial facilities, which means one or more *plan checkers* have to look at the documentation provided by you, the installer, and verify that the electrical and structural components meet the minimum requirements. Consequently, you need to provide enough information to the reviewers so they can make informed decisions about the system. After they've gone over the proposed system, they'll either issue the permit or ask for more information before letting you proceed. In the next sections, I explain what you need for the mechanical and electrical permits that apply to commercial PV systems.

## Mechanical permits

Before issuing a mechanical permit for a commercial PV system, a building department wants to guarantee that the building will be able to handle the addition of a PV array on the roof. The process for this type of mechanical permit is a bit more complicated than it is for a residential PV system.

On commercial projects, the structural integrity of the building needs consideration very early on. Many commercial buildings are built only to the minimum structural requirements, which means adding a PV array on top of them isn't possible. If you're looking to install a PV system on a commercial building's roof, ask the client to pay for a full structural analysis before you go too far in the design and permitting processes. It can be very disappointing for all involved to kill a project because you discover that the building isn't strong enough to hold the array.

Only a structural engineer can perform a full structural analysis. This engineer needs a certain amount of information from you regarding the PV installation, so you'll have to do some initial design work. Generally, though, the amount of design work you need to do (specifically, analyzing the site and estimating the total array size) is minimal and time well spent upfront. (If for some reason you can't provide the drawings, the engineer can always do a site visit, but using the drawings is the preferable method.)

If the structural engineer gives you the go-ahead for a roof-mounted array, you can then move on to the actual permitting process with the jurisdiction's building department. Just like you need to prepare some minimum electrical drawings, you need to draft some mechanical and structural documents as well. A structural engineer can handle the building's structural elements, but you need to take care of the system-specific mechanical drawings. At a minimum, these drawings should include

- **A racking plan:** This drawing shows the entire roof with a racking system (minus the modules). It's helpful to see how the array is held in place (with ballast or penetrations; see Chapter 16) and where the rack will be located in relation to the other equipment on the roof.
- **Racking details:** This drawing typically shows a close-up of one or two rows of modules to give the person looking at the plans an understanding of the way the racks and modules will be installed. Racking details often include a specific detail for the roof penetrations or ballast methods for clarity.

I suggest using computer-aided drafting (CAD) programs to create these drawings. A number of CAD programs exist, and the right one for you depends on the number of features you require. All of these programs have a steep learning curve, though, so you may want to find someone in your area who can take your ideas (and hand-drawn sketches) and turn them into professional drawings while you get up to speed on the CAD software of your choice.

REMEMBER

If the array is going to be ground-mounted or top-of-pole mounted, you don't have to worry about the building supporting the array. You do, however, need to satisfy the local building department's request for documentation regarding the racking system you plan to use (specifically, the manufacturer's instructions). PV arrays mounted elsewhere than the roof can be very large, so you may also need to have a geotechnical engineer verify that the racking manufacturer's instructions are valid for the soil type in which you plan to install the array.

## Electrical permits

The basic rules and computations I show you in Chapters 11 and 17 still apply to the electrical design and installation of commercial PV systems, but because of the number of modules and possible system variations, the whole process can become overwhelming . . . until you know exactly what you need to obtain the necessary electrical permits. Before issuing an electrical permit for a commercial application, the building department wants to ensure that the system components will be installed correctly and that they'll operate safely with the electrical systems already in place.

REMEMBER

The electrical diagrams for commercial projects should include all the electrical details that you'll use during the installation so you can accurately represent the entire project to anyone who needs to review it. Also, by putting all the information down in the following format, you'll be more prepared and better able to accurately plan the installation. Here's what the electrical drawings you prepare for commercial projects should include, at a minimum:

- **A title page:** This drawing includes all the names and contact information for everyone involved in the job. It's a great place to include a short narrative and/or description of the purpose and scale of the PV project. (No, this part isn't really a drawing, but it's still considered part of the drawing set.)
- **A roof or ground plan:** This is a plan view showing the building or ground area where the PV system will be installed (see the previous section for the scoop on plan views). This drawing can also point to the general locations of equipment and electrical components such as the PV disconnects, the utility service (for grid-direct and utility-interactive, battery-based systems), and the building's main distribution panel.
- **One-line drawing:** This drawing calls out all the individual electrical components used and their locations in the system (refer to Figure 14-1 for an example of a one-line drawing). This is a great place to indicate the conductor and conduit sizes and specific information about disconnects and combiner

boxes; it's also a good spot to show exactly how the PV system will interconnect to the utility (for grid-direct and utility-interactive, battery-based systems only).

- **Electrical details:** This drawing will most likely vary from job to job because the details you need to show on one job may not be required on the next (or they may need to change). However, some typical electrical details include the method of grounding the array, the point of utility interconnection, and PV string configurations.

# Not Just Pretty Pictures: Creating Drawing Sets

*Drawing sets*, the individual drawings that you put together for the building department and your installation crew, can help you in a few ways if they're prepared in a professional and organized manner:

- **When you present them to a building department, you automatically look well-organized and professional.** This appearance can help make the processes of obtaining permits before the installation (as I describe earlier in this chapter) and inspection after the installation (which I describe in Chapter 18) go much more smoothly.
- **You can also use these drawings during the installation process.** If you're working with an installation crew, these drawings provide them with a set of clear directions so they can spend their time building, not trying to figure out what you were thinking with your design.

REMEMBER

When you're making your drawings of the PV system, there's really no such thing as too much information. By documenting the entire design and detailing the system construction, you create a certain level of quality control early on that can be checked and verified in the field during the commissioning process (flip to Chapter 18 for the scoop on commissioning). As you create a few drawing sets, you quickly become aware of any information that needs to be included (or in the rare case, what can be excluded).

In the following sections, I provide some specifics for a few of the drawings. *Note:* The exact format and layout of your drawings will ultimately be up to what you feel is best.

TIP

A number of drawing programs are available to you. One of the most popular is *AutoCAD*. Many engineering firms use this program, so if you want to share electronic files with an engineer but you don't want to invest in *AutoCAD*, I suggest you at least find a program that can talk to *AutoCAD*.

## Calling out components clearly

In all of your drawings, establish a consistent method for calling out and specifying the exact components used in the design. I find it easiest to include this information as part of the electrical one-line drawing that I describe in the earlier "Permits for electrical components" section. You show all the equipment on that drawing anyway, so specifying the part numbers and any special requirements here is a good idea and not terribly difficult (refer to Figure 14-1 to view a sample one-line drawing).

REMEMBER

If any equipment noted on your drawings can't be substituted with similar pieces, state somewhere that the designated equipment must be installed.

## Depicting equipment locations

The location of all the major pieces of equipment should be shown on one or two drawings as well. This visual helps the building department understand the installation better and allows the inspector to decide whether any components should be moved.

Roof plan views are great places to include this information. In residential systems, the equipment locations can generally be shown very easily on the overall site plan (which I mention in the earlier "Permits for mechanical components" section). In commercial systems, you may need to point to the locations on the roof plan and site plan and then include details of the equipment locations on other drawings (such as a details drawings page).

## Showing conductor-sizing calculations

Including the conductor-sizing calculations in a drawing is a great idea that can help clarify your design process. Many plan checkers and inspectors may not fully understand the requirements in Article 690.8 of the *NEC*®, but including your calculations can help them understand those requirements. For residential and commercial projects, if you can include these calculations on the one-line drawing, great. If you have to place the calculations on a separate notes sheet, that's fine. (I describe the notes sheet in the next section. For the step-by-step process of sizing conductors, turn to Chapter 13.)

TIP

I find it especially helpful to include *NEC*® references when showing conductor-sizing calculations to a jurisdiction that doesn't have a lot of PV-specific experience. You can also easily include the voltage-drop calculations in this same area. Doing so is helpful for rebate programs that review the drawings for accuracy and overall efficiencies.

## Jotting down job notes

REMEMBER

I like to include an area on one sheet (or on a completely separate sheet) that includes site-specific notes. Each job has some differences, and those differences should be called out so that everyone working on the project is on the same page. Your site-specific notes can include any special requirements for the installation crew, special considerations for the product staging locations, or even the specific electrical calculations. The important thing is to write them down so you can reference them in other drawings to make sure all parties are aware of the requirements and don't overlook anything by mistake. (***Note:*** I personally consider a notes section or sheet a mandatory part of my drawing sets, but if you don't have one, the building department won't deny your permit application.)

### HELPING HANDS

As the PV industry has matured and grown, a few organizations have been able to make positive and substantial changes for the industry as a whole. They're out there working with local, regional, and national organizations to help PV installers and designers streamline their jobs.

One such organization is the Solar America Board for Codes and Standards (Solar ABCs; `www.solarabcs.org`). This group has tackled many issues with a major effort in standardizing permitting for PV systems, especially residential installations. In October 2009, Solar ABCs released a very informative document titled "Expedited Permit Process for PV Systems." This document contains a wealth of information and even includes samples for residential drawings like the ones I cover in this chapter. If you plan to install residential PV systems, I suggest you use this document and even give a copy of it to the local building department.

Another noteworthy group is the Interstate Renewable Energy Council (IREC; `www.irecusa.org`). One resource it works on and makes available to everyone is the Database of State Incentives for Renewables and Energy Efficiency (DSIRE). This database allows you to view the incentives for renewable and energy-efficiency measures all the way down to a local level. DSIRE is a great resource that's used by many in the PV industry; find it at `www.dsireusa.org`.

IN THIS CHAPTER

- » Staying safe on the job site
- » Boosting yourself with ladders (the right way, that is)
- » Bringing up rooftop safety concerns
- » Keeping electricity in its place
- » Being aware of battery safety

# Chapter 15

# Staying Safe Anytime You Work on a PV System

When you're installing, maintaining, or fixing a PV system, you need to have your wits about you at all times. If you aren't constantly aware of your surroundings and all the hazards you're exposed to, you can get seriously hurt, very quickly. Yes, safety on a job site is all about common sense. But if I've learned anything from reading the newspaper each morning, it's that the old saying is true: Common sense isn't all that common. That's why I devote this chapter to showing you PV-specific safety issues, including those that involve general job site safety, ladders, roofs, electricity, and batteries.

On-site safety is also important because you typically don't install a PV system all by yourself. Usually you install a system with the assistance of at least one other person. And if you're in a larger company, some people may design systems for others to install.

TIP

If you want to become fully aware of all the on-the-job hazards — as well as the proper ways to deal with them — I suggest you attend worksite safety classes through the Occupational Safety and Health Administration (OSHA). This federal agency offers various levels of classes that can help you become aware of OSHA requirements and how to meet them. For more information, head to `www.osha.gov` or the state contractors board Web site.

# Getting a Grip on General Construction Site Safety

A PV installation has many of the same hazards that any construction site has, including the chance for a number of individuals (or even a combination of different trades) to be working on the site at any given time. This fact requires your awareness of the different hazards and appropriate preparation. In this section, I cover several common safety issues surrounding PV installations.

## Identifying job-site obstacles and putting on protective gear right away

Every job site you'll work on will have a number of obstacles to avoid. You'll encounter skylights, plumbing vents, and attic fans on rooftops — not to mention all the equipment you're going to install. On the ground, you'll have to deal with stairs, pathways, and more. When you show up at a job site, evaluate it for hazards before you start working. Identifying issues before they ever come up is your best defense against getting hurt.

REMEMBER

On top of the obstacles you identify when you show up are the ones you add in the form of racking, PV modules, and other system gear as the day progresses. So take inventory of the obstacles when you show up and unload for the day, but keep in mind the new obstacles that'll show up as you work.

By identifying the potential hazards ahead of time you can also figure out the proper personal protective equipment (PPE) you need. At a minimum, you should wear the following whenever you're on a job site:

- **A hard hat:** Although not the most comfortable thing in the world, a hard hat protects your head from falling objects.
- **Eye protection:** Never install or perform maintenance on any part of a PV system (particularly the batteries in a battery-based system) without good-quality eye protection. Always wear a standard pair of safety glasses whenever you're on a site. (I like to have indoor and outdoor versions to help shade my eyes appropriately.) You should also always have a pair of goggles on hand for working on batteries. Sure, safety goggles may make you feel like a self-conscious teenager in chemistry class, but they're an absolute necessity.

- **Gloves:** Make a serious investment in a good-quality pair of all-around work gloves because you'll be handling sharp (and sometimes very hot) pieces of metal. If you're going to be working with batteries, buy some gloves that are acid resistant to help keep your hands protected from the strong acids present in batteries (whether sealed or flooded).
- **Footwear:** The shoes you choose to wear depend on the environment in which you're working. If you're in a new-construction industrial setting, steel-toe boots may be the most appropriate choice. If you're installing a ground-mounted PV array, a sturdy pair of standard work boots or hiking boots may be better. And if you're up on a residential roof, a pair of work boots with soft rubber soles is your best bet.

Some specialty pieces of PPE may include high-voltage gloves for the times you're using a meter inside boxes and face shields for guarding against arc faults. You need the gloves anytime you want to put your hands inside a box (such as junction boxes, combiner boxes, disconnects, and inverters) to check voltage or current. PPE for arc fault protection is necessary whenever you're working in any of the electrical panels in commercial situations.

## Safely working alone and with others

Working alone is never the first choice in construction — too many problems can sneak up on you that require a second person in order to maintain proper safety — but chances are you've been in the situation where waiting for a second person (or even getting a second person on-site) just isn't feasible.

REMEMBER

If you're working on a PV system alone, be aware of your limits. Realize that if you try to perform a task that really needs a second set of hands and you get hurt, you won't save any time or money; in fact, just the opposite will occur. Never look at any job as more important than your personal safety.

REMEMBER

When working with others on a job site, you need to have a different level of awareness. Here are some guidelines to follow:

- **Never assume where people are or what their next move will be.** If you move behind someone, announce your presence.
- **Use directions with clear references.** This tip is handy when you're working with someone to accomplish a task together, like carrying a large, heavy object across a roof. If you're facing each other and you tell your partner to turn left, what will he do? Move to his left, which is your right? If you say "move west," "turn clockwise," or something similar, your partner will better understand what you want him to do.

## Taking in tips for tool safety

Because you use a number of tools on any given job, make sure those tools are properly maintained and used for the safety of you and everyone else on the job site. Some tools, such as saws, have parts (think blades) that wear out. It's your responsibility to make sure these parts are replaced as needed because accidents quickly occur when tools aren't properly maintained.

WARNING

Never use the tool on hand to simply get the job done (for example, using a wrench as a hammer). Always take the time to find the right tool, even if you have to climb off the roof to get it. This practice will save you in the long run. Using the wrong tool inevitably results in you breaking something (either the tool or the component you're working on), and because most components in PV systems are expensive, breaking them isn't in your best interest.

You also need to respect the safety features built into the equipment. Defeating a safety feature, like bypassing the cutting guard on a saw, is a bad idea that *will* get someone hurt. You never know when someone will pick up your tool to use "for a second" and not see that the safety feature is bypassed.

## Limiting your exposure to the elements

As a PV installer, you're going to be exposed to some of the most extreme conditions possible. You'll show up first thing in the morning and get to work on wet, slippery surfaces. As the day wears on, you'll be on a south-facing roof with temperatures exceeding 120 degrees Fahrenheit (49 degrees Celsius). And then you get to go crawl in the attic space to pull some conduit and run wire. Given these facts of the job, it's your responsibility to limit your exposure to these extreme conditions as much as possible.

REMEMBER

One way to limit your exposure to excessive sun and other elements of nature is to be disciplined about stopping and taking breaks. I know how easy it can be to keep working "just to get this one thing done," but if you do that, you're going to overextend yourself. Having to take a break because you injured yourself is never as fun as taking a break just because you choose to.

You also should invest in a good-quality wide-brimmed hat, wear lots of sunscreen, and drink more water than you can imagine. If you're working in the middle of the summer, plan to start and end early too. Roofs and attics are too hot by midday for you to continue working safely because you won't be able to think clearly and will therefore make mistakes. ***Remember:*** The work isn't going anywhere, and you can't get it done from the emergency room.

## Stowing a first-aid kit on the job site

You need to have a proper first-aid kit at every job site. You can purchase one at any safety-supply store or through a reputable safety-supply catalog. Just be sure to buy one that's stocked with treatments for electrical burns, instant ice packs, and bandages for cuts. If you want, you can buy a kit made especially for electrical injuries in addition to a general first-aid kit.

TIP

I encourage you to require everyone working with you on the PV installation to receive first-aid training and certification. This training only takes a few hours and will make everyone more confident and secure. You can find first-aid training in your area by contacting the local American Red Cross office; visit `www.redcross.org` to find yours.

# Looking at Ladder Safety

If you've decided to work with PV systems, you've decided to include ladders in your life on an almost-daily basis. Therefore, you have to become familiar with and willing to practice ladder safety every time you pull one off your rig. The following sections show you some of the major considerations regarding ladder safety, but they're not all-inclusive. Investigate the OSHA requirements in the state where your client is located to see how they affect ladder use. The OSHA Web site (`www.osha.gov`) has links to regional and state chapter offices that you can contact.

## Selecting your stash of ladders

Ladders are classified in three ways, according to their use and complexity:

- **Stepladder:** This is a freestanding ladder that's designed for use in the fully open position. Stepladders come in various lengths and can be used to access points above one's head, like when you're installing conduit along the top of some wall-mounted equipment.

  WARNING

  Never use stepladders to move from one level to another, like from the ground to a roof. They aren't designed for this use, and they may not support you properly if you try to use them this way.

- **Straight ladder:** Straight ladders give access to different levels as well as elevated points along a vertical surface, but areas accessible from a straight ladder are limited by the ladder's height. You can use a straight ladder when you need to access a roof surface from the ground level.

» **Extension ladder:** A form of straight ladder that allows the user to increase the overall length by moving a "fly" section along the "base" section of the ladder, extension ladders allow you to access different levels. You can use an extension ladder when you need to access a roof surface on a multistory building.

Another ladder classification to know is the *load rating*, which indicates the maximum weight a ladder can handle (every ladder has one of these ratings). Ladder manufacturers come up with their load ratings by accounting for the weight of the user and any additional weight, such as tools the user is carrying up the ladder. The load-rating classifications are as follows:

» Light duty (Type III), load rating 200 pounds

» Medium duty (Type II), load rating 225 pounds

» Heavy duty (Type I), load rating 250 pounds

» Extra heavy duty (Type IA), load rating 300 pounds

» Special duty (Type IAA), load rating 375 pounds

For any ladder that you intend to use on a construction site, use the higher grades — IA and IAA. Using these grades of ladder ensures that your ladders will be able to handle nearly any load that you need them to on-site. Ladders with reduced load ratings may only work in certain situations.

The next factor to consider when selecting the right ladder for a job is the ladder height. The number-one thing to keep in mind here is that when height is listed on a ladder, you don't get to use all of it. Depending on the classification and overall height, the last few rungs of a ladder are typically off-limits. Always look at the ladder manufacturer's instructions for the allowable access heights. Table 15-1 shows the standard maximum wall and roof heights allowed for various straight and extension ladder lengths.

**TABLE 15-1** **The Working Heights for Various Ladder Lengths**

| Ladder Size | Maximum Working Height for Working on a Wall | Maximum Working Height for Accessing a Roof |
|---|---|---|
| 16 feet | 13 feet | 9 feet |
| 20 feet | 17 feet | 13 feet |
| 24 feet | 21 feet | 17 feet |
| 28 feet | 24 feet | 21 feet |
| 32 feet | 29 feet | 25 feet |

Stepladders come in a variety of sizes as well. Just refer to the label on the side of one to determine its maximum working height.

You should seriously consider buying fiberglass ladders only; fiberglass is non-conductive and considered the best material for ladders used in conjunction with electrical work.

## Properly setting up any ladder

After you have a ladder of the correct type and length on-site, you need to set it up properly. Before you get into the setup process, make it a habit to check the ladder for damage every time you use it. You never know when something has broken, and halfway up to the roof is no time to identify a problem.

When setting up a stepladder, verify that the legs are fully extended and that the braces are down and locked. All four legs need to be firmly in place on the working surface to keep the ladder from moving when you climb on it. ***Note:*** Stepladders are designed for use with the legs extended, so don't use them to lean against a vertical surface. You won't get the proper support in this position, and the ladder may slide out from under you.

When using straight and extension ladders to access roof surfaces, proper setup is critical. If you set up the ladder too steep, you run the risk of having the top falling from the roof. If you don't set it up steep enough, the bottom can kick out and fall out from under you. By taking some simple precautions, you can greatly reduce the possibility of ladder accidents (I won't say eliminate because even safety precautions can't save you from ill-advised actions). Here are some guidelines:

- When setting up a ladder for access to a roof, you want to have the ladder rise 4 feet for every run of 1 foot. This is typically called the *4-to-1 rule.* Now before you get too worried about having to bring a protractor with you to every job site, here's a quick and easy way to verify the proper angle: After setting up the ladder, place your toes so that they touch the base of one of the ladder's legs. Then reach your arm out at a 90-degree angle from your body with your fingers extended. If your fingers barely touch the ladder rungs, you have the correct angle.
- You also need to make sure you have the ladder extending past the roofline at least 3 feet. This distance allows you to have a firm handhold when stepping back onto the ladder from the roof. Each ladder rung is about 1 foot apart, so when you're setting the ladder up, make sure three rungs extend past the roof's edge.
- Secure the top and bottom of the ladder to properly protect everyone who's using it. You can stake the ladder to the ground or attach it to the building with a strap. Another method is to have a person hold the bottom of the

ladder when anyone is climbing up or down (this option isn't always realistic though, so staking should be your first choice).

If the working surface allows, you can drive a metal stake into the ground and attach the bottom of the ladder to the stake. With this method, the first person up straps the ladder to the building by taking a proper nylon strap wrapped around a ladder rung and then attaching the strap directly to the building with a method that's appropriate for the building material (wood screw, masonry bolt, or metal screw).

Figure 15-1 illustrates the proper ladder setup.

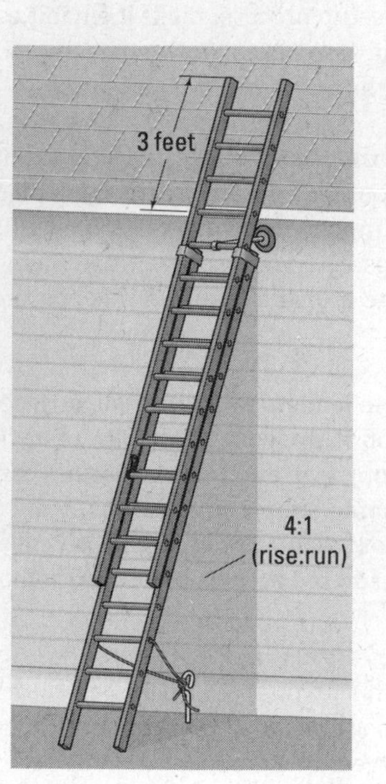


**FIGURE 15-1:** The right way to set up a ladder.

# Raising the Issue of Rooftop Safety

Probably one of the most dangerous places you'll work is the one you'll have to get on regularly — a rooftop. As soon as you set foot on a roof, you're fully exposed to the elements. You'll also likely be working with others and needing to navigate around multiple obstacles — all while properly installing the PV system. On top of all that, you must remember that you're performing electrical and mechanical work up there — as if one of the two wasn't dangerous enough by itself. In the following sections, I introduce you to some of the major safety considerations associated with working on a roof.

## Restraining yourself with fall protection

Fall protection is mandated whenever you're doing work where the potential to fall is more than 6 feet — work that includes installing a PV system on a rooftop. For PV installations, *fall protection* consists of a fall-arrest harness that keeps you from free-falling more than 6 feet or coming into contact with the next level down (see Figure 15-2a). The harness has a ring on the back that a safety line attaches to; the other end of the line is anchored to the roof (check out Figure 15-2b).

You can find multiple anchoring systems for various roof types. Some are intended for installation beneath the roofing material and become a permanent fixture of the roof (however, they're only visible from the roof's surface), whereas others can be removed and reused on future jobs. Numerous safety shops and suppliers can help train you on the proper applications and methods for using the equipment. Just do an Internet search for "construction safety equipment" or look in your local Yellow Pages to find a number of suppliers.

Fall protection at job sites has received a lot of attention and is especially scrutinized by OSHA thanks to the number of work-related injuries that have happened due to falls. OSHA inspectors visit job sites to confirm that all employees are properly trained and using the required PPE, including fall-protection gear. The fines associated with OSHA violations aren't minor, so proper use of PPE is your best approach (pleading ignorance isn't an option with OSHA). Head to www.osha.gov to find resources that can help you determine the requirements you'll be subjected to.

## Storing your tools

Of course you need to bring numerous tools with you to the roof when you're installing a PV system, but loose tools can be a major accident waiting to happen up. Although it's very courteous for you to yell "Headache!" to your buddies on the ground as your drill slides down the roof, it's more courteous if you don't let the drill fall in the first place. Follow these guidelines:

- Be careful to lug only the tools you need.
- Use tool belts that buckle around your waist and have enough storage to hold the tools you need on you.
- Invest in some quality tool bags that have straps and D rings attached to them so they can be properly secured. Such tool bags are handy because you need numerous tools and components on the roof, but you can't carry them all on your waist.

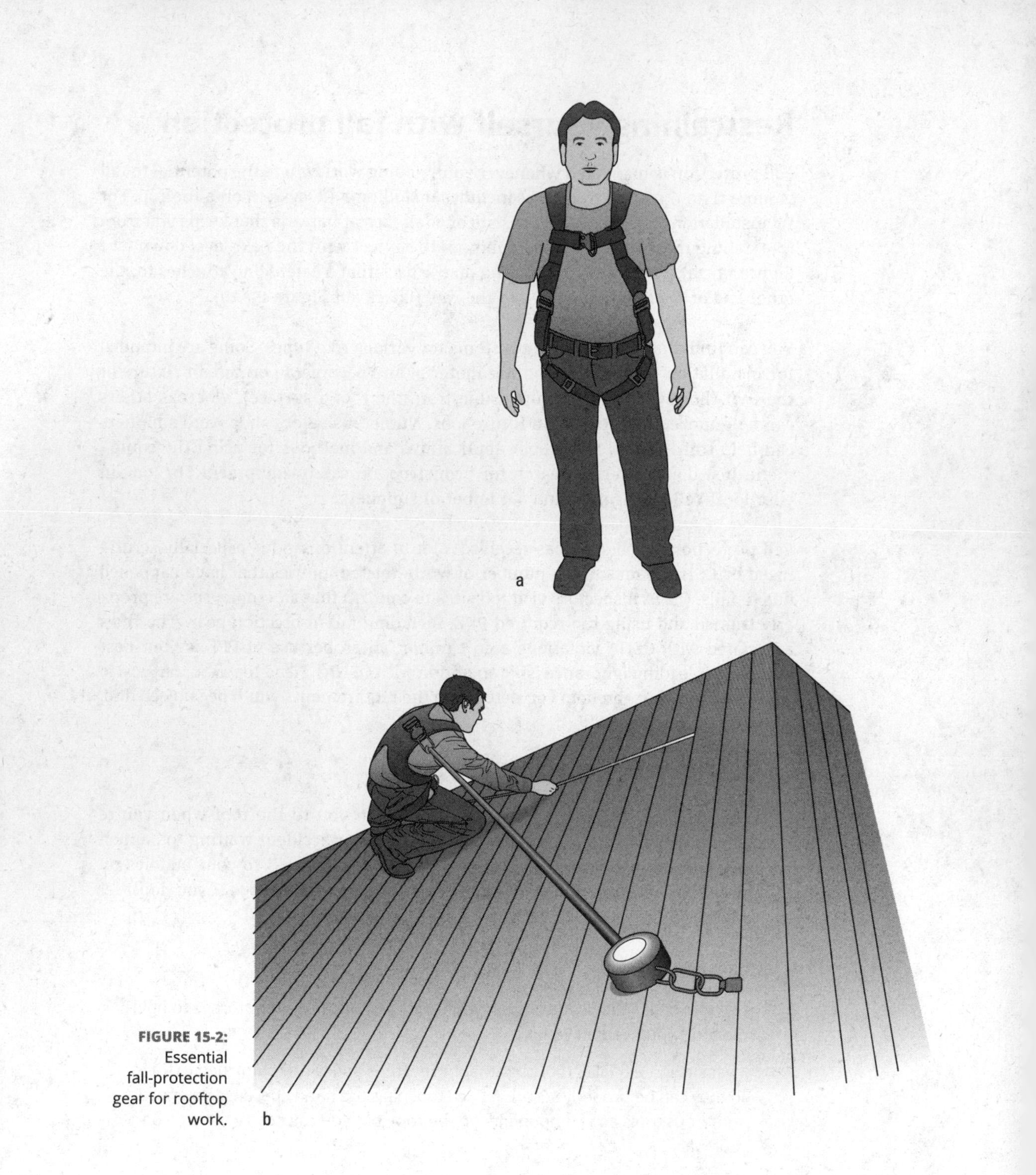


**FIGURE 15-2:** Essential fall-protection gear for rooftop work.

## Maintaining safe walkways

In many areas of the United States, *walkways* (clear space along the edges of PV arrays) are now required along the perimeter of any PV array. This requirement is primarily for the safety of firefighters, but it has the added benefit of helping ensure your safety. A firefighter may never need to use the walkway, but you probably will while performing system maintenance.

Check with the building inspector at the jurisdiction in your client's area for any walkway requirements. If none exist, I suggest accounting for at least 18 to 24 inches of walkway along a minimum of one edge of the array (but preferably two for easier access).

# Examining Electrical Safety

Electricity is a very real and important safety hazard. I've seen too many people start working with PV systems and take the attitude that it's only solar and it can't hurt you like "regular" electricity can. Don't fall prey to this myth. Electricity can injure or kill you regardless of the source. Also, even though PV modules start at a low voltage, your job is to wire them up so they provide voltages that are higher than standard AC voltages, like the 120/240 VAC in most homes tied to the grid. If you don't think these are reasons to keep safety in mind at all times, then you're in for trouble down the road.

When you install, maintain, or fix a PV system, you're subject to multiple electrical risks. In the following sections, I note the general shock hazards to watch for and provide safety pointers on working with circuits. (Chapter 17 has the scoop on how to install all the electrical elements in a PV system.)

## Staying aware of general shock hazards

You should always be aware of the potential for being shocked when working with electrical equipment. You should also keep in mind that when you're working with PV systems, multiple sources of power are present — that equals numerous hazard possibilities. Here are the major shock hazards you need to think about at all times:

» **Power tools and extension cords:** Inspect this equipment before using it and immediately replace any tools or cords that appear damaged. In particular, make sure all the equipment and cords are properly grounded. Never use a cord that has a missing or damaged grounding connection. If you do, you'll raise your chances of getting shocked.

» **PV modules:** These babies don't have on/off switches. When you take a module out of the box and expose it to sunlight, it'll start sending out current if given the proper path. To avoid being shocked, always plug in the PV source circuits last (see Chapter 18) and wear high-voltage gloves when you put your hands in a box to make any electrical measurements.

Currents as low as 100 milliamps (0.100 A) can disrupt the heart's normal functions. The PV modules you'll be working with will have individual outputs of 5 to 10 amps — enough to cause serious injury and even death.

» **AC circuits:** These circuits pose their own electrical hazards. You'll probably need to make connections inside the main distribution panel (MDP) or subpanel. If power is present in these panels, you run the risk of touching *conductors* (wires) that'll send current running through your body. When you need to work inside the MDP or a subpanel, turn off the power to ensure you aren't exposed to any dangers. The loads in the panels will shut off, but your client should be willing to reset a few clocks in exchange for guaranteeing your safety.

» **Rooftop dangers:** The majority of PV arrays are installed on the tops of homes and offices, so odds are good you'll be working on rooftops when installing your PV system designs. Pay special attention to the presence of overhead power lines and always look up before setting up and working on your ladders. Speaking of ladders, avoid using ones that are made of conductive materials; instead, invest in a properly rated fiberglass ladder. (I cover ladder and roof safety earlier in this chapter.)

## Working with circuits

Working on the different circuits in a PV system exposes you to a special set of electrical hazards. Why, you ask? Because PV circuits can never be turned off, and the AC power provides yet another shock hazard. This simple fact means that voltage (and the shock hazards that come along with it) is a constant presence. So be smart about your approach when dealing with circuits of any kind and never let your guard down. Also, take some time to review the guidelines that I provide in the next sections for making connections and disconnections in a system.

### Following a specific order of connection

You can keep yourself fairly protected during the initial system installation process by being careful to make the electrical connections systematically and in a particular order so as to keep the voltage contained and away from you.

REMEMBER

First things first when you start installing any PV system: Make sure all the disconnects and overcurrent protection devices (either circuit breakers or fuses) you install are in the off position. This check keeps any circuits from becoming unintentionally connected during the installation process.

As you install the PV array (that's the PV source circuit in electrical terms), you generally want to make the series connections as you mount the modules. If the system contains a roof-mounted array that consists of multiple strings, you may have to make the series connections as the modules are attached to the rack because you'll have limited access to the PV wiring as the other modules are installed (the modules will be only a few inches off the roof surface, and the wiring may be blocked by additional rows of modules). Almost all PV modules now come with quick-connect plugs that allow for series connections to aid the installer in wiring the modules in series strings.

You can make these series connections safely as you mount the modules, but the final connections require special consideration. Imagine a row of ten PV modules, each one wired in series with the next. The positive connector of the first module is connected to the negative connector of the second, the positive of the second is connected to the negative of the third, and so on. As soon as this string is mounted and wired as a single series string, you have a negative wire connected to nothing on the first module and a positive wire connected to nothing on the tenth module. To get those two connections to the combiner box or junction box, you must install a *home-run cable*. This cable has a quick-connect plug on one end so you can connect it to the quick-connect plug on the module. The opposite end of the cable is bare so you can place it inside the combiner box or junction box and connect it to the proper terminal.

When connecting the home-run cables, work from the combiner box or junction box back toward the PV array. Connect the bare conductor to the proper terminal inside the box before connecting the quick-connect end to the module. This will help keep you safe and minimize your exposure to shock.

REMEMBER

To make sure all the conductors in the system are safe, you should only make this home-run connection at the module when you're ready to *commission* (turn on) the system — which is after everything else has been installed and wired (see Chapter 18). If you do this, all the other conductors on the PV side of the system won't be energized during installation, which means you can work safely inside all the combiners, disconnects, and inverters.

REMEMBER

For all the other PV-side circuits you install in a PV system (specifically, the PV output circuits and inverter input circuits), use the same general process. Start wiring at the end of the system that's not hot (meaning there's no voltage present) and work toward the portion of the system that has the voltage source (either from the inverter toward the grid in a grid-direct system or from the inverter toward the battery bank in a battery-based system). Always turn off breakers and disconnects inside distribution panels so you can get your hands inside the boxes without exposing yourself to danger.

## Working on the AC circuit

In both grid-direct and battery-based systems, the AC circuit you must work with is the inverter output circuit (although you can have more than one of these circuits if the system you've designed calls for multiple inverters). The exact final connection point for the AC circuit varies with the system type, but the safety considerations are universal. You want to make sure that any panels — main distribution or sub — you're working in don't have any power present. (You can do so by performing the same locking and tagging procedure I describe in the next section.)

TIP

You can also buy special lock kits that are designed to go over circuit breakers and let you verify that no one will turn on the breaker behind you and apply power to a panel you thought was "dead."

## Disconnecting properly

All PV systems have at least one AC and one DC disconnect that can be used to stop any current flow. Before conducting even minor maintenance or troubleshooting work on PV systems that are operating and producing power, check that all the associated disconnects are turned off.

REMEMBER

If you ever need to expose yourself to the circuits of a PV system (whether it's operating or not) to conduct checks or perform routine maintenance, take it slow and think your steps through before opening the boxes. If you have to work on an inverter or troubleshoot a PV array, your first step should be to approach the system and turn off all the power sources to that equipment. This means the AC and DC disconnects for a grid-direct system and the generator and/or utility, battery, and PV circuits for battery-based systems.

Regardless of the system type, your next step is to lock and *tag out* (place identification tags on) the disconnects so no one can come up behind you and flip the switches back on. The lock prevents anyone other than the keyholder (you) from turning on the system. The tag (which is attached to the lock) lets people know that the system shouldn't be turned on and provides your contact information so people can contact you if necessary. Figure 15-3 shows an example of a locked and tagged DC disconnect.

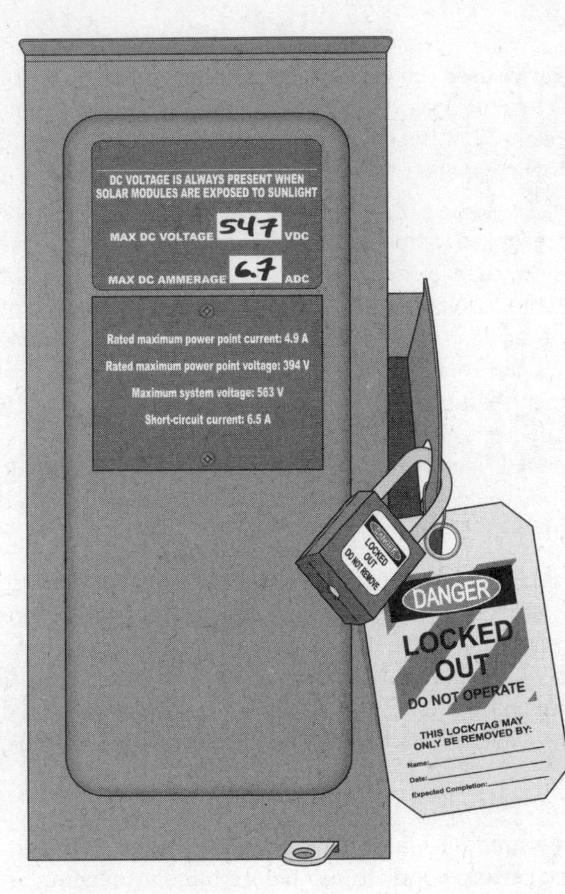


**FIGURE 15-3:** A locked and tagged disconnect.

REMEMBER

After you've turned off and locked all the disconnects, go back to your truck and have some coffee. Not thirsty? Check your e-mail instead. Whatever you do doesn't matter as long as you take a few moments to step away. Sure, you may feel like jumping right into the next task, but every inverter and charge controller used in a PV system has capacitors inside it. What's more — those capacitors do a pretty good job of holding onto the voltage that was applied to them for a few minutes. So even if you've turned off the disconnects, voltage can still be present on both sides of the DC disconnect. (In fact, your disconnects should be labeled per *National Electrical Code*® [*NEC*®] requirements that such a hazard exists. However, you may one day have to work on someone else's installation. In this case, you can't trust that the system has the proper labeling, so your best bet is to work smart.)

WARNING

Even though you've used the disconnects to shut off the current flow from the PV array and inverter, the array is still producing voltage. Consequently, all the PV circuits that are ahead of the disconnecting means will still be hot. Never disconnect at the following spots:

- If you've been called to troubleshoot a PV array, or if you see something you want to quickly change on the roof, it can be tempting to grab that home-run cable I describe in the previous section and simply unplug it to disconnect the array. However, you should never disconnect the modules in this location while the modules are *under load* (meaning current is flowing from the modules to the batteries or inverter). This point is so important to remember that the connectors on the backs of the modules are marked "Do not disconnect under load." If you were to pull these connectors apart under load, you'd be lucky if you could walk away from the situation and describe the sight and sound of the arc you pulled.
- Another tempting place to disconnect the PV strings is inside a combiner box, using the small fuse holders installed to pull the fuse out and break the circuit. These fuse holders, like the module connectors, aren't meant to break the circuit under load. If you try to do this anyway, you'll pull an arc that can sustain itself until the sun goes away, which can create not only a shock hazard to you but also a fire hazard because the plastic fuse holder could catch fire.

REMEMBER

When you're in a situation where something isn't right, take your time and examine what you're getting yourself into before blindly touching and pulling wires, opening fuse holders, or pulling apart module connectors. Check for the presence of current and voltage by using your digital multimeter the way I describe in Chapter 3. By making these checks before pulling out the fuses or pulling apart the PV module connectors, you ensure that you aren't needlessly placing yourself in harm's way.

# Charging Ahead with Battery Safety

In addition to presenting electrical hazards, batteries also present chemical reaction hazards because they're just heavy plastic containers of acid. Even getting them on-site and in place presents its own set of problems. Before you jump into working on batteries, check them over to make sure there aren't any cracks in the case or problems with the posts (like loose fittings).

The first thing people think of when I mention batteries and safety is a *fault situation*, which is when the positive terminal of a battery is in direct contact with the negative terminal of the same battery thanks to the presence of a metal object lying across both terminals. When a fault situation occurs, the chemicals inside the battery can easily move electrons from one side to the other, creating very high currents. (I'm sure you've heard at least one story of your buddy's grandpa who dropped a wrench across the terminals of a battery and welded the wrench to the battery. Although this story is probably repeated and embellished more than any fishing story, it serves as a very real and important reminder of one of the dangers of working with batteries.)

When installing batteries, you have to use the right tools and maintain a sharp awareness of what you're doing to eliminate (or at least greatly reduce your exposure to) that risk. Make sure all the tools you use in battery installations are properly insulated to minimize the potential of shorting the batteries. (Even though electrical tape is amazing, wrapping the handle of your wrench with it doesn't count.) Tools with two layers of plastic coating are perfect.

You also have to take special care when handling batteries because the vast majority of them contain lead and sulfuric acid. Protect yourself from the hazard of chemical reactions with the right personal protective equipment (PPE). I explain the PPE you should have in the earlier "Getting a Grip on General Construction Site Safety" section, but you may also want to invest in an apron to avoid tiny holes in all of your clothes from the battery acid.

In addition to wearing the right PPE, be sure to remove any and all jewelry you have on whenever you're working near batteries. I once witnessed the results of a wedding ring completing a battery circuit. Luckily, the finger (and the person attached to it) survived. The ring, on the other hand, was cut off and resembled a high school art project more than a wedding ring.

Batteries used in PV systems are generally very large and heavy. Always use proper lifting techniques and move the batteries carefully. Most batteries have carrying handles on the top, but that doesn't mean you should try moving them yourself. I suggest grabbing a partner to help you move these beasts around. With two people involved, you save yourself from back pain and also avoid the potential of spilling any acid out the top (it's easier to gently lift and set down batteries with two people).

IN THIS CHAPTER

- » Taking stock of the racking options available for PV arrays
- » Loading up the roof
- » Sealing the roof before you attach an array
- » Keeping the array on the ground with the right support

# Chapter 16
# Assembling the Mechanical Parts

For many people moving into the field of PV system design and installation, the electrical side of the installation process is a little different from their current knowledge base but close enough to what they know that they can pick up that part of the process rather quickly. The mechanical side, on the other hand, can be very overwhelming for a number of people; the idea of going onto someone's roof, drilling a hole in it, and then being held responsible for any future issues is a bit much for some folks. Yet in order to be successful in the PV world, you have to feel comfortable with the mechanical aspects of the job. Why? Because the reality of PV installations is that there's just as much — and often more — detail involved in the structural and mechanical components of a PV system (in other words, the mounting of a PV array and all that goes with it) as there is in the electrical components.

Not to worry, though. The mechanical aspect may pose a challenge or two early on, but you can get the hang of it and have well-installed PV systems. In this chapter, I show you the major racking options for use on roofs or out in fields. I also get you familiar with some calculations you need to make when considering the effects of a PV array on a building, all the way from the dead loads imposed by the array to the live loads that'll be changing constantly. Finally, I introduce you to some of the latest methods for properly sealing the holes you inevitably have to drill into someone's roof as you attach the array, and I explain options for supporting ground and top-of-pole mounting.

# Surveying PV Array Mounting Methods

During the site assessment that I describe in Chapter 5, you have to identify where the PV array will be mounted. The location you choose generally dictates the type of mounting structure you must use. Multiple options exist for each racking type, which is why I spend the next several sections reviewing the major racking solutions available for your installations. ***Note:*** Although the general concepts for the major racking manufacturers are the same, the execution during installation varies, so be sure to read and follow the manufacturer's instructions when installing a racking system. You should also use stainless steel hardware to match aluminum racking systems so that the materials don't deteriorate over time.

Keep in mind that you can use a commercially available racking system and modify it for your specific application. For example, some PV pros have used a racking system designed for ground-mounted arrays to mount a PV array as an awning on the side of a building. You need to work in conjunction with the racking manufacturer on such projects, though, because it may need to evaluate applications other than the designed installation.

## STICKING WITH STORE-BOUGHT RACKING FOR A WHILE

When you know what the racking solutions are and what the components consist of, you may be tempted to make your own racking. However, I strongly encourage you not to. Fashioning your own racking may seem simple on the surface, but when you dive into the details, it often proves otherwise. Commercial racking systems have been fully designed and engineered to meet the criteria for the specific environment in which the array will be installed. Not only that, but the manufacturers supply information regarding the proper methods for attaching the rack to the structure (or ground) to meet various wind loading requirements and to stay within the limits of the racking material itself. When you use your own racking material, you have to verify that the materials you're using and how you're installing them meet building code requirements.

Find a commercially available racking system and use it for at least a year or two. When you're first getting started, you don't need to add another level of difficulty to your already tough job. After you feel like you really know the ins and outs of installations, then you can revisit the homemade rack idea. You may have great ideas on ways to improve the rack you've been using after you get some time in with that system.

## Roof mounting

The most popular place to locate a PV array is on the roof of a building, especially in grid-direct systems. In many situations, the roof may be the only option for mounting a PV array. My own house is a perfect example of this. Thanks to a small lot, neighbors' trees, and historic-district limitations, the roof was my only option for mounting my PV array.

The roof of a home or office building has a number of advantages when it comes to mounting a PV system:

- It generally has the best access to the solar resource.
- A roof-mounted PV array allows you to use otherwise "unused" space and save the yard space.
- It's in close proximity to the existing electrical system, helping to reduce the overall cost.
- Roof-mounted installation has a minimal effect on a site.

WARNING

Of course, the roof isn't always the absolute best location for a PV array. Here are some of the disadvantages of putting a PV array on top of a building:

- The array's orientation is dictated by the building.
- The modules operate hotter (therefore less efficiently) due to their proximity to the roof.
- The array's size may be limited due to the available area.
- When the roof needs replacing, the array will also need to be removed (the roof should be in good condition to begin with so this won't need to happen for more than 15 years).
- The array may be fairly inaccessible, in which case snow and ice can't be cleared from it.
- *Roof penetrations* (the holes you need to make to mount the array on the roof) lead to potential roof leaks. (I discuss the best methods for sealing holes in the later "Sealing roof penetrations with flashing" section.)

In the following sections, I break down roof mounting into three categories: flush-to-roof racking, tilt-up racking, and ballasted racking.

## Flush-to-roof racking

For residential PV installations, the most popular method of attaching modules to a roof is a *flush-to-roof racking system.* As the name implies, this racking system places the modules parallel and very close to the roof's surface (see Figure 16-1). This racking system results in an array that's aesthetically pleasing and minimizes the effects of *wind loading*, the uplift effects of wind on the back of an array and building. (*Note:* Flush-to-roof racking isn't as popular for commercial applications because it isn't well suited for the flat roofs that are commonly used on commercial buildings.)

**FIGURE 16-1:** Flush-to-roof racking.

REMEMBER

Despite the vast number of commercially available flush-to-roof racking systems, each one with its own unique features, all of these systems function the same way. A basic flush-to-roof racking system consists of

- Footings for holding the entire system to the roof
- Rails to support the modules
- Clamps that hold the modules to the rails

The manufacturers of flush-to-roof racking systems offer several footing options so you can raise the array off of the roof's surface anywhere from 4 to 10 inches. The standard offering is a short foot that keeps the array very close to the roof's surface. (Another option: posts that can be installed and sealed from leaks with the help of standard roofing products.) After you secure the footings to the roof, you can install the rails in preparation for the modules.

The components that really make the flush-to-roof racking system a winner are the manufacturer-provided, top-down mounting clamps. As you can imagine, if you place a PV module flush to the roof with limited access to the back of it, the easiest way to secure it to the rail system is from the top of the module (not the mounting holes at the bottom of the module).

Top-down mounting clamps work in conjunction with the rail system supporting the modules. When the module is on the rail and in place, the top-down clamps are placed on the edge of the module and then tightened to the nut or bolt inside the rail system. (Because the attachment method occurs from the top side of the modules, this mounting system is also sometimes referred to as *top-down mounting.*)

WARNING

When working with roof-mounted systems in residential applications, many PV installers are tempted to overcome the less-than-ideal orientation by tilting the array off of the roof's surface, resulting in an array that points closer to the optimum location. Tilting the array results in more racking parts because you need legs to raise the racking system off of the roof surface. I encourage you to avoid installing PV systems in this manner for two reasons:

» **Aesthetics:** Although I think a PV system is a beautiful sight, not everyone shares that opinion. If you install a PV system that's the structural equivalent of Frankenstein (and people do), the message to many folks is this: See how ugly solar is? That bad reputation doesn't help promote the cause of solar energy.

» **Structure:** Tilting up arrays on residential roofs requires increased engineering to ensure the arrays stay put. This additional engineering costs more upfront, particularly in the materials used (the additional rack pieces and increased number of footings) and the amount of time you have to spend on the job site. Usually this added cost isn't made up for by increased energy production, which means all the effort you put into changing the array's orientation (and all the client's money!) actually has a negative effect on the array's financial value.

## Tilt-up rack mounting

Another common way to mount PV arrays to roofs is with a tilt-up rack mounting system. This is generally the same type of system used in ground-mounting applications (which I touch on later in this chapter); it's merely adjusted for rooftops.

In the preceding section on flush-to-roof racking, I tell you not to tilt the array off the roof's surface. So what makes me change my tune here? The simple fact is that tilt-up systems make the most sense when the PV array will be mounted on "flat" roofs. (Technically, the term *flat roof* isn't accurate; such a roof should be called a *low-pitch roof*, but *flat* is an accepted term.) Because of the lack of a built-in tilt, the racks have to be tilted up to increase energy production.

## TAKING CARE WHEN MOUNTING AN ARRAY ON A METAL ROOF

Thanks to their long life spans, light weights, and limited maintenance requirements, metal roofs, particularly the standing-seam variety, are the preferable roofing choice for many people. Standing-seam metal roofs have seams in them every 16 inches that stick straight up to provide the sheeting with some rigidity. These seams also give you a point where you can attach a special clamp called the S-5! clamp. (The manufacturer [S-5!] makes clamps that fit nearly any standing-seam profile and even offers specific clamping options for PV systems.)

When you're trying to mount an array on a metal roof with standing seams, be sure to keep the following points in mind:

- The metal sheets are designed so they can expand and contract along the roof as temperatures vary. If you don't know where these expansion and contraction points are in the roof, you run the risk of placing your PV array right on top of them and not allowing the roof to function as it should. This can cause major damage to the roof through buckling of the sheeting.
- Another point of warning is how the metal sheeting is attached to the roof. Most roofers don't expect anything to go on top of the metal roof. Consequently, they install the metal sheets using the minimum number of attachments required. But if you add PV modules on top of that roof and use S-5! clamps to hold the array down, now the only thing holding the roof and the PV array to the structure are the attachments originally made by the roofer. ***Tip:*** Try to work with the original roofer to make sure the PV array and metal roof will stay in place in the face of large windstorms.

To keep the manufacturing costs to a minimum, many racking manufacturers modify their flush-to-roof racking systems to accomplish the tilt-up feature. The rack is secured to the roof with the same type of footings, and the modules are supported with the same rails; however, the backs of the modules are tilted off of the roof surface with a piece of aluminum that allows the array to be tilted at the desired angle. Many manufacturers have telescoping legs that allow the installer to choose from a range of tilt angles (and even allow for seasonal adjustments in the array tilt).

TIP

When you're dealing with a tilt-up mounting system, sealing any roof penetrations is extremely important because water doesn't flow off a low-pitch roof as fast as it does on a sloped roof. In order to maintain the building owner's roofing warranty, you have to use a sealing method that's appropriate for the roofing material (see the later "Sealing roof penetrations with flashing" section for details). Most roofing manufacturers also require that roofers in their network are the ones doing this work in order to make sure the work is done properly and without opening anyone up to potential liabilities. Therefore, when working on a low-pitch roof that requires penetrations, I strongly suggest you enlist the services of a proper roofer who can help you with the work and maintain any roofing warranties. Start by working with the building owner to find the original roofer. If that doesn't work, you can contact local roofing contractors to see who can do the work and maintain the warranty for the roof you're working on.

Rooftop arrays that use tilt-up racks must be spaced so that they don't create shading from one array to the next. You determine the proper layout during the design process (described in Chapters 11 and 12). You can use either trigonometry or a drawing program to determine the minimum amount of space needed between the rows.

## Ballasted racking

Because of the overwhelming desire to keep water out of buildings, many people, both PV pros and building owners, like to use ballasted racking systems on flat roofs instead of penetrating the roofs and attaching tilt-up racking systems. Ballast racks use the weight of the array and additional weight (generally concrete blocks) to keep the array in place; otherwise, they're built just like tilt-up racking.

In addition to the row spacing I mention in the preceding section, here are a few considerations you need to make before using ballasted racking systems:

- **Increased weight on top of the roof:** Because of the extra weight ballast racking puts on a roof, you must have a structural engineer evaluate the system. The customer generally pays for this expense, and it adds time on the front end of the design process.
- **Limited array tilt:** Because the weight of the system must hold everything down, you may have to reduce the array tilt to keep the wind uplift to a minimum so the system doesn't blow away.
- **Potential water damming:** You need to allow rainwater to flow to the designated drains easily. If you block water paths, you may accidentally create a situation where water can leak into the building.
- **Roof layout has to avoid peaks and valleys:** When you use a ballasted system, the rack is in direct contact with the roof, so it can't cross any peak or valley (which is tough considering no roof is truly flat). Consequently, you need to account for all the roof features during your site survey and be very accurate when documenting them.

## Ground mounting

When your client has sufficient space on his property, a ground-mounted PV array is a terrific first option for any PV system type (grid-direct or battery-based). Ground-mounted systems avoid most of the disadvantages of the roof-mounted systems mentioned earlier. They can typically be oriented in whatever direction you need to maximize system performance. Sometimes you may need to follow a property line; other times you may have to avoid shading issues. Yet more often than not, you have some freedom in selecting where to place the array. Ground mounting also lets you avoid issues that can arise with roof penetrations and allows for easier access during maintenance.

REMEMBER

Another advantage of ground-mounted systems is the fact that they operate at a cooler temperature than arrays mounted on rooftops, slightly increasing the array's power output. And as I show you in Chapter 6, the cooler an array operates, the more power you can get from the modules.

Ground-mounted racking systems, like the one shown in Figure 16-2, are often identical to tilt-up racking systems except in how the racks are held down. In ground-mounted systems, either concrete-encased poles or ground screws are common (more on these methods in "Supporting Ground and Top-of-Pole Mounting" later in this chapter). However, some racking manufacturers allow you to support the rails from the bottom up with a system known as *bottom-up rails*. With bottom-up rails, the rail sections have predrilled holes that match up to the holes on the bottoms of the module frames. To secure the modules on the rack, you just use a nut and bolt assembly. In racks that have plenty of access on the backside of the array, the bottom-up system may be preferable to top-down mounting so that you don't have to set up ladders or scaffolding to access the front sides of the modules.

WARNING

Though ground-mounted PV systems have many advantages, they aren't without their issues. You need to keep the following in mind when working with this mounting option:

- **The footings:** Most people are more comfortable digging in the ground than drilling holes over where people sleep at night, but you still need to dig those holes correctly. I go though the major considerations regarding ground-mounted array footings later in this chapter.
- **The required ground work:** Besides the footings, you have to do a fair amount of ground work for the electrical portion of the job. Specifically, you have to dig a trench from the array location to the inverter and main distribution panel location. And because you're burying electrical wires, that wire run and all the components you use must comply with the *National Electrical Code®* (*NEC®*).

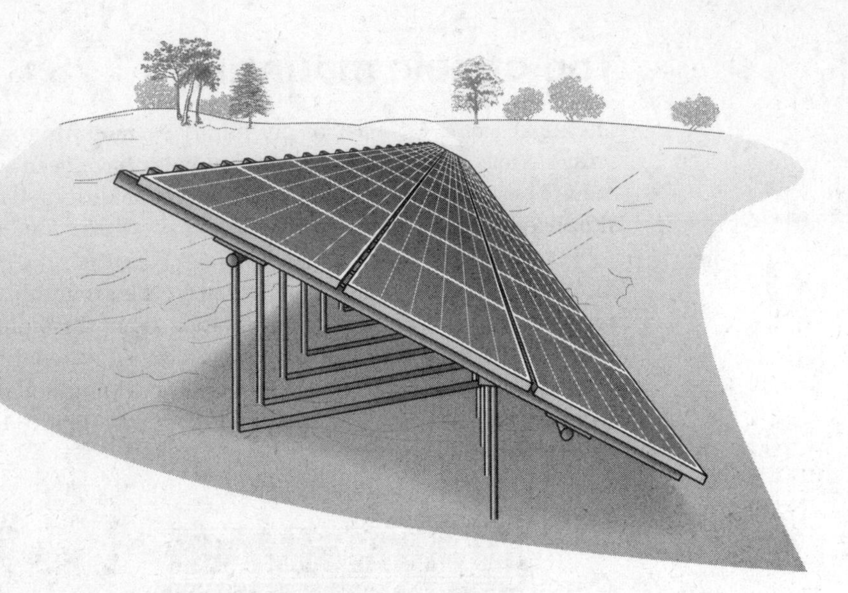

**FIGURE 16-2:** Ground mounting.

- **The accessibility of the modules:** One last consideration is the accessibility of the modules to anyone who wants to approach the system. Here's why you need to think about this point:
  - **The *NEC*® has rules you must follow for any wiring on PV systems that's considered readily accessible.** The *NEC*® makes you keep any exposed wiring and connectors out of the reach of unqualified people to protect those who may not realize what dangers they're exposing themselves to because they didn't read Chapter 13 of a certain PV design and installation book with a pretty black and yellow cover. As far as the *NEC*® is concerned, the easiest way to meet this requirement is to make the wiring and connectors not readily accessible. You can do this by fencing in the array, placing all the wiring and conduit in raceways, or raising up the system so the wiring and connectors are no less than 8 feet from the ground. Each one of these solutions has its own set of challenges, so I can't give you one easy answer. On the positive side, racking manufacturers have been developing solutions to use in conjunction with their systems to aid PV installers in solving the problem of meeting this *NEC*® regulation.
  - **You don't want thieves to be able to walk away with the array.** PV modules are valuable items, and some ill-intentioned folks have caught on to that. If you place an array in a location where anyone can get to it, you and your client should at least consider installing some basic antitheft devices. These can be as simple as specialty hardware that requires specialized tools to remove or as elaborate as camera systems and alarms. I personally like the idea of fences with locked gates; this option solves the *NEC*® problem (see the preceding bullet) and helps deter potential thieves.

## Top-of-pole mounting

The most iconic (at least in my mind) PV mounting system is the top-of-pole array. Arrays mounted on the tops of poles have been used for many years now; they're best for any PV system (grid-direct or battery-based) and can be relatively simplistic in design and application.

In top-of-pole systems, a hole is dug in the ground based on the wind loading, soil type, and array size. The pole is then set in this hole and surrounded by concrete. The racking system itself is a bottom-up system (described in the preceding section), which means you can set up scaffolding underneath the array location and have full access to the bottom of the array. Figure 16-3 shows a typical top-of-pole racking system.

**FIGURE 16-3:** Top-of-pole mounting.

As with ground-mounted systems, top-of-pole arrays operate at cooler temperatures, allowing for increased power levels. In addition, top-of-pole arrays give you complete freedom in determining the array tilt and orientation. Why? Because the racking system is connected to the pole at a single spot, which means you can point the array in any direction and pick (or even seasonally adjust) the array tilt to increase the array's energy yield.

TECHNICAL STUFF

Top-of-pole arrays also open up the possibility of using a tracking system. Tracking PV arrays have the ability to track the sun from east to west each day and make small adjustments daily to account for the changes in the sun's altitude angle. Tracking systems can help increase the overall energy output of an array, but they do introduce a moving part into your PV array where there wasn't one before. So deciding whether or not to include a tracker requires some analysis on your part.

TIP

PV arrays mounted atop poles require a fair amount of ground work that includes setting the pole and trenching the *conductors* (wires) back to the rest of the power system. They also require you to find a way to safely elevate yourself when installing them. The easiest way is to use scaffolding set up around the base of the pole. The scaffolding allows you to move freely beneath the array and provides you with full access to the racking system.

When shopping for top-of-pole mounts, be aware that your options may be limited by the overall square footage of PV modules that can be installed on a single pole. The exact power rating of an array mounted on a single pole varies with your PV module selection, but the maximum size is approximately 3.6 kW for any one pole. This value means that you must either install multiple poles or investigate other mounting solutions if you're working with larger arrays. If you're installing multiple pole mounts, be sure to accurately account for shading between the two poles. After all, it can be quite frustrating to take the time required to install two or more pole mounts only to see one shading another.

As for mounting the PV array, the same rules that apply to ground mounting also apply for top-of-pole mounts, at least in terms of conductor and connector accessibility. Although with pole mounts, raising the modules above the 8-foot mark to make them inaccessible (in the eyes of the *NEC*®, anyway) can be easier.

TIP

Another consideration to make in terms of the pole and array height is the amount of snow received at the site. When calculating the array height, keep in mind that if any snow accumulates on the array, as soon as the sun comes out, the dark PV modules will warm up, and the snow will begin to slide off. This means that you need to have the bottom edge of the array high enough above the ground so the snow can slide off completely. If you don't, the snow will build up and block the PV array's solar access, causing energy production to be greatly reduced or even eliminated and a lot of weight to be added to the bottom half of the array.

## Building-integrated mounting

One type of PV mounting that gets a lot of attention is the one that incorporates the PV modules directly into the building materials (instead of simply placing the modules on top of a roof). This is known as *building integrated photovoltaics* (or BIPV for short). BIPV isn't as popular as the other racking methods I mention earlier in this chapter, but it's gaining more traction. BIPV can take many different forms and use any PV module technology (either crystalline or thin film; see Chapter 6 for details). Systems have been installed where the roofing material is a PV array and where the PV array was integrated into the curtain walls installed along the face of the building. Some of the most well-known BIPV products are those that are installed into either a new or existing roof, not necessarily replacing the roof but rather becoming part of the roof.

REMEMBER

BIPV is best suited for new construction. Although it's mainly used in grid-direct systems, it can also be used in battery-based systems. For residential applications, PV manufacturers are producing small modules that are intended for installation on roofs that'll also use tile or concrete shingles. The modules are manufactured in the same dimensions as standard roofing products so they can be installed in the appropriate location and then surrounded by the traditional roofing tiles. (If you're considering installing one of these products, you need to evaluate the wiring methods you'll have to use and contact the manufacturer for specific installation requirements.) The end result is a roof that now produces electricity. Because the PV modules become part of the roofing system in building-integrated mounting, you can see why this is difficult to do if the roof already exists.

BIPV can also be used on flat commercial roofs that use a membrane roof. The manufacturer takes up to six individual modules (each module is approximately 15 inches wide by 16 feet long) and incorporates them into a large membrane sheet (which is made of the same material as the existing roof). These modules on the now-large (approximately 10 feet wide by 20 feet long) sheet are all wired together, leaving you with one very large module. These large sheets can then be placed on a flat roof and "welded" to the roof using standard roofing practices, thereby creating a lightweight PV array that's fully integrated into the existing roof.

WARNING

BIPV does have some downsides, as you can see from the following:

- Because BIPV isn't as common as any of the racking systems used with traditional framed modules, a BIPV system can be more expensive. Part of that cost increase is due to the material itself. Another part of the added cost is the required coordination of different building trades. For example: If the array is part of the window glazing, you need to closely coordinate all the individuals responsible for each part of the installation.
- Because BIPV is part of the building, the PV array (usually, anyway) isn't facing the optimum orientation. For a flat roof, for example, the PV array isn't tilted up at all; you therefore have to accept system losses due to this lack of tilting. On top of those losses, you must also consider how dirty the array will become over time. Flat roofs aren't clean locations, and even in very wet climates, they don't get washed off very well. Have your client plan to monitor the array's energy output and get up on the roof periodically to clean it.

# Considering Loading When You Mount an Array on a Roof

Every time you want to attach a PV array to a structure, you need to look at the effect the array will have on the building as a whole. The last thing you want is for the building to fail because you didn't consider the *loading* (the addition of weight to the roof) and the existing structure. The type and amount of loading depends on the roof-mounting system you choose to use and where the array is placed on the roof. For example, a flush-to-roof system that stays a few feet in from a roof's edges will have greatly reduced wind loading effects compared to a tilt-up array mounted near the peak of a roof.

As a PV system designer and installer, you must follow various building codes while designing and installing any roof-mounted arrays. The sections that follow review those codes and some common loading calculations you need to perform when designing a PV system with a roof-mounted array. (***Note:*** In the following sections, *loads* and *loading* don't refer to the electrical loads discussed elsewhere in this book; instead, they refer to structural considerations.)

## Following building codes

The *jurisdiction* (the city or county) you're working in has a set of building codes that it uses as a guideline for how much weight can be placed on a roof. The basis for the building codes in most locations in the United States is the International Building Code (IBC). The IBC contains multiple references to the American Society of Civil Engineers' (ASCE) publication titled *Minimum Design Loads for Buildings and Other Structures (ASCE 7-05)*.

REMEMBER

Just like jurisdictions can follow the *NEC*® as much as they want, the IBC is merely a guideline. Many jurisdictions have additional requirements beyond the requirements listed in the IBC. To find out the exact requirements where you're working, talk with the inspectors and someone in the local building department's permit office.

## Accounting for additional dead load

*Additional dead load* is the weight of the PV modules and all the associated racking materials installed on the roof. For most homes built in the last 30 years, the addition of a PV system is well within the structural limitations required. However, if you're dealing with homes built prior to the 1970s, as well as homes built in locations where building codes took a little longer to be adopted, you should enlist the help of a structural engineer to verify that the roof framing can support a PV array. (*Note:* Even in relatively old homes, you can often spend a little time adding the proper amount of support. My home was built in 1929, which meant it wasn't ready for PV. I had to add some additional bracing to support the existing rafters, but doing so didn't require any major efforts.) Finding a structural engineer familiar with PV systems is nice but not an absolute requirement. When it comes down to it, the structural engineer doesn't care too much about what the equipment on the roof is; she just wants to make sure all the loads are accounted for. You can generally find some good engineers to work with by asking some of the local general contractors who they use regularly.

REMEMBER

When you look solely at the additional dead load of a PV system, the weight is generally close to 3 to 4 pounds per square foot. As long as this *dead load value* (which is the overall array and racking weight divided by the square footage of the array) is less than 5 pounds per square foot, you're okay. Under normal conditions, this amount of weight is so far below the requirements for modern building codes that it rarely causes any issues.

TIP

Here's an interesting fact to keep in mind: The 3-to-4-pounds-per-square-foot weight is less than the weight associated with adding a second layer of composition asphalt shingles, which is a standard allowance in roof load calculations. So if your client's roofing material is one that allows for multiple layers (like composition asphalt) and there's just a single layer of roofing material present, then adding a PV array will be less of a load than another layer of shingles. In other words, you should be just fine.

In addition, by placing a PV array on the roof of a home, you're removing the areas where people can walk on the roof — a live load that all roof systems account for (I explain live loads in the next section). By removing the possibility of people walking on the roof (at least in the section where the PV array is), you can exchange the new dead load for that live load consideration. Doing so allows you to install flush-to-roof-mounted PV arrays on roofs that may not have allowances for the addition of future roofing materials on top of existing ones. (Note that ballasted and tilt-up racking systems still need additional considerations due to the added weight and wind loading.)

Although the overall additional dead load for a PV array is relatively small, the issue for many building departments is that you're taking that distributed weight and transferring it to a few points, making the building handle additional *point loading* (transferring the weight of a distributed area to a single location). What many jurisdictions want to see is a minimum number of attachment points to help distribute the additional weight of a PV array over a number of roof-framing members. For flush-to-roof systems, the most common way jurisdictions calculate the minimum number of attachments required is to take the gross weight of the PV array (including the modules, racking, and any other material installed on the roof) and divide that by the number of proposed attachments. If the amount of weight per attachment exceeds a specific value (generally 45 pounds per attachment), the jurisdiction may require additional roof penetrations to help distribute the weight or an engineering review of the roof's structural members. Ballasted and tilt-up racking systems need an engineering review anyway to make sure the point loading doesn't exceed the roofing structure's capabilities.

The numbers presented here work fairly well across the United States, but of course you can't have a single answer for all places. Locations such as hurricane-prone Florida and earthquake-prone California may have additional requirements, which is why you need to talk with the city or county's building department and make sure you design your system to its requirements.

Also, as soon as you start looking at installing a PV array on a commercial building, you can't make any assumptions. During the design process, engineers commonly specify materials and methods that are *code minimum*, meaning there may not be any additional structural capabilities in the building to support a PV array. Consult with a structural engineer early in the process to verify whether the building can physically handle a PV array. I've been involved in projects where the sales process went too far before an engineer revealed that the building was below code requirements without any PV and that adding anything to the roof would be a disaster.

## Looking at live loads

Any load that isn't constant (like the weight of the array is) is considered a *live load*. Live-load specifications vary for every location and jurisdiction. Because you're most concerned with record weather events, the live-load specifications may seem extreme for 99.9 percent of the time, but you need the array and building to stay put 100 percent of the time. In the following sections, I point out several live-load considerations you must make.

### Wind loading

The amount of wind loading your array is exposed to is highly dependent on the array's location. You can find a map of the United States in the IBC (Figure 1609 if you really want to look it up) that shows the basic wind speeds for use in designs; you can also visit `www.windspeedbyzip.com` for wind-speed data. The best way to find out the wind speeds in your client's location and make sure you use the exact numbers that the jurisdiction bases its calculation on is to ask the building department what it uses. The IBC uses multiple variables in defining the basic wind speed, including the characteristics of the surrounding terrain and the building height.

Your client's exact wind loading is important to look at because it affects your choice of the appropriate array location and mounting method. If you want or need to place the client's PV array close to the edge of a building, the array will likely need additional fastening to the roof (through extra penetrations or extra ballast material) due to the increased wind loading at the roof's edge. The IBC helps you go through the calculations, but it doesn't take you through them step by step. A great resource to walk you through the entire process is the February/March 2010 issue of *SolarPro* magazine, available for download at `solarprofessional.com`. If you still feel uncertain about the calculations, consult a structural engineer to verify the design.

## Snow loading

Snow loading can be one of the most difficult live loads to predict and account for in your design. For flush-to-roof systems, the snow loading on a building should've been taken into account with the original building design; the addition of a PV array should therefore have little impact, meaning you don't need to design for any additional snow loading when you're dealing with a modern roof (one constructed within the past 30 years).

For rooftop arrays mounted with tilt-up racking, snow tends to drift and accumulate around the backs of modules (where they're raised), creating increased point loading on the roof. This extra point loading can last for months in some locations — not a good thing. In addition to this drift, as snow accumulates on the tops of modules, it slides off the faces of the modules to the roof's surface as soon as the sun comes out, creating even more point loading. With these systems, you need to work with a professional engineer to make sure the racking system and building will hold up during the times when the snowdrifts occur.

Because ballasted systems are also tilted up, they're subject to the same snowdrift problems. However, these systems may pose even more problems due to the extra weight holding the array down to begin with. Not to beat a dead horse, but this situation is where a structural engineer earns her keep.

For all PV systems, you should also think about where the snow is going to go when it begins to melt. Snow-covered PV arrays still heat up very quickly, which means all that snow can slide right off them in a flash, posing a risk to anyone standing below the modules. If you're installing a PV array in a snowy area, try to keep the array away from the edge of the roof, or install snow guards to reduce the risk of personal injury due to falling snow chunks.

## Seismic loading

Section 1613 of the IBC insists you account for seismic loading, which is the pressure exerted on an array during an earthquake. Any racking system that has a positive attachment method (one that penetrates the roof), such as flush-to-roof mounting and tilt-up mounting, generally meets these requirements. If you're using a ballasted racking system, you may need to address additional considerations as required by the local jurisdiction and the IBC.

# Properly Attaching an Array to a Roof

Whenever you're going to attach a PV array to a roof, you need to make sure that the attachment method is appropriate, that it'll hold correctly, and that you aren't allowing water to enter the building. In the next sections, I introduce you to the hardware used for making the connection and the options available to you for proper weather sealing.

## Making attachments with lag screws

Attaching flush-to-roof and tilt-up racking requires you to penetrate your client's roof — in other words, you have to drill holes; there's no way around it (unless of course you opt for a ballasted racking, as described earlier in this chapter). The most common way of attaching either type of racking to a roof is with a lag screw that's connected directly to a roof-framing member, which is either an engineered truss or a roof rafter.

Another very common method for attaching a lag screw is to get inside the attic space and attach a 2-x-4 or 2-x-6 wood block between two of the roof-framing members at the location where you want to make the roof penetration. This technique lets you create roof penetrations at the exact location you want, regardless of the roof-framing location. Another benefit is that the point loading you're creating on the roof system is now being distributed to two rafters or trusses rather than just one. This second method is preferable because its added strength allows you to avoid the roof framing.

TIP

The exact dimensions of a lag screw are often determined for you based on the racking manufacturer's requirements and the wind loading considerations for your client's location (I cover wind loading earlier in this chapter). However, this fact doesn't mean you shouldn't verify the lag screw's ability to keep the racking system attached to the roof. Lag screws are characterized by their pullout strength in various types of wood by the number of pounds per embedded inch they can withstand. For example, a 5⁄16-inch lag screw (a common diameter for roof penetrations) has a pullout strength of 266 pounds per inch in Douglas fir. Because you usually won't know what the exact type of wood a client's roof is made of, I suggest you assume a wood type of spruce/pine/fir. This wood type has the lowest pullout strength (205 pounds per inch for a 5⁄16-inch lag screw) in order to cover all possibilities. You may be required to produce this information for the local jurisdiction, so be sure to have an idea of the pullout strength ahead of time.

When installing lag screws, you first need to drill a pilot hole that's 60 to 75 percent of the lag screw's diameter. Because the lag screw will follow the path your pilot hole drills, make sure the hole is perpendicular to the roof framing by using a rafter angle square (also known as a Speed Square) beside your drill. If the pilot hole isn't square, the lag screw won't be able to set fully, and the roof penetration may begin to leak after a few years.

If required by the local jurisdiction, you can determine the maximum uplift force that'll be imposed on the array with the help of calculations outlined in *ASCE 7-05*. Using the tables in *ASCE 7-05*, you can determine the minimum required lag screw length. Keep in mind, though, that the lag screw has some materials to get through before it actually reaches the roofing member, and the tables are based on the lag screw being embedded into that wood. Take into account the depth of the footing that the lag screw is going through, the roofing material, and the plywood roof decking material when calculating the overall lag screw length.

TIP

The calculations to determine the lag screw depth and footing spacing are most critical for systems that use tilt-up racking that exposes the array to much greater wind loading. Unfortunately, the calculations and all the considerations are beyond the scope of what I can present here, so I suggest either teaming up with a structural engineer or doing more research on this subject by examining the following sources:

- International Building Code
- The North American Board of Certified Energy Practitioners (NABCEP)'s study guide for Certified PV Installers
- *SolarPro* magazine, specifically the February/March 2010 issue

## Sealing roof penetrations with flashing

Recently, the methods used for sealing roof penetrations have come to the front of the PV industry's collective mind. Why? Because the majority of construction-related litigation in the United States is based on water intrusion. Given that many business can't survive a major lawsuit, performing due diligence on your sealing methods during every system installation can only serve you well in the long run.

I know a lot of PV pros out there have used nothing more than aluminum L feet (a piece of aluminum that looks like an *L* and measures about 3 inches in width and height) screwed directly on top of the roofing material with just enough roof caulking in between the two to keep the water out (even I'm guilty of installing numerous systems like this). I also know that a number of installers who use this method don't see the need to change their methods now. Fortunately, you can get off to a good start with the knowledge of how to seal a roof penetration the right way.

One very specific IBC requirement you should be aware of is 1503.2, which requires that flashing be installed around roof penetrations in order to keep moisture out of the building. What's *flashing*, you ask? It's a physical barrier that extends past any roof penetration to keep water from reaching the hole in the roof. Figure 16-4 shows you a couple examples of flashing.

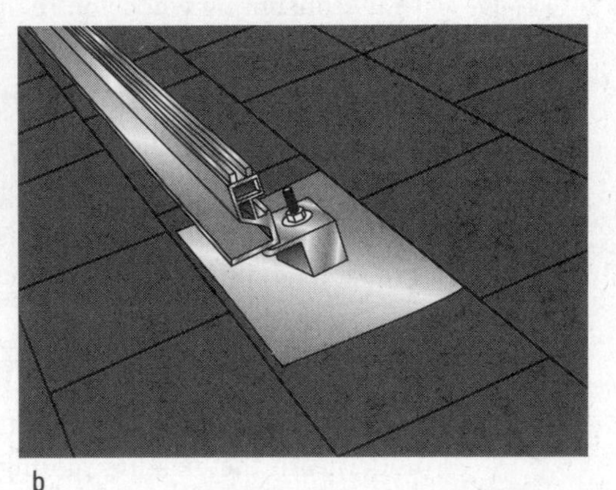


**FIGURE 16-4:** Different types of flashing.

To meet the IBC code requirement of flashing all roof penetrations, you need to use either a post-type footing (which is provided by the racking manufacturer) or a block-type mount with integrated flashing (which comes from either the racking manufacturer or an independent third-party company). I describe these types of flashing in more detail in the following sections.

## Post-type footings

The post-type of roof penetration, shown in Figure 16-4a, is a great flashing option, especially when you're installing a PV array on a brand-new roof or on a building that's getting reroofed. It's ideal for these scenarios because you can set the posts down on the roof before the roofers come in and start their work (or in the case of a reroof, after the original roof has been removed and before the new one starts to go down). When the roofers show up, they can seal the roof as they normally would; they just have to work around a bunch of posts. These posts are sealed the same as any other roof penetration the roofers deal with every day, so they shouldn't be new to the roofers.

For the best results, coordinate the placement of post-type footings with the roofer beforehand. The last thing you need to do is surprise the roofing crew one morning with a roof full of posts that it wasn't expecting. Trust me, you don't want to wind up on any roofer's bad list. In fact, why not partner up with a good roofer for all of your roof penetrations? Roofers are much more knowledgeable about proper roofing techniques and materials, and the roofer's job will be easier and higher quality in the end, so the relationship will be mutually beneficial.

You can install post-type footing on an existing roof for use with either flush-to-roof or tilt-up racking systems, but this is more difficult and typically requires you to cut into the existing roofing material for each post. Doing so is very time-consuming when you're up there on a hot (or cold) roof.

## Block-type mounts with integrated flashing

The block-type mount, or flashed penetration, shown in Figure 16-4b is another option for flush-to-roof and tilt-up racking systems. It's good for use in new construction or on existing roofs, and you can find a number of different options available for just about every roofing material used today.

Block-type mounts with integrated flashing are constructed with a mounting block attached to the top of the flashing material to prevent water from penetrating the flashing. To install this flashing on a roof, take the flashing and slide it up underneath the existing roofing material so that the top part of the flashing is under the two rows or shingles directly above the roof penetration, keeping water from reaching the hole in the roof. You can then mark the hole location and drill a hole in that spot. Place a small amount of roofing caulking to seal the hole you drilled and insert a lag screw through the integrated block (be sure to tighten it into the roof-framing member). Your first few block-type mounts may take some time to get the hang of, but before your first row of footings is set, you'll have the process down. A great tool to invest in is a roofing pry bar, which helps you set the flashings properly without tearing the roof apart in the process.

Different variations of this flashed penetration are available for nearly every type of roof covering (composition asphalt, wood shake, and tiles). They're all more expensive initially than simply attaching an L foot directly to the roof deck, but I consider them cheap insurance against bigger problems and the proper way to make the connection as defined by building codes.

# Supporting Ground and Top-of-Pole Mounting

For PV arrays atop ground and top-of-pole mounts, you need to think about how you're going to secure the racking system into the ground. Often you can work directly with the racking manufacturer to help determine the specifications for the foundations. Many of them have engineered their systems and can share that information with you. They're generally willing to dictate the diameter and depth of the hole(s), as well as the required spacing for ground-mount systems. (Spacing considerations need your attention only when multiple poles are going in the same area. Top-of-pole mounts use just a single pole.)

For locations that have difficult soil conditions (such as bedrock close to the surface or very sandy soil), or if you want to raise the array higher than standard heights, you may need to have a geotechnical engineer evaluate the design to determine whether any additional reinforcements are needed. I suggest you find someone who's familiar with the local conditions by searching the Yellow Pages or talking with general contractors in the area.

The most common material used for the vertical pole support is steel tubing. The sizes range from 2 inches all the way to 8 inches in diameter, depending on the exact racking method.

When installing ground-mounted systems, site preparation is vital. The reason ground mounts need extra attention over top-of-pole mounts is that without the proper preparation, a ground-mounted array can quickly become cockeyed during installation. Even a small ⅛-inch deviation in the beginning can have huge effects just 50 feet down the line. The most accurate way to prepare the site for a ground-mounted system is to spend some time with a surveyor's transit and leveling tool. Doing so helps you mark the coordinates for all the footings you need to install and ensure that they're in a straight line.

You have two options for securing the posts into the ground via the footing:

- Historically, the most common method is to dig an appropriately sized hole, place the pipe in the hole, and fill in the remainder with concrete. This is a very effective and secure method, but it can be rather labor intensive. In order to do it right, you need to place the pole in the ground accurately and then secure the pipe in a vertical position so that when you pour the concrete, the pipes stay in place. (For detailed top-of-pole installation steps, check out issues 108 and 109 of *Home Power* magazine, available at homepower.com.)

- Another method that's gaining popularity is the use of helical piers. These systems essentially drill the pipe into the ground. The pipes have blades along their bottoms that make them look like a very large screw. They're screwed into the ground with the help of a tractorlike vehicle that has a special attachment installed for drilling the piers into the ground. These systems don't require any digging or concrete, but they do call for specialized equipment and expertise if you want to install the piers accurately. (***Note:*** At this time, helical-pier systems are limited to ground mounts because the piers aren't large enough for most top-of-pole installations.)

IN THIS CHAPTER

» Placing electrical equipment in the right spots

» Reviewing some wiring basics

» Examining the two components of grounding

» Connecting to the grid

# Chapter 17

# Integrating the Electrical Elements

After you analyze your client's site, specify the equipment needed for either a grid-direct or battery-based system, size all the components, and assemble the mechanical parts (as I explain in Chapter 16), you need to install the electrical elements of the system safely and according to *National Electrical Code*® (*NEC*®) requirements. To do so, you have to know Code references as they pertain to PV systems specifically and electrical systems generally.

The electrical portion of PV installations tends to be highly scrutinized by inspectors and Code officials (such as plan reviewers in the building department) reviewing projects. In addition, many inspectors have now seen enough PV systems to know what to look for and can identify problem spots (see Chapter 18 for the full scoop on the inspection process).

In this chapter, I dive into some Code-related topics to know during your installations. I cover equipment locations, wiring, grounding considerations, and the requirements you're responsible for when connecting to the utility.

# Location Is Everything: Knowing Where to Place Electrical Equipment

As you install a PV system, you're bound to find a number of potential places to locate the (many!) different pieces of electrical equipment. Your charge is to evaluate the different possible locations and decide on the best spot. This spot must be compliant with the *NEC*®, adhere to the manufacturers' requirements, and make sense for the system owner and his situation. The following sections focus on the Code requirements for the electrical equipment locations, specifically making sure the equipment is properly rated and grouped together to allow for maintenance and emergency shutdown.

REMEMBER

You must evaluate the site-specific requirements on a per-job basis.

## Manufacturers' requirements for equipment locations

A statement found early in the *NEC*® — Article 110.3(B), to be exact — mandates that all equipment must be installed and used in accordance with any instructions for that equipment. Consequently, to be compliant with the Code, you must install all electrical equipment according to the various manufacturers' instructions. So if the installation manual says you can install the equipment only on even-numbered days and you install it on an odd-numbered day, you're in violation of the Code. Granted, that's a silly example, but it illustrates the importance of knowing a manufacturer's requirements. I've seen installations that failed inspection because the inverters weren't spaced properly according to manufacturer instructions; in each case, the inspector demanded that the misplaced inverters be moved.

REMEMBER

Avoid placing any electrical equipment in direct sunlight (with the exception of the PV modules, of course; I explain how to install the modules in Chapter 16). For the inverters, this is good practice because an inverter's ability to produce power is reduced as it grows hotter. Equipment such as overcurrent protection devices (OCPDs) are also affected by increased temperatures (you may need to derate them per the manufacturer's instructions if the temperatures get too high). Other equipment, such as disconnects and *combiner boxes* (boxes that place all the PV strings in parallel), may not normally be negatively affected by high temperatures, but when in direct sunlight for multiple hours each day, the labels you use on them may degrade with time and be difficult to read. For conduit runs, you may not be able to avoid locations in direct sunlight. If this is the case, make sure the conduit is properly rated for outdoor installation by referring to the manufacturer's spec sheet.

REMEMBER

Any electrical component you install with electrical terminations inside of it can be called an *enclosure* (some examples include disconnects, combiner boxes, and inverters). Make sure you know the National Electrical Manufacturers Association (NEMA) enclosure ratings before installing any electrical equipment in a PV system because the NEMA rating of the enclosures you install affects where the equipment can be installed. The most commonly seen NEMA enclosure ratings for PV equipment are

- **NEMA 1, indoor installations only:** This enclosure offers no protection from dust or water entry. (Nearly every battery-based inverter is rated NEMA 1 — in fact, I can't think of any exceptions.)
- **NEMA 3R, outdoor installation, typically on a vertical surface only:** This enclosure type offers protection from falling rain. (Most grid-direct inverters — but not all — have at least a NEMA 3R rating.)

WARNING

  The reason I specify vertical surfaces for NEMA 3R enclosures is due to the listing process. When manufacturers have their equipment tested, the standard for NEMA 3R is that the equipment is mounted on a vertical wall. Therefore, unless specifically stated otherwise, you can install enclosures rated at NEMA 3R only in the vertical position. It can be tempting to install a NEMA 3R enclosure on a roof at the same angle as the roof (I've even seen this done), but the problem is that the enclosure wasn't designed for that application. If installed on the same angle as the roof, it may not be able to properly protect the electrical connections inside from rain and ice.

- **NEMA 4, outdoor installations on horizontal or vertical surfaces:** These enclosures can withstand direct spray from a hose and splashing water.

## Locations for disconnecting means

When I refer to *disconnecting means*, I mean that, in the most generic sense, you need to install disconnects such that when you go to work on an inverter, you can safely disconnect both the AC and DC sides of the inverter from all sources of power with approved methods at the inverter location.

REMEMBER

The *NEC*® requires the PV disconnecting means be installed in a readily accessible location, either on the outside of the building or after it enters the building, at the nearest point of entry. But placing the disconnecting means properly can be a difficult task within most PV systems because many PV output circuits (which run from the combiner box to the DC disconnect) enter a building up at the roof location, and no "readily accessible" location exists up there for placing a disconnect either inside or outside of the building. Thankfully, an exception applies to this Code section that refers you to Article 690.31(E). It allows you to install the PV source circuit or PV output circuit inside a building without first installing a disconnect so long as the *conductors* (wires) are inside metallic conduit until they reach the disconnect.

TIP

Most grid-direct inverter manufacturers now include a disconnect switch that can be integrated into their inverters, allowing you to disconnect the DC power source right at the inverter (and sometimes even the AC power source with the same switch). These disconnects are a separate piece that mates to the inverter during installation. The inverter can then be removed safely, leaving the disconnect on the wall and allowing for the installation of a new inverter. (See Chapter 10 for more on disconnects integrated into inverters.)

Many inverters that include an integrated disconnect make provisions for the installer to also use the disconnect as a combiner box. This means you can bring down multiple PV source circuits from the roof to the combiner/disconnect and place them all in parallel at the inverter location. In fact, the 2011 version of the *NEC*® will require a disconnecting means at the point of PV source circuits placed in parallel. Using the integrated combiner/disconnects from the manufacturer can help you easily meet this requirement.

REMEMBER

For those inverters that don't have an integrated disconnect, you need to install AC and DC disconnects at the inverter location. This process involves mounting at least one disconnect next to the inverter to disconnect the PV source circuits. If the inverter is installed at the main distribution panel (MDP), the circuit breaker used to make the connection to the utility can serve as the AC disconnecting means. If, however, the inverter is mounted at a location other than the MDP, you must install an additional AC disconnect at the inverter location. If the inverter and AC disconnect happen to be located at the utility meter location, you may satisfy both the *NEC*®'s and the utility's requirements for disconnects with the same disconnect (you find out more details about utility disconnects later in this chapter).

The *NEC*® also says that a disconnect "shall not be required at the PV array location," which means that for ground- and pole-mounted arrays (see Chapter 16), you can run PV source circuits without installing a disconnect at the array. However, most PV installers consider it a best practice to install a disconnect there because doing so provides you with an easy place to disconnect the PV circuits. (Besides, when the 2011 version of the *NEC*® is available, you'll need a disconnect if you combine the circuits at the array.)

## Combiner boxes and junction boxes and wiring, oh my!

Making the transition from module conductors in free air to the conductors you run through the client's building (or underground) to the disconnecting means has to happen in some sort of enclosure, either a combiner box or a junction box. (A *junction box* is an enclosure that looks very similar to a combiner box on the outside. Inside, it uses terminals to transition the PV source circuit wires [USE-2 or PV wire; see Chapter 10] to the building wiring [THWN-2]. These terminals don't place any conductors in parallel though, so you have the same number of conductors entering and leaving the junction box.) This enclosure typically lives at the array location, thereby reducing the length the exposed conductors have to run.

REMEMBER

Until your client's jurisdiction is using the 2011 version of the *NEC*®, you don't have to have a disconnecting means adjacent to a combiner box. So if you want, you can install a combiner box at the array location and bring a single PV output circuit down to the inverter. Generally, however, running the individual strings to the inverter and using a combiner box at the inverter location is considered a safer method. If that's the method you choose to use, then you must still incorporate a junction box at the array location to transition the outdoor wiring to indoor wiring that you can run in conduit.

TIP

Make sure the junction or combiner box you choose has the proper NEMA ratings for the environment you install it in (see the earlier section titled "Manufacturers' requirements for equipment locations" for details). Also, you may need to consider the heat effects on the fuses if you decide to mount a combiner box on a rooftop or in a location where the temperatures will be greater than the manufacturers' ratings (see Chapter 13 for more on this).

# Working on Wiring

The actual act of wiring up a PV system isn't all that complicated if you've taken the time to plan for and set up your installation. Of course, this doesn't mean the process is without its own set of issues and difficulties. In the sections that follow, I point out some of the installation-specific details you need to keep in mind. (If you're curious about the different types of conductors and conduit that are typically used in PV installations, head to Chapter 10; for the how-to on sizing conductors and conduit, see Chapter 13.)

## Seeing red (and green and white): Color-coding

When it comes to wiring PV systems, one of the most common mistakes I've seen is someone mixing up the color of the conductors used. The *NEC*® has a very specific set of requirements for color-coding based on the function of the conductor in the circuit.

REMEMBER

The *NEC®* requires the following color-coding for wiring in PV systems:

- » Equipment-grounding conductors (EGC) must be green, green with yellow stripes, or bare wire.
- » Grounded (or neutral) current-carrying conductors must be white or gray.
- » Ungrounded (or hot) current-carrying conductors can be any color other than those mentioned for equipment-grounding and grounded conductors. However, the most common colors are black and red.

To help you remember what colors of conductor go where, think of the 120/240 VAC wiring system in your home. When you look at the wires coming into the outlets around your house, you have one black wire, one white wire, and one bare copper wire (which is sometimes green instead). The black wire is the one that's connected to the circuit breaker in the MDP or subpanel, the white wire is connected to the neutral busbar, and the bare (or green) wire is connected to the ground busbar. (In that same MDP, the neutral busbar is connected directly to the ground busbar.)

If you look at a 240 VAC load, like your electric clothes dryer, you see a red wire in there too. In this case, the red wire and the black wire are connected to a two-pole circuit breaker. In this wiring scheme, the black and red wires are known as the ungrounded (or hot) current-carrying conductors, and the white wires are known as the grounded (or neutral) current-carrying conductors. The bare or green wires are the equipment-grounding conductors; they don't carry current in properly installed and operating circuits.

So what does all of this mean to your PV system? As you well know, a PV module has two conductors on the back of it. As with all DC circuits, one side is positive (marked with a plus sign), and the other is negative (marked with a minus sign). Because DC electricity flows in a single direction (see Chapter 3 for more on DC electricity), the designation of the positive and negative sides, or *polarity*, is important to start with. Although all the PV module conductors are marked positive and negative, the best way to define the different parts of the PV circuits really isn't with the terms *positive, negative,* and *ground.* This terminology isn't incorrect, but you can do better. Most people associate red wires with the term *positive* and black wires with the term *negative.* When you install a PV system, you're going to have a positive and a negative conductor, but one of those conductors will be connected to ground (just like the neutral conductor is in the MDP at your house).

TIP

If you can get yourself into the habit now, remove the words *positive, negative,* and *ground* from your vocabulary and start referring to them as the *grounded current-carrying conductors, ungrounded current-carrying conductors,* and *equipment-grounding conductors.* This way you can easily follow the color-coding required by the *NEC®*.

- For a PV system that has the negative connected to ground across the ground fault protection (GFP) device, the negative becomes the grounded current-carrying conductor (just like the neutral in the AC system) and is white or gray. (Chapter 10 has more info on ground fault protection.)
- The positive is the ungrounded current-carrying conductor (just like the hot conductor on the AC side) and can be either red or black. I suggest using red rather than black because even though you may have banished the word *positive* and its association with red wiring in DC systems from your consciousness, you may be the only person in the jurisdiction who has done so. If someone comes up behind you, opens up a box, and sees a red, white, and bare (or green) wire, that person (assuming he knows the wiring is for a DC system) will probably associate the red with *positive.* He may still have to think about the white wire, but you've done your best to help him out.
- As in all electrical systems, the equipment-grounding conductors are green, green with yellow, or bare copper. This standard is consistent with all the wiring on the AC side.

## Managing wires on PV modules

The quick-connect plugs installed on the backs of PV modules are a sort of double-edged sword. On one hand, they greatly reduce the amount of time you have to spend making *module interconnections* (the electrical connection from module to module). On the other hand, they introduce an issue that wasn't present before: the need for wire management on the backs of the modules. Before the quick-connects, the wiring was most commonly done by connecting conduit from one module to the next. Although this task was time consuming, it automatically protected and managed the wires. Nowadays, most modules can't accept conduit because the electrical boxes integrated onto the modules have no way to accept conduit connections. Consequently, you have to find other ways to keep the conductors safe and secure.

REMEMBER

The point of wire management is to provide a safe and reliable method of keeping the conductors in place for the life of the PV system. When installed correctly, that system should last more than 25 years, which means you need to protect the conductors from damage from all the elements, especially damage that may be caused if the conductors are blowing around in the wind, in contact with the roof surface, or being pulled down as snow and ice melt along the backs of the modules.

You can use a couple different methods to secure the exposed conductors running along the backs of the modules:

- **Specially designed clips:** Multiple companies manufacture specially designed clips that you attach to the PV module and then snap the conductors into. You can buy these cable clips in different sizes to accommodate different wire types (such as USE-2 or PV wire; see Chapter 10 for more on these conductors).

- **Plastic cable ties:** Also called *zip ties,* plastic cable ties are very easy to use and commonly available, but I don't like to rely on them. You can buy plastic cable ties that are UV-resistant (usually the black ones), but I've seen these so-called UV-resistant ties fail in just a few years of being out in the sun. Even though most plastic cable ties aren't in direct sunlight, I just don't trust them to last more than 25 years. I suggest purchasing stainless steel cable ties with nylon coating instead.

TIP

A combination of cable ties and cable clips is generally your best bet for long-term reliability.

## Protecting wires with conduit

Chapter 3 of the *NEC*®, "Wiring Methods and Materials," specifies the requirements for properly protecting conductors with conduit. In the earlier "Locations for disconnecting means" section, I note the need for metallic conduit for some roof-mounted PV arrays. Another common approach is to use PVC conduit for underground and exposed outdoor locations. (See Chapter 10 for an overview of the common conduit types and acceptable locations; refer to Chapter 13 for details on conduit sizing.)

Regardless of the type of conduit used or the type of circuit (AC or DC), you and a partner can pull conductors through conduit the same way. When pulling the conductors, use *fish tape* (a stiff, yet flexible, metal "rope") and push it from one end of the conduit to the opposite end. After it appears on the other end, tape your conductors to it and pull the fish tape and conductors through the conduit.

TIP

I also suggest pulling a nylon string through the conduit along with the conductors. If you ever need to pull any new conductors through the conduit, the nylon string serves as the fish tape, saving you the hassle of trying to push more fish tape though a conduit that already has conductors inside it.

TIP

If you want extra help getting the conductors into your conduit, you can buy lubricant, affectionately called *slime*, to help slide the conductors through the conduit. Slime is cheap and helpful anytime you have to pull the conductors.

## Bonding Yourself to Grounding

Grounding is the most confusing, debatable, and fun topic in the PV world. In fact, grounding is so much fun that the pages in my copy of the *NEC®* that deal with PV grounding are falling out because I spend so much time flipping them back and forth in sheer joy.

There are two distinct components to grounding: equipment grounding and system grounding. These components have some common features, but they also have different installation requirements. So when you're thinking about grounding, be sure to think about which component of it you're considering.

- *Equipment grounding* is the act of connecting the pieces of equipment together electrically.
- *System grounding* is the act of connecting the equipment to the conductors buried in the ground.

The information I present in the following sections is designed to get you familiar with the components of grounding and help you understand the differences between equipment and system grounding. Entire books have been dedicated to these topics, so I don't expect to make you an expert on the subject here. However, after reading these sections, you should be able to properly address the grounding issues you'll face most of the time.

### Equipment grounding

The point of equipment grounding is to make sure all the electrically conductive materials are kept at the same voltage potential as ground. So if a conductor accidentally touches any equipment (a disconnect or PV module, for example), the current has a low resistance path to ground, allowing the safety equipment to activate and reducing (but not eliminating — never forget that) shock hazards to anyone who may touch that piece of equipment. The equipment-grounding conductor (EGC) runs right alongside the other conductors in the PV circuits, as described in Chapter 10.

The *NEC®* specifically calls out that "exposed non-current-carrying metal parts of module frames, equipment, and conductor enclosures shall be grounded." The requirement to ground this equipment is present for all PV systems, regardless of the system voltage, which means you must always install equipment grounding for your PV installations.

In the following sections, I show you how to make these connections as well as how to size them properly.

## Making the equipment ground connection

To properly ground all the conductive pieces of equipment, you need to install what's known as an *equipment-grounding conductor* (EGC). The *NEC®* defines an EGC as "the conductive path installed to connect normally non-current-carrying metal parts of equipment together and to the system-grounded conductor or to the grounding-electrode conductor, or both." (I cover the system-grounded and grounding-electrode conductors in the later "System grounding" section.) In short, this verbiage means you need to connect all of those metal pieces of equipment together and then connect that mass of metal to the system ground.

For items such as combiner boxes, disconnects, charge controllers, inverters, and any metallic box holding electrical equipment (such as the battery box), the easiest way to install an EGC is with a properly sized conductor that's connected to a ground lug inside the boxes. (A *lug* is a device that allows you to terminate a conductor connection.) All equipment manufacturers have a very specific location and type of lug that you can use to connect the ground, so if a ground lug isn't in the box, buy the lug specified by the manufacturer.

Using a conductor to connect each module together can be a difficult task. PV modules often have a place on their frames that's intended for an EGC; usually this spot is in the middle of the long edge of the frame. The problem is that manufacturers often don't supply hardware for you to actually make the connection. And to make the situation even worse, a module's instructions generally give you little (if any) direction on how to connect an EGC.

This lack of clear guidelines has caused PV installers to seek outside help with connecting modules. Currently, two methods are used (see Figure 17-1):

- **Ground lug:** By attaching a ground lug that's rated for outdoor use (and direct burial) to the back of each module and connecting a conductor to each lug in the array (see Figure 17-1a), you can place the lug at the location provided by the manufacturer (although this location may not be in a convenient place based on your chosen mounting method). However, attaching ground lugs is a time-consuming and detailed process.
- **Grounding clip:** By placing a specific grounding clip between the module and the racking system, you can *bond* the modules to the mounting rails (in other words, you can establish electrical continuity). A grounding clip pierces both the module and the racking system at the same time, bonding the two together (see Figure 17-1b) and allowing you to use the mounting rails as the EGC, eliminating the need to run wire all along the back of the array. You need

to connect all the rails together to maintain a continuous grounding path, but you can do this pretty easily before installing any of the modules. After ali the modules have been installed and bonded to the rails, you can connect a single EGC to one of the rails and bring it into the junction box (or combiner box) with the grounded and ungrounded current-carrying conductors from the array.

WARNING

Using grounding clips isn't as labor intensive as using ground lugs, but it isn't a foolproof method. You absolutely have to make sure that you place the grounding clips in the appropriate locations (as specified by the manufacturer of the grounding clips you're using) and that the grounding clips are engaged properly (meaning they're in full contact with the module and rack). If you install grounding clips incorrectly, then the module won't be properly bonded to the rail. Also, grounding clips are only rated for a single use. So if you tighten them to the rail and then want to move something, you must replace them.

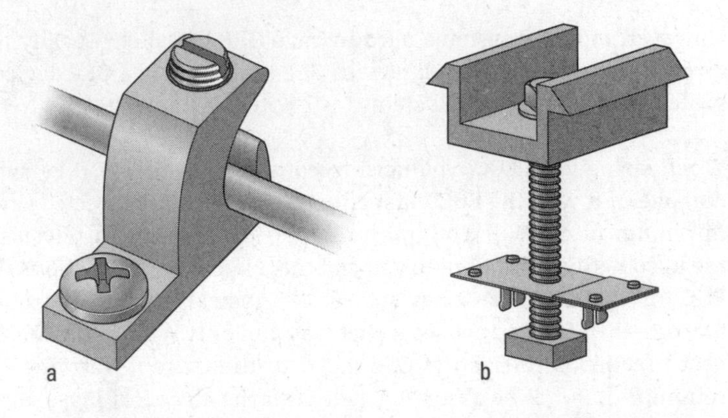


**FIGURE 17-1:** Equipment-grounding options for PV modules.

REMEMBER

Both the ground lug and grounding clip methods allow you to install a continuous ground path for the PV modules. Make sure that if any module is ever removed from the system, the path to ground for all the other modules isn't interrupted. If you use the lug method, this generally means you have to use a continuous conductor. If you use grounding clips, then as long as all the modules are properly bonded to the rail (which is the path to ground), the ground path won't be removed until the racking system is removed.

## Sizing equipment-grounding conductors

An EGC is sized based on the requirements set forth in the *NEC*®. For PV systems, the size is primarily based on the OCPD used in the circuit (I explain how to size these devices in Chapter 13). This sizing methodology also assumes that a ground fault protection (GFP) device is installed in the system. In Article 690.45, the *NEC*® refers you to a table to determine the size of your EGC based on the amperage rating of the OCPD protecting the circuit. All you have to do is find the OCPD rating and move across the table to find the minimum EGC size required.

As I explain in Chapter 13, not every PV system requires OCPDs. When dealing with one that doesn't, you must size the EGC by assuming that the OCPD amperage rating is equal to the PV-rated short circuit current and using that value in the table in Article 690.45 of the *NEC*®. You can never use an EGC smaller than 14 AWG, though, regardless of the module's short circuit current value.

The *NEC*® has special provisions for PV arrays that are installed in systems where no GFP is present. In these scenarios, you must size the EGC to have an ampacity rating of at least two times the conditions of use ampacity for the current-carrying conductors (I touch on conditions of use in Chapter 13). In this scenario, you need to refer to the conductor ampacity tables (*NEC*® Tables 310.16 and 310.17) to find the right size for the EGC. This requirement is in place to account for the possibility of current flowing continuously though the EGC in a fault condition.

However, most PV systems need to have GFP installed. Luckily for you, all grid-direct inverters come standard with GFP built in, and a GFP device is easy to integrate into battery-based systems (as I note in Chapter 10).

When sizing your EGC, you need to consider whether it'll be subject to physical damage. If it will, the EGC must either be physically protected (in conduit) or be a minimum of 6 AWG. The interpretation of "subject to physical damage" may mean something different to you and me. I know of jurisdictions that consider any EGC on the back of an array subject to physical damage, but other jurisdiction-powers-that-be may not see it that way and only require the *NEC*® minimum. The best practice is to size the EGC equal to the current-carrying conductors (as a minimum). But to be absolutely safe, call the electrical inspector in the jurisdiction where your client is located before installing the system. Sure, this phone call may seem like an extra hassle, but just think of the time it'd take to remove the wire you originally installed and replace it with another if you didn't call to find out the jurisdiction's preferred EGC size beforehand.

## System grounding

All the EGCs you run will be connected to system grounding. The system ground is composed of a grounding electrode (GE) and a grounding-electrode conductor (GEC).

- The GE is the conductive object in direct contact with the earth. It's often a *ground rod* — a copper rod that has been driven 8 feet into the ground (some jurisdictions require two ground rods driven a minimum of 6 feet apart and also that the two be bonded together; if you're adding new GEs, make sure you know what the jurisdiction requires).

Another commonly used grounding electrode is called a *Ufer ground.* This is where the steel reinforcement in the concrete foundation of a building is used as the contact with the earth.

- The GEC is the conductor that connects the grounding electrode (either a ground rod or a Ufer ground) to a point where all the other grounded conductors can be connected. Typically, this point is a 6 AWG or 4 AWG wire that's connected to the grounding electrode on one end and the ground busbar inside the MDP on the other end. (A *ground busbar* is where all the EGCs are terminated; it's in direct contact with the metal MDP enclosure.)

Article 690.47 is the main *NEC*® reference for PV system-grounding requirements; a number of changes were made to this article in the 2008 version of the *NEC*®. I walk you through the commonly accepted and generally approved methods for connection and sizing in this section.

## Making the system ground connection

For utility-interactive PV systems (either grid-direct or battery-based), the PV system grounding is connected to the existing system grounding installed for the utility system. You can make this happen in a couple different ways. ***Note:*** The method you choose depends on the location of your inverter and its relation to the existing grounding-electrode system.

Figure 17-2 helps illustrate the two common ways to make the system-grounding connection in a utility-interactive system:

- **A ground rod (or two) connected to the inverter system grounding by a 6 AWG GEC:** Keep in mind that you must connect the existing GE (from the utility) to this new grounding-electrode system you just installed. The new and existing GEs must be connected together with a GEC that's the same size as the existing GEC from the utility. (See Figure 17-2a.)
- **A GEC that's run from the inverter to the existing GE:** As long as the GE from the inverter is continuous and sized based on the *NEC*® requirements (I cover the sizing in the next section), most jurisdictions accept this grounding method (shown in Figure 17-2b). You can even use this GEC to bond the equipment installed on the AC side of the system as it runs through the equipment. You can't break this GEC, but you can crimp EGCs to the GEC inside the equipment to bond the equipment to the GEC. This method provides the low-resistance bond to the grounding system that the Code wants you to get.

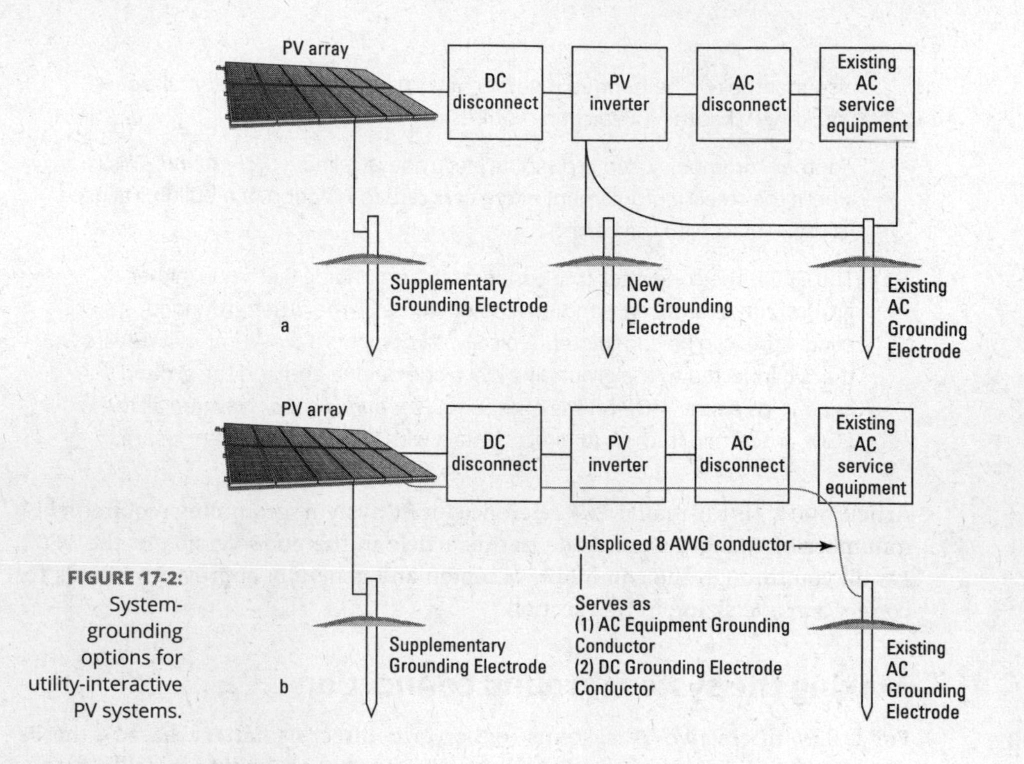


**FIGURE 17-2:** System-grounding options for utility-interactive PV systems.

For stand-alone, battery-based systems, you must install a brand new GE (because there won't be one from the utility). You then have to connect two GECs, one originating from the AC MDP and the other from the DC wiring enclosure (where the inverter connects to the battery bank). These GECs will be either 6 AWG or 4 AWG, depending on the GE.

## Sizing system grounding

As with the EGC that I describe earlier in this chapter, the GEC must be sized based on the requirements as listed in the *NEC*®. Article 690.47 references the grounding Code article to determine the correct size. In short, the GEC has to satisfy the requirements for the AC side (Article 250.66) and the DC side (Article 250.166). Article 250.66 includes a table to reference, and Article 250.166 outlines the different possibilities and the requirements for each. For utility-interactive systems, whether grid-direct or battery-based, a GEC that meets the AC portion of this requirement is already installed by the original electrical contractor who installed the electrical system; this GEC has already met the Code requirements. Stand-alone, battery-based systems, on the other hand, require a new GE and GEC be installed as described in the preceding section. In all three system types, the GE is installed as close to the MDP as possible. When sizing a GEC to meet the DC portion of the requirement, the size should be no smaller than an 8 AWG conductor.

REMEMBER

Of course, the final size of the GEC depends on the location of the GEC and the existing GE. For example, if the GEC is going to be exposed to physical damage (meaning it'll be run outside without conduit), it needs to be at least 6 AWG. The size of the conductor provides it with some physical protection. If you can run the GEC from the inverter to the GE without exposing it to damage, an 8 AWG GEC for the DC side may suffice. These sizes often work well for residential and small commercial systems.

# Connecting to the Utility

For utility-interactive systems (both grid-direct and battery-based), you need to make an interconnection to the utility grid. This interconnection point can be accomplished in a variety of ways, but regardless of your point of interconnection (or the type of utility-interactive system, for that matter), you must meet a number of *NEC®* requirements. The next sections highlight the requirements you're most likely to encounter on a regular basis and explain the different types of interconnections you can make.

## Determining the utility's requirements

Anytime you're installing a utility-interactive system, the utility will want to know that you're connecting to its grid with a PV system as opposed to having a PV system connected to its grid without its knowledge (known as *guerrilla solar*). By now, most utilities in the United States are familiar with the process and have some standard paperwork available, typically a net-metering agreement. Here are the standard contents your average net-metering agreement should contain (the agreement in a particular jurisdiction may include additional points):

- The system owner agrees to some fundamental equipment requirements to ensure the safety of the utility's workers.
- The utility outlines how it'll calculate the energy sent to the utility and the amount sent to the home or business (this is generally referred to as the *true-up period*).
- Special utility-mandated safety requirements are noted. (The number-one piece of safety equipment required is a visible and lockable disconnect, which may be the same disconnect that you install to comply with the *NEC®*; see the earlier "Locations for disconnecting means" section for more on this).

TIP

If you're installing PV systems in a specific utility's area, I suggest requesting a copy of the net-metering agreement before you ever talk to customers. That way you know what the utility requires before you even set foot on someone's roof and can be thinking through all those requirements during the site survey (which I describe in Chapter 5). Having a list of the utility's requirements ahead of time can help you avoid any surprises (like unplanned disconnects) at the end of the job when they'd be very expensive for you to fix. (That's right. You're on the hook for mistakes like this because it's your responsibility to know the requirements.)

REMEMBER

Many utilities want to see a disconnect they can walk up to 24/7 and disconnect the PV system from the utility grid. Figure 17-3 illustrates the common location for a utility-required disconnect (which is before the inverter interconnects to the utility and next to the existing utility meter). The idea is that if a utility worker is working on the lines, the disconnect gives him the ability to disconnect the PV array from the grid and eliminate the possibility of being shocked (or killed) from a PV system. The reality of the situation is that any properly installed inverter will shut down as soon as the utility goes away (see the anti-islanding discussion in Chapter 9). Some of the largest utilities have eliminated or relaxed their requirements, indicating a higher level of comfort with PV technology and the safety equipment already installed. If the utility you're working with does require the disconnects, you may be able to start a dialogue to eliminate this requirement by doing your research on these other utilities.

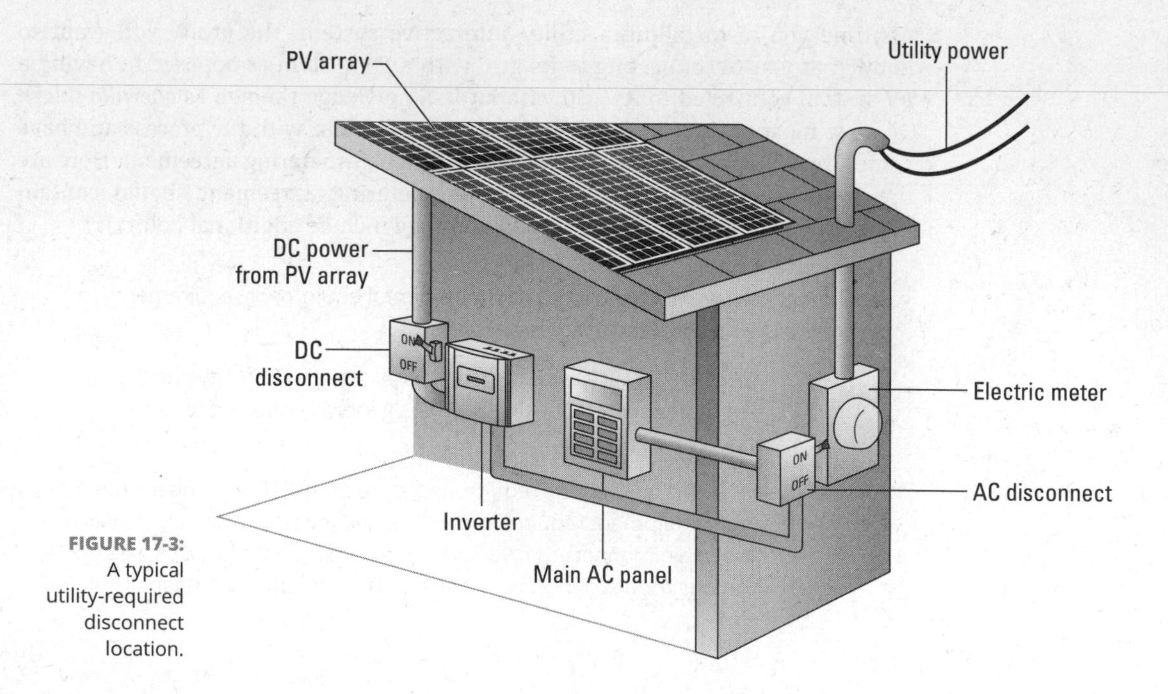


**FIGURE 17-3:** A typical utility-required disconnect location.

## Making a load side or line side connection

To meet the *NEC*® requirements for connecting to the utility, you have to refer to *NEC*® Article 690.64 (if you're working with the 2011 version of the *NEC*®, it's Article 705.12). The following sections set the requirements PV system installers are held to, depending on the interconnection point: load side or line side.

- In residential applications, you typically achieve an interconnection to the utility by placing a circuit breaker inside the MDP and connecting the inverter output circuit to that breaker. It's the easiest method of connecting the inverter because adding a breaker to an MDP is a straightforward process. Connecting to the utility this way is known as a *load side connection* because you're connecting to the load side of the main OCPD in the MDP.
- If adding a breaker to the MDP isn't an option (because the MDP is full or the PV array output is too large), you can make a *line side connection,* which is when you connect the inverter's output circuit to the utility before it reaches the main OCPD in the MDP. This approach is common in commercial applications.

If you want (or need) to connect to the line side of the main OCPD, you should discuss the options with the local electrical inspector. This method requires you to make connections that the inspector may not be familiar with, and walking through the options beforehand instead of after you've installed the system can save you a lot of trouble.

***Note:*** If you need to interconnect with the utility at this point, you should be aware of the following main Code requirements:

- You need to have a disconnect and OCPD immediately adjacent to the point of interconnection. This disconnect should be *grouped* (located) with the other disconnects for the facility.
- The disconnect should be rated for at least 60 A, and you must use a fused disconnect to meet the OCPD requirement. However, the fuses can be less than the 60 A specified, based on the inverter output current.
- You can't have more than six disconnects to shut off all the power sources. This requirement means that if you're in a commercial facility that has six disconnects already installed, you can't add a seventh one for the PV system; instead, you must find a way to disconnect all the power sources with fewer than six switches.

For the more common systems that are installed on the load side of the main OCPD, the *NEC*® has seven different requirements you need to meet. Addressing and meeting all seven requirements may seem daunting at first, but in reality, all seven requirements are achievable for the majority of your installations. I note the requirements in the following sections.

## Placing each inverter on its own circuit breaker or disconnect

Each source interconnection must be made on a dedicated circuit breaker (or fusible disconnect). This requirement is in place to ensure that all the inverters are connected to their own OCPD. If they are, then they're protected adequately and can be properly isolated.

## Using a circuit breaker with the correct rating

The busbars inside the MDP (or subpanel) you connect the circuit breakers to have an ampacity rating just like all the conductors in your system. This provision in the Code allows the sum of the circuit breakers supplying power to the busbar (in other words, the main OCPD from the utility and the PV breaker) to exceed the busbar rating by 120 percent.

This means that, for example, you can have an MDP in your home that has a busbar rating of 200 A (you can find this information on the label inside the main cover) and a main OCPD (main breaker) rating of 200 A (both of these are very common ratings for the equipment installed in many homes) and still install a PV system on that panel because of the 120-percent rule.

To determine the maximum PV breaker you can put on this MDP, find the busbar rating, multiply it by 120 percent, and subtract out the existing main OCPD size. For the 200 A busbar and 200 A main OCPD example:

$$200\text{ A} \times 120\% = 200\text{ A} \times 1.2 = 240\text{ A}$$

$$240\text{ A} - 200\text{ A} = 40\text{ A (the maximum inverter breaker size)}$$

You can only operate that PV breaker at 80 percent of its rating continuously, so the maximum inverter output current is limited to 32 A (40 A × 80%). You determine the maximum inverter output current value in Chapter 13.

## Working with ground fault protection integrated into an overcurrent protection device

If the main OCPD has a ground fault protection (GFP) device built into it (which is typical for larger commercial systems but unheard of for residential systems and rare for small commercial systems), the main OCPD must be listed for current flowing backward through it. This provision is why line side connections are so common for larger commercial systems. Finding a main OCPD with integrated GFP listed for current flowing the "wrong" way is a difficult task to say the least.

## Making your mark on panels with multiple power sources

REMEMBER

You have to properly mark any electrical panel (whether MDP or subpanel) that has both utility and PV sources to alert any emergency personnel that this electrical system has multiple power sources. Also, if you connect the inverter to a subpanel, you must place a label on the MDP as well, indicating the presence of the PV array and where to find the disconnecting means.

## Ensuring that circuit breakers can be backfed

The circuit breakers you use must be suitable for *backfeeding* (having current go to the utility rather than a load). Some circuit breakers only work with current flowing from the utility to the load. If you put current the other way (from an inverter to the utility), the circuit breakers may not be able to properly protect the conductors and equipment.

TIP

This backfeeding requirement really only affects you if the circuit breakers you want to use are labeled with *line* and *load* directly on the breaker. If you don't see these designations, you're okay.

## Understanding that fastening kits aren't required

Some inspectors have required PV installers to install fastening or "hold-down" kits for the PV circuit breakers. The thought process behind this mandate was that if the breaker comes off the MDP busbar, it'll still be "hot" and have the potential to cause shock and fire hazards. Of course, this scenario can't actually happen due to the anti-islanding requirements of today's inverters. A provision can now be found in the Code that specifically states that these fastening kits aren't required.

## Locating the protection

The final provision/requirement dictates where the PV circuit breaker is installed. If you exceed the busbar rating with the utility and PV breakers (which you will 95 percent of the time or more), the PV breaker must be installed on the opposite end of the busbar from the utility breaker. Why? Because if the currents are flowing in opposite directions, no place on that busbar will ever be exposed to more current than its rating allows.

REMEMBER

You also have to label the PV breaker with a warning that no one can ever relocate the breaker in the panel. This label's purpose is to ensure someone doesn't come in behind you and move the breaker without realizing the possibilities for overloading the busbar.

IN THIS CHAPTER

» Performing a pre-commissioning inspection

» Following a commissioning sequence

» Having an installed system inspected

» Maintaining a PV system

# Chapter 18

# Commissioning, Inspecting, and Maintaining a PV System

After the PV array and all the associated equipment have been installed, the entire system is ready to be checked out and turned on; this start-up process is officially known as *commissioning the system.* But before you get too excited and start flipping switches, you need to make sure that everything has been installed correctly and that no obvious issues are present that can be resolved sooner rather than later. In this chapter, I show you how to conduct these preliminary checks and introduce you to the main points for commissioning a PV system. When in doubt, consult the equipment manufacturers to make sure you aren't going to cause damage to the equipment (and yourself!).

After commissioning, you need to have the local building department send an inspector to look over your system; this inspection officially completes the system installation process. I delve into inspection issues in this chapter.

PV systems are low-maintenance systems, but they're not no-maintenance systems. Your client must interact with the system, particularly if it's a battery-based one. However, the responsibility of system maintenance isn't your client's alone. If you can, I suggest including some ongoing maintenance (typically an annual check) as part of your installation contract. Performing an annual check allows you to catch potential problems early and gives the client the benefit of having you review her system each year. I cover maintenance specifics at the end of this chapter.

REMEMBER

Always keep the main goals of system commissioning, inspection, and maintenance in mind: verifying a quality installation, establishing a safe system, documenting the array as it has been installed, and ensuring proper long-term performance. When you do find problems, don't be in a hurry to fix them. Instead, take small steps to address the problem(s) so you don't create more issues than you originally had.

## Making a List and Checking It Twice: Preparing for Commissioning

The first step in *commissioning* (or starting up) a PV system is to spend some time upfront making sure the system is installed correctly and safely. In other words, you need to inspect your own system (note that this self-inspection is separate from the inspection performed by one of the local building department's representatives; see the later "Arming Yourself for Inspection Issues" section for more on that). In the following sections, I explain how to inspect the electrical and mechanical elements of a PV system before beginning the official commissioning process.

TIP

When possible, bring in someone who hasn't been involved with the actual installation to check your work. Someone who has a fresh set of eyes may be able to spot a problem that you overlooked. Also, plan to have at least one person from your installation crew on hand so she can assist with any changes that need to happen immediately.

### Mechanical elements? Check!

You need to ensure that the mechanical portion of the system is what you expect and that it can keep the array in place. The next sections help you know which questions to ask as you go about examining the entire mounting system. (Chapter 16 has details on assembling the mechanical elements.)

### Footings

Always check out the *footings* (the method used to keep the array in place). Ideally, you should conduct this portion of your pre-commissioning inspection before the entire array is installed. If a problem is discovered too late, fixing it can be very difficult and expensive. Save yourself the hassle and your client some cash by asking these questions:

- For rooftop arrays, are the footings installed in the proper locations and with the proper spacing?
- Was the correct hardware used in the installation of the footings?
- For ground-mounted arrays and top-of-pole arrays, were the holes dug to the proper dimensions? Was the concrete poured high enough? Is the spacing between poles correct?

### Racking

You need to check the racking system (which is composed of rails and clamps) as part of your inspection of the system's mechanical elements, but that can be a challenge, particularly if you're dealing with a rooftop array. Then again, accessing ground-mounted and top-of-pole arrays can be just as tricky after all the ladders and scaffolding have been removed. Use caution and remember to ask the following questions:

- Are the modules properly connected to the racking system with the correct hardware?
- Are the modules grounded to the racking system the way they should be (with grounding clips or an equipment-grounding conductor attached to each module, as described in Chapter 17)?
- Is the racking system properly attached to the footing system?

## Electrical elements? Check!

As you examine your PV system, you have to look at all the electrical components to verify that everything was installed correctly. You don't have to stick your hands in any boxes and confirm proper voltages, but you do have to look everything over to see whether it looks right. In the sections that follow, I provide checklists of all the electrical elements you need to inspect before you begin the commissioning process.

REMEMBER

If you catch a mistake in your work, stop and think it through. And before you go to make changes, verify that the system is in a safe state for you to start playing with your tools; see Chapter 15 for details on how to stay safe whenever you work on a PV system.

## PV modules

It may sound silly, but your electrical inspection should start with the PV modules. Take time to make sure that the array installed matches your one-line drawings (see Chapter 14 for more on these) by asking these questions:

- Is the correct number of modules installed?
- Is the manufacturer and model number the same as it is on the plans drawn up for the permitting process (see Chapter 14)?
- Are the modules wired in the correct string configuration? This configuration can be series, parallel, or series-parallel (see Chapter 3).

## Batteries

If batteries are installed, they need special attention. Take time to evaluate the installation and future maintenance of the batteries by asking the following:

- Are the batteries properly vented to the outside? All batteries release hydrogen, although sealed batteries release very little of it. (Check out Chapter 7 for an introduction to the different types of batteries.)
- Are the batteries installed in a proper enclosure? The batteries should be protected from damage and anyone who may hurt themselves in their presence.
- Are the batteries installed and wired properly? The system will require a very specific battery voltage, and if the batteries aren't wired right, the system won't work.

## Charge controllers

Whenever you're dealing with a battery-based system — whether utility-interactive or stand-alone — you need a charge controller to make sure that the batteries are properly charged from the PV array. Before applying power to the charge controller, be sure to ask these questions:

- Is the charge controller installed in a proper location? It's typically mounted very near the inverter and battery bank.
- Are disconnects and overcurrent protection devices present on both the input and output sides of the controller?
- Are the correct wire gauges attached to the controller?

## Inverters

**Because the inverter is such a major component of any PV system, if any inverter issues exist, they can be total showstoppers. Ask these questions when inspecting an inverter prior to commissioning:**

- Does the inverter's output voltage match the utility grid voltage? Some inverters only connect to a single utility voltage, whereas others allow you to field-select the voltage. Either way, the inverter and utility voltages need to be verified. (See Chapter 11 for more information.)
- Is the inverter installed according to manufacturer recommendations? Inverters are often installed without the necessary clearances for proper cooling and access. Make sure yours isn't one of them.
- Is the inverter mounted in the best location? Is it installed in the direct sun? Is there airflow for the inverter to dissipate heat? All inverters are negatively affected by heat, so try to keep them as cool as possible.
- Does the inverter's voltage window match up with the installed array correctly? This syncing of voltage windows really should've been addressed during the design process, but it's worth examining in the field anyway just to be safe. (Flip to Chapter 11 for details.)

## Conductors

**When checking the *conductors* (wires), make sure you have visual access to a few points and ask the following questions:**

- Is the correct conductor type used for the installed environment? For example, is the conductor used along the array appropriately rated for exposure to high temperatures and sunlight? Does the conductor used in the conduit have the appropriate ratings for the locations in which it's installed? (See Chapter 13 for more on sizing conductors.)
- Is the wire gauge specified in the plans what's actually installed? (A conductor's *gauge* is its physical size.)

» Are there signs of potential wire damage from installation? If any conductors look like they may have been pinched or appear damaged from being pulled through the conduit, note that so you can investigate.

» Is the color-coding correct? The grounded current-carrying conductor should be white or gray, and the ungrounded current-carrying conductor should have insulation that's any color other than white, gray, or green (but preferably it's red). (Chapter 17 has information on color-coding.)

» Is the PV source circuit wiring properly supported? These conductors are generally limited to the wiring along the backside of the PV array (see Chapter 17 for details), but it may include other areas.

## Conduit

Your PV systems will almost always contain some amount of conduit, so be sure to ask these questions when conducting your pre-commissioning checks:

» Is the correct conduit used? Some locations may require metallic conduit, whereas others may allow PVC. (Chapter 10 has info on conduit types.)

» Is the conduit supported correctly? To determine whether or not it is, you need to know when the conduit needs support. If you don't know this, you at least need to be able to look up the specific *National Electrical Code®* (*NEC®*) article that deals with the conduit installed.

» Have expansion fittings been installed? Conduit, especially PVC, expands and contracts with changes in temperature. Article 352.44 of the *NEC®* requires you to allow for expansion and contraction by installing specialized fittings. Otherwise, the conduit can be compromised, and the wiring may not be fully protected.

## Disconnects

Disconnects go on both the AC and DC sides of PV systems. And for many grid-direct systems, the inverter also includes at least the PV disconnect. When you're looking your system over in preparation for the system commissioning, ask yourself the following:

» Are the disconnects installed per Article 690.14 of the *NEC®*? Can you disconnect the PV array and AC power out of the inverter, and are these disconnects grouped together?

» If the utility requires a visible, lockable disconnect, is it installed correctly according to the utility's requirements?

REMEMBER

- Are the correct conductors being disconnected in each disconnect? The *NEC®* says that you "shall not disconnect the grounded current-carrying conductor." See Article 690.13 in the *NEC®* for more info.
- Are the installed disconnects properly rated for their environment and for the voltage and current values they'll be carrying?
- Are the proper labels installed on the disconnects? If you have any question about the purpose of a disconnect, you can't expect someone who's not at all familiar with the system to know what she's dealing with. (I describe proper disconnect labels in Chapter 10.)

### Overcurrent protection devices

Often, overcurrent protection devices (OCPDs) are installed on both the DC and AC sides of the inverter. Prior to flipping the on switch, you not only need to check that the correct OCPDs are installed but also that they can properly protect the system. Take a few moments to ask the following:

- Are the current ratings correct for the circuit? An incorrectly sized OCPD (either too big or too small) will cause problems. (Flip to Chapter 13 for information on sizing OCPDs properly.)
- Do the OCPDs have the correct voltage ratings? This is crucial, especially on the PV source circuits. The OCPDs must have the proper DC ratings, or else they won't be able to properly protect the system when they're really needed.
- Are the OCPDs in good condition? A fuse or circuit breaker that's faulty on day one can lead to a number of hours in troubleshooting. To verify whether an OCPD is able to pass current when the system is turned on, perform the continuity test I describe in Chapter 3.

## Start 'Er Up: The Commissioning Process

After you examine the system to verify that all the mechanical and electrical elements have been installed correctly and safely (see the previous sections), you can get to the real fun of *commissioning* (turning on) the system. Take this process one step at a time so you don't skip any crucial portions. The following sections cover the major aspects of the commissioning process.

REMEMBER

After you commission the system and confirm that all is well with it, you must shut it down until after the system has passed the local building department's inspection (I fill you in on this later in the chapter).

## Putting safety first

The first step in commissioning a PV system is making sure you and the entire system are safe. How do you do that? In Chapter 15, I go through the order of making the PV connections to guarantee that you stay safe during the installation process. Just follow these simple rules once again during the commissioning process to ensure the safety of yourself and the members of your installation crew. The number-one strategy? Always work on circuits that aren't live and use proper wiring sequences to make sure none of the circuits are live. You also need to confirm that all the disconnects are in the off position and locked out (I show you how to do this in Chapter 15).

For grid-direct systems, don't connect the PV source circuits to the combiner box or junction box *or* make the AC interconnection to the utility grid until the system passes the commissioning list I present later in this chapter.

The same PV and utility (when available) interconnection rules apply for battery-based systems as they do for grid-direct systems. After all, battery banks are just another source of power that you need to wire in the correct sequence in order to avoid possible shock hazards. The connections between the battery bank to the inverter(s) should be made at the inverter and inverter disconnect first before being made on the battery bank. This way you never have live conductors in your hands waiting to cause trouble.

Regardless of the type of system you're dealing with, you should always use a disconnect to make the final electrical connection. The first time you connect power to an inverter and the capacitors accept a charge, there's often a spark. If you use a conductor to make this final connection to the power source, you may put yourself at risk for a serious shock. If, however, you use a disconnect to make that final connection, the spark will be contained within the disconnect, and you'll never be exposed to the shock hazard (or even see it).

Here's one last thought on safety: Be sure to keep ladder safety and precautions for fall hazards in mind during the commissioning process. After all, you're probably going to take multiple trips up and down a ladder and spend some time up on a roof. Check out Chapter 15 for a thorough review of these safety considerations.

## Gathering the gear you need

As with any job, you need to make sure you have the correct tools for the work. Fortunately, the tools used to commission a system are similar to the ones you use to install it. I suggest you gather the following gear:

- **A commissioning form, pen, and clipboard:** I provide a sample commissioning form for your viewing pleasure in the next section.

- **A 10-in-1 screwdriver:** This handy tool should have everything you need to start up the system.
- **A digital multimeter (DMM):** Ideally, your DMM has a clamp DC current meter (see Chapter 3).
- **High-voltage safety gloves and arc-flash safety glasses:** Use this safety equipment whenever you're measuring voltage and current values.
- **Infrared thermometer:** You use this device to measure the temperature of the modules when they're operating to verify their performance.
- **Irradiance sensor:** You can buy an irradiance sensor for about $150. That may seem pricey, but this tool is useful in measuring the irradiance the array is seeing. You can then use this irradiance value to verify the array's power output.
- **Proper fall protection for any fall hazards you may be exposed to:** This is the same fall protection gear you use when installing a system — a body harness and safety ropes.

REMEMBER

In addition to collecting and carrying this gear with you, use it! The number of people I see with safety gear sitting next to them always amazes me.

## Commissioning different types of systems

After you visually check and confirm the installed equipment, you're ready to make the final electrical connections and check the system's operation. To do so, you must make electrical measurements before and after the disconnects are turned on. As you find out in the sections that follow, the steps differ for grid-direct systems and battery-based systems.

REMEMBER

Always try to commission a system on a nice, sunny day when you have plenty of time to run the tests. By turning the system on in good weather, you can test the system when it's operating close to its maximum values. I also recommend that you take your time. Rushing this process to save 30 minutes isn't going to do anybody any good in the long run.

TIP

Be sure to record all the data you collect so you have an accurate record of the system's status on day one. You can create a checklist similar to Figure 18-1 that you can use on all of your installations. This form gives you places to record important system information, such as the array specification, voltage, and inverter power output values. By using the same form each time, you ensure that the proper checks are made (so you can sign off on the system) and documented for future reference (so you have an official record of the system's status). This information will serve as your base of reference if you're ever called in to troubleshoot any issues in the future.

**General Information**

Site address ______________________

Your name ______________________

Date ______________________

Time ______________________

Currently producing (Watts) ______________ (total from all inverters)

System peak Watts ______________ (STC rating)

Inverter manufacturer ______________________

Inverter model number ______________________

Inverter serial number ______________________

Currently producing (Watts) ______________________

Utility meter number ______________________

Module manufacturer ______________________

Module model number ______________________

Actual grid voltage (Measured at point of interconnection) L1-L2: ________ L1-N: ________ L2-N: ________

**Array Information**

| | Array A | Array B | Total |
|---|---|---|---|
| Array true bearing/azimuth (degrees) | | | — |
| Array tilt (degrees) | | | — |
| Inverter quantity | | | |
| Module quantity | | | |

**Component Numbers and Inverter Production**

Performance Verification

Back of module temp: ________

Solar irradiance: ________ (in module plane)

Measure temp & irradiance at the same time as reading inverter outputs

| | Location and quantity | Model number |
|---|---|---|
| AC disconnect | | |
| DC disconnect | | |

**Inverter 1 PV String Measurements**

| | Number of modules | $V_{oc}$ | $I_{mp}$ |
|---|---|---|---|
| String 1 | | | |
| String 2 | | | |
| String 3 | | | |
| Continue on additional sheet if necessary | | | |

**Inverter 2 PV String Measurements**

| | Number of modules | $V_{oc}$ | $I_{mp}$ |
|---|---|---|---|
| String 1 | | | |
| String 2 | | | |
| String 3 | | | |
| Continue on additional sheet if necessary | | | |

Notes:

**FIGURE 18-1:** A sample system-commissioning form.

## The process for a grid-direct system

REMEMBER

In this section, I explain the general process for commissioning a grid-direct PV system with the help of a member of the installation crew. However, the following steps should never override the commissioning process outlined by the manufacturer's manual that came with the inverter:

1. **Visually inspect the entire system (see the earlier "Making a List and Checking It Twice: Preparing for Commissioning" section).**
2. **Confirm that the ground fault protection fuse in the inverter is good and return it to its place before applying any power to the inverter.**

   A simple continuity check (like the one I describe in Chapter 3) will prove whether the fuse is good. If the fuse is blown before you ever turn on the inverter and you don't know it, you'll spend more than a few hours running in circles looking for a problem that may not exist.
3. **Lock out and tag all the disconnects for the AC and DC conductors in the PV system.**

   Refer to Chapter 15 for a description of this process.
4. **Go to the PV array and open the junction box or combiner box.**

   Put on your high-voltage gloves and safety glasses and then use your DMM the way I show you in Chapter 3 to verify that the voltage and current levels in each circuit equal zero. After you know that none of the strings have voltage or current present, open any fuse holders and make sure the fuses aren't in them.
5. **Proceed to the array and connect the home-run cables from the array to the junction box or combiner box.**
6. **Return to the junction box or combiner box, check the strings individually for voltage and polarity, and record the voltage values for every string.**

   REMEMBER

   I explain how to conduct a polarity check in Chapter 3. If after conducting this check you find that you have more than a couple of volts difference between strings, you need to investigate the series string connections to verify that you didn't make a mistake along the way.
7. **Insert the fuses (when present) and close the fuse holders.**

   In systems with only one or two strings, you may not have fuses installed (turn to Chapter 13 for more on the requirements for fuses).

   Because you've locked all the disconnects below the array, the system won't turn on and put you at risk. All the PV conductors after the junction box or combiner box will have voltage present, but until a load is introduced, there won't be any current flow.

8. **Back at the utility point of interconnection, have an installation team member make the connection to the utility.**

   The standard way of doing this task is to connect the inverter output circuit to the dedicated circuit breaker in the main distribution panel (MDP).

9. **Apply AC power by turning on the AC disconnecting means.**

   The AC disconnecting means is the breaker that the inverter output circuit was just connected to. If an external AC disconnect is used with the PV system, check the voltage levels in that disconnect and turn it on to send AC voltage to the inverter.

10. **At the DC disconnect, verify that the proper DC voltage and polarity is present at the disconnect terminals and then turn on the DC disconnect to connect the inverter.**

11. **Watch the inverter for any error codes.**

    The inverter should turn on and start producing power after the five-minute start-up sequence. If you happen to see an error code, refer to the inverter's manual (or call the manufacturer's technical support line) to determine the exact cause for it.

## The process for a battery-based system

The process for commissioning a battery-based system (either utility-interactive or stand-alone) is much the same as the process I outline for commissioning grid-direct systems in the preceding section.

After completing Step 7 in the preceding set of instructions, proceed with the following steps (unless of course the inverter's manual specifies a different commissioning sequence; always follow the manufacturer's instructions to ensure proper operation):

1. **Check the voltage and polarity from the battery bank to the battery disconnect(s).**

2. **If the utility is present, have an installation team member make the interconnection to the utility.**

   This part is the same as for grid-direct systems. A breaker in the MDP connects the inverter to the utility.

3. **If the voltage and polarity values are correct, apply DC power from the battery to the inverter by turning on the inverter's DC disconnect.**

4. **Connect the charge controller to the battery bank by turning on the charge controller's output disconnect.**

Refer to the charge controller's manual and make any adjustments to the controller that are necessary for proper operation.

5. **At the PV array disconnect(s), verify the voltage and polarity from the array to the disconnect(s).**
6. **Connect the PV array to the charge controller when you're satisfied with the voltage and polarity values.**
7. **Connect the AC power to the MDP (and subpanel in a utility-interactive system).**

   Each panel will be connected via a dedicated circuit breaker.
8. **Verify the voltage to and from the MDP (and subpanel) before flipping the disconnects.**

REMEMBER

After you commission a battery-based system, you must spend some time configuring various set-points on the inverter according to the manufacturer's instructions and the particular requirements for your client's system. The factory default set-points are a good place to begin, but ultimately you need to change at least a few set-points in order to optimize the system.

## Verifying that the system is working

After the system has been fully commissioned and is operational, you can take a few measurements and make a few calculations to determine whether the array is operating satisfactorily or whether there may be an issue. When taking these measurements, you should always have at least two people on-site helping you out. One person should be at the array, and the other person should be at the inverter location.

- The person at the array measures and records the temperature of the modules. To do this, you use an infrared thermometer to take a reading on the back side of the module. Try to get three or four readings so you can average the values. The person at the array is also responsible for measuring the irradiance that the array sees. This involves taking the irradiance sensor and (without shading the modules) pointing it in the same direction as the array. The display will read out the number of watts per square meter ($W/m^2$) that the array sees at that moment.
- The person at the inverter records the power output as displayed from the inverter and measures the voltage and current outputs of the inverter output circuit. This is most easily accomplished with two meters, one clamp meter that's constantly reading current and a second one that's constantly reading voltage.

Because the amount of current (and therefore power) is dependent on the irradiance value, you need to record the irradiance and power output values at the same time for the most accurate result. If you're working on a nice, clear day, the irradiance values will change very little, giving you more time. If the day is cloudy or partly sunny, you need to quickly record the values.

I've found that walkie-talkies work very well in this situation so that the person at the array and the person at the inverter can stay in contact and record their numbers simultaneously.

Next come the calculations. To figure out how the array is performing compared to how you expected it would, follow these steps:

1. **Adjust the array's maximum power voltage for the operating temperature.**

   I explain how to do this in Chapter 11.

2. **Adjust the array's maximum power current or the measured irradiance value.**

   Refer to Chapter 6 for the how-to.

3. **Multiply the temperature-adjusted voltage by the irradiance-adjusted current to find the corrected PV array power output.**

4. **Multiply the corrected PV array power output by the total system efficiency after taking losses such as the module nameplate power tolerance, wiring losses, and inverter efficiency into account.**

   Grid-direct systems tend to operate at 90-percent efficiency, and battery-based systems are closer to 85-percent efficiency.

5. **Compare the value you calculate in Step 4 with the value read off the inverter's display.**

   For another check, compare the inverter's display to the value you get by multiplying the voltage and current readings from the inverter output circuit.

All of these values should be reasonably close (within 5 to 10 percent). Of course, a number of variables can sway these numbers, so don't get too caught up in the discrepancies if the numbers don't match up perfectly.

If you calculate dramatically different numbers than what you see on the inverter or measure with your meters, investigate the problem. Maybe part of the array is shaded or a fuse has popped in the combiner box. Always take a slow, systematic approach when troubleshooting. Don't change too many things at once, or else you may not know what fixed the problem (or what made it worse). In this situation, I like to work from the inverter toward the array to try and find the problem. Doing so allows you to look at the whole system and then systematically break it down into the smaller components.

# Arming Yourself for Inspection Issues

After the system has been installed and commissioned, you need to call for an inspection (or inspections). The purpose of an inspection is so the local building department can verify that the components of the PV system are installed safely. In many jurisdictions, you have the option of being present during the inspection, as long as the building department's inspector has full access to look at the installed components. I recommend that you be present for your first installation or two; this way, you can answer questions and generally educate the inspector about the system. After a few good installations and inspections, you'll feel pretty comfortable with the process, and your presence at the inspection will become truly optional.

I must warn you. No two inspections work quite the same way. I've been a part of both inspection extremes, one where the inspector did little more than look up at the array and another where the inspector asked me to set up my ladder and spent 30 minutes on the roof with me going over every detail.

The sections that follow cover a few of the most common inspector concerns and comments about PV installations.

REMEMBER

One common complaint from PV installers is that inspectors have unreasonable requirements for the systems. During the inspection process, your previous interactions and questions with the inspector can greatly aid you both. When you think about it, inspectors are viewed as "experts" on a number of different types of electrical and mechanical systems; although they may not be true experts on PV systems, they typically have years of industry-related experience as well as knowledgeable questions and observations. As an installer, you can help guide and teach the inspectors you work with about PV systems. Just remember that people in general (and especially people in authority) don't like to be told how little they actually know. My suggestion is to work with the inspector and help her understand PV systems better in order to work out solutions before throwing up your hands in disgust and running to the inspector's supervisor with complaints.

TIP

Having the attitude that the inspector is against you only leads to trouble; stay positive and cooperative, and you'll have much better results.

## Not having "a neat and workmanlike manner"

One of my favorite *NEC*® requirements is the statement that "electrical equipment shall be installed in a neat and workmanlike manner." This statement appears in Article 110.12 is as a blanket statement for all electrical installations. Thanks to this statement, an inspector can look at your installation and fail you based on sloppy installation techniques.

TIP

Even though the Code doesn't define or give specific examples of "a neat and workmanlike manner," a note in Article 110.12 references a book you can buy to see industry-accepted practices for good workmanship: *NECA 1-2006 Standard Practice of Good Workmanship in Electrical Contracting (ANSI)*, published by the National Electrical Contractors Association.

TIP

Inspectors give more concrete examples and areas of noncompliance in their inspection reports, which they share with you before they leave the job site. An inspector's report includes all the Code references for the items you need to fix. Studying the reports from a few early installations will help you in future installations, especially when you're working in the same jurisdiction.

## Forgetting about aesthetics

Yes, beauty is in the eye of the beholder, and no, your PV array isn't being inspected for beauty. But you don't want to install an unsightly PV system. If you do, you may be hit with a charge of noncompliance with the "clean and workmanlike" clause I mention in the preceding section. Installing an eyesore of an array also discourages others from seriously considering solar power. I've seen PV arrays mounted on homes where even I, the guy who thinks PV is one of the prettiest sights in the world, have to shake my head in disgust. You want the neighbors to see the array and want one too, not wonder how much their property values went down because of that mess next door.

TIP

To keep your systems from falling victim to this aesthetics problem, always take a step back and picture the installed system. For roof-mounted systems, keeping the array in the same plane as the roof (if at all possible) is a great practice. If you need to lift the array off the roof, keep the lift minimal. A giant erector set on a client's roof is just asking for trouble.

## Failing to manage conductors on the array

Keeping all the conductors along the back of a PV array together and properly secured can be a more difficult job than it sounds like it should be. Very few modules allow you to connect conduit to them (for installations mounted close to a roof, this would be nearly impossible anyway), so you have to find a way to strap those conductors and protect them. Chapter 17 notes ways to support the backs of the module conductors so you can stay out of trouble.

## Neglecting to label the system

Article 690 in the *NEC®* has numerous requirements for labeling a PV system. Many of the necessary labels are standard ones, and some even come preinstalled on the equipment you're using for the system. The problem arises when you don't have all the proper labels installed. I introduce the idea of labels in Chapter 10, but in this section, I delve into the specifics.

REMEMBER

Some of the common labels that you must install include the following (refer to Article 690, Section VI of the *NEC®* for more):

- **A warning indicating that the inverter is supplying only 120 VAC (Article 690.10):** This label addresses a safety issue when a specific wiring method called multiwire branch circuits is present, which is why it's only applicable for some battery-based systems.
- **Descriptions of the purpose of all PV system disconnects (Article 690.14):** This label gives anyone who needs to disconnect the PV system the proper information regarding which disconnects to turn off.
- **A warning indicating a disconnect that may have all terminals energized when the switch is in the off position (Article 690.17):** This label applies to the PV source circuit disconnects for grid-direct systems and the inverter input circuit for battery-based systems.

REMEMBER

According to Article 690.53 of the *NEC®*, several specific details are required on PV disconnects:

- **The rated maximum power point current (at STC):** This is the number of strings in parallel multiplied by the modules' $I_{mp}$ value. For example, if you have 3 strings in parallel, each with an $I_{mp}$ of 5.4 A, the label needs to indicate a rated maximum power point current of 5.4 A × 3 = 16.2 A.
- **The rated maximum power point voltage (at STC):** This value is equal to the $V_{mp}$ for each module multiplied by the number of modules in the series string. So if you have 10 modules in each string of your array and each module has a rated $V_{mp}$ of 36.2 V, this label indicates a rated maximum power point voltage of 36.2 V × 10 = 362 V.

- **The maximum system voltage:** This is the temperature-adjusted open circuit voltage (see Chapter 11 for details) for the modules multiplied by the number of modules in a series string. For instance, if your array has 10 modules in series and you determine that the cold temperature-adjusted voltage is 48.7 V, this label needs to list the maximum system voltage as 48.7 V × 10 = 487 V.
- **The short circuit current for the array:** A note associated with this label requirement directs you to the calculation of maximum circuit current. What you're supposed to label isn't the listed short circuit current of the array but rather an adjusted $I_{sc}$ value. To get the right content for this label, multiply the number of strings in parallel by the modules' listed $I_{sc}$ and multiply that by 1.25. So if you have 3 strings in parallel with each string listed at 5.8 A, your label reads 5.8 A × 3 × 1.25 = 21.75 A. Chapter 13 has more about this calculation.
- **The maximum rated output current for the charge controller:** You can pull this number right off of the charge controller's spec sheet. Of course, this label is only necessary if you're installing a battery-based system that requires a charge controller.

REMEMBER

Here are some labeling guidelines specific to different types of PV systems:

- For utility-interactive systems (either grid-direct or battery-based):
  - Label the point of interconnection (typically a breaker in the MDP) as a power source with the nominal system voltage and rated current output (Article 690.54). This labeling requirement is repeated in Article 690.64 with an additional requirement for systems that exceed the panel's busbar ratings where the point of interconnection occurs. (I discuss this requirement in Chapter 17.)
  - If the utility service disconnect (the main breaker in the MDP) and the PV system disconnect aren't in the same location, you need to install a plaque indicating the location of both disconnects (Article 690.56).

    TIP

    In this situation, I find that a simple plan view of the building (with a north arrow to indicate direction) with the disconnects called out works best. This drawing gives anyone who needs quick access an easy-to-understand picture rather than a short novel on where to find the disconnects.
- For stand-alone, battery-based systems, you need to install a plaque that lets emergency personnel know that a stand-alone system is on-site and how to access the disconnecting means (Article 690.56). You also need to have labeled all the disconnects.

TIP

Have this plaque ready and a spot to install it in mind, but wait until your inspector shows up to install it so you two can agree on the location. This way the inspector gets to have a say; as a result, you'll surely pass this portion of the inspection.

TIP

Because most PV system labels aren't installed until the end of the job, they can easily be overlooked. I suggest planning and buying your labels early in the installation process so you don't forget about them. You may also want to discuss what the local inspector wants to see in terms of label materials and the method you use to attach the labels.

# Surveying System Maintenance

One of the best features of a PV system is the minimal amount of maintenance required. Because a PV system has no moving parts, you don't have to regularly maintain parts that will ultimately wear out and fail. Yes, the inverter will likely need to be replaced once during the life of the system (make sure your client is aware of this). And if batteries are installed, the battery bank will need to be replaced one or two times over the system's life as well. Beyond that, though, you (and your client) should expect the array to operate for years and years with little maintenance.

Note that I said *little* maintenance, not *no* maintenance. There are several maintenance tasks that you should consider performing on a regular basis for your clients; I outline them in the following sections.

REMEMBER

Most system owners are paying a lot of money to have PV systems installed, and they have the right to expect the systems to perform for them. On top of that, many rebate programs put the installer on the hook for warranty issues for up to five years. To me, all of this together means that spending 45 to 60 minutes each year on-site to verify that the system is performing as expected should be considered in the initial sale. *Note:* If you've installed batteries in a client's system, you need to plan on more than the annual maintenance check for optimal performance (see the later "Maintenance on a higher level: Taking care of battery banks" section for details).

TIP

As long as someone is monitoring the system output on a regular basis, any problems can be detected and dealt with rather quickly. It's when the system isn't being monitored that problems arise. The question then becomes: How long has there been a problem, and who's at fault for not catching it? As I explain in Chapter 9, one monitoring feature some inverters now offer is the capacity to send an alert via e-mail or text message when a problem occurs. Depending on the monitoring sophistication and options the system owner signs up for, this alert can be a notification of low power production, or it can be an actual error code indicating a problem with the array. Very often, these more advanced features can notify multiple people (including you) when an error occurs. If you can call your client to notify her that you're aware of a system issue before she is, you'll look like a hero.

## Mechanical maintenance

The systems used to mount the PV array should be maintained on an annual basis; you should check the module-to-rail connections and rail-to-footing connections (when they can be reached) using torque wrenches and nut drivers. When installed correctly, the racking system should be able to hold the array in place for many years. Of all the PV systems I've inspected annually, only a handful of mounting connections required retightening.

## Electrical maintenance

When performing maintenance on the electrical components of a PV system, you shouldn't have to spend a lot of time crawling along the roof and opening disconnects and inverters to check the wiring. Some visual checks and a few checks at terminals should be enough to make sure all the electrical components are functioning properly, as you find out in the next sections.

### Wiring connections

REMEMBER

Of all the electrical components that may need attention, you should tackle the connections made inside the junction box or combiner box first. This box is often located right at the array location and exposed to extreme changes in ambient temperature — changes that can cause the terminals holding the wires in place to loosen slightly over time. Use a torque wrench to make sure the terminals are tightened to the correct value (which is dictated by the manufacturers) and ensure that the connections aren't creating a *high-resistance connection* (which is when the terminal holding the conductor is loose). You can correct a high-resistance connection by tightening the terminal to the specified value with your torque wrench.

### Module wiring

The conductors used to connect the modules in series are extremely prone to damage and also deserve to be checked each year. As the array spends years in the sun, these wires can begin to rub against the racking system or the roof itself, creating weak points and possible safety hazards (such as shock and fire hazards).

By visually inspecting the module wiring annually, you can catch any potential problems early and provide corrective measures (such as verifying that the conductors are held securely to the modules and rail).

TIP

For many systems, getting back to the wiring after the array has been installed and functioning for some time can be a difficult task. Clearly, preventing the wires from ever getting to this point will pay off in the long run when you have to perform maintenance. Take the time needed to properly support and manage the conductors as described in Chapter 17. Note that you may need to remove some modules for better access. Although removing modules is a pain, it's worth the effort to eliminate shock and fire hazards in the long run.

### PV cleanliness

One area that will be variable is the amount of time the PV array can go before requiring a cleaning. Some system owners let the rain clean the array and hardly notice any reduction in energy output; other owners may need to wash the array multiple times a year. The amount of power an array can produce is dependent on the intensity of the sunlight striking the array, so if the array is dirty, the irradiance will be reduced.

REMEMBER

You can clean the array as part of your annual maintenance schedule for the client, but you should have your client keep an eye on the array and its power output levels and clean it whenever the power output values seem low. Reassure your client that cleaning the array doesn't really take a lot of effort. All she has to do is go to the array first thing in the morning (before it has a chance to heat up so there's no risk of cracking the glass with a sudden blast of cold water) and use a garden hose to spray the modules. If necessary, she can use a soft rag to clean off anything stuck to the glass (like bird droppings). Advise your client to never use an abrasive cleaner, though, because the glass shouldn't be scratched.

## Maintenance on a higher level: Taking care of battery banks

REMEMBER

Whenever batteries are involved in a PV system, the system owner has taken on an additional level of maintenance. As an installer, you can offer to come by more than once a year to help maintain the battery bank, but the system owner needs to be more involved with this type of system regardless of the number of times you come by. The exact schedule for maintaining the batteries depends on your client and how she uses the batteries.

» For sealed batteries used in a utility-interactive system, I recommend maintenance be conducted every six to nine months.

» For stand-alone systems where the batteries are cycled regularly, I suggest no more than three months between regular maintenance tasks.

Here, I describe the most important maintenance tasks for battery banks.

## Keeping the proper torque

*Torque* refers to the tightness of the conductors at the battery terminals. Because the battery terminals are typically made of lead, which is a very soft metal, the connections to these terminals can loosen over time. If they become too loose, a high-resistance connection results, and that can lead to multiple problems. Use a wrench to fix the torque when necessary.

REMEMBER

Check the connections on the battery terminals on a regular basis. A good time is the same day when you equalize the batteries (for flooded batteries) or when the batteries are exercised (in utility-interactive systems only). You'll be there doing some work anyway, so you may as well make a party of it. The person in charge of the battery maintenance can tighten the connections, and you can check the torque values as part of your annual maintenance visit.

## Exercising a battery bank in a utility-interactive system

In a utility-interactive, battery-based system, the battery bank sits fully charged the majority of its life. If the battery bank isn't exercised (in other words, if the bank doesn't run some loads), when the day that your client really needs power from the batteries rolls around, the bank won't perform. So, the battery bank needs to be forced to run some loads two to three times a year. Doing so helps keep the battery bank happy so it simultaneously lasts a little bit longer and gives you and your client the ability to judge the exact amount of capacity it has for the times when the grid fails.

*Note:* I like to see the client perform this task because it keeps her more involved in her system and aware of the system's limitations. Of course, you may need to remind her to do this on a regular basis.

REMEMBER

Batteries deteriorate over time, so checking the battery bank's ability to run loads a few times a year becomes more important as the battery bank ages and gets closer to retirement.

## Watering flooded batteries

For battery banks that use flooded batteries (typically found in stand-alone systems), adding water to the bank is simply a part of life. As a battery charges and discharges, hydrogen (a key component of water) is released, which makes the acid solution stronger, causing the fluid level inside the battery to drop. (I go through the specifics of battery charging and discharging in Chapter 7, so head there for the full scoop.)

REMEMBER

Your client needs to add water to the batteries to keep the acid solution at the right concentrations and to keep the batteries' lead plates submerged in fluid. The exact rate at which water must be added depends on a number of factors that are specific to your client's system. The system owner should check the water levels at least once monthly, especially as she gets used to living with a battery bank. After she's more comfortable with the battery bank and feels like she's in tune with the watering intervals, she can increase the time between checking the batteries.

TIP

I strongly encourage you to check and add water as part of your annual maintenance plan. However, don't let the client think this task is your job. She should be responsible for watering the batteries at other times of year. Be sure to clearly establish this expectation early on because if there's a misunderstanding, replacing a battery bank can be an expensive proposition.

## Equalizing flooded batteries

Another maintenance item that must be performed for flooded batteries is the *equalization charge*, which is a purposeful overcharge of the batteries to help break off any lead sulfate that has accumulated on the plates and stir up the electrolyte. In a stand-alone, battery-based system, you almost always need the assistance of an engine generator to create an equalization charge. Why? Because the PV array typically won't have enough current for long enough periods to effectively equalize the battery.

If your client is comfortable with the idea of performing the equalization charge and can be properly trained, then she should be the one to equalize the battery bank. If, however, she isn't willing to do this task or you don't feel confident that she can fulfill it safely, you need to go on-site to perform the equalization charge for her.

REMEMBER

As with battery watering (which I cover in the preceding section), the exact time frame between equalization charges varies among systems. For systems using flooded batteries (typically stand-alone systems), the batteries should be equalized every three to four months. You should consult with the battery manufacturer to determine the exact time frame for your client's situation, though, because no two systems are alike. The manufacturer will also recommend a voltage to equalize to.

WARNING

Always check with the battery manufacturer's recommendations before equalizing a battery. The various manufacturers have very specific requirements for equalization voltage values that must be strictly followed. One constant, though, is the need to house the battery bank in a well-ventilated room because batteries release a lot of gas during the equalization process. Another constant is the need to add water after performing an equalization charge — not before; if you have too much water in the batteries before equalizing them, you run the risk of having acid spill out over the top.

# 5 The Part of Tens

IN THIS PART . . .

Welcome to the Part of Tens! Here, you'll find some of the highlights for long-lived, successful PV systems.

Chapter 19 covers ten ways to avoid costly changes when it comes to keeping your installed PV systems up to code, and Chapter 20 goes through the items you can use to keep systems operating optimally so that they provide years of hassle-free electricity for your clients.

IN THIS CHAPTER

» Supplying the proper structural support and keeping things dry

» Making sure your conductors are rated properly, grounded, and more

» Giving your system some space and attaching the right labels to it

# Chapter 19

# Ten Ways to Avoid Common Code Mistakes

When installing PV systems, you have multiple options for both the mechanical and electrical portions of the installation. Yet regardless of your chosen method, you must keep in mind all the appropriate codes to ensure a safe and secure installation.

- On the structural side of installations (which I describe in Chapter 16), the International Building Code (IBC) is the governing document. It's the basis for local building codes, but it isn't the only structure-related set of regulations you must deal with; the local jurisdiction will likely have additional structural guidelines that you must follow.
- On the electrical side, the *National Electrical Code®* (*NEC®*) is the primary document that will be referenced by the local authority having jurisdiction (AHJ). I go through many (but not all) of the *NEC®* sections that apply to the electrical part of PV installations in Chapter 17. You, the installer, must be familiar with these Code sections and install your systems to meet *NEC®* requirements.

This chapter is a quick reference to the most commonly encountered problem areas as they relate to complying with the *NEC®* and the IBC. By avoiding these mistakes from the beginning, you can save yourself a lot of time (because you don't have to waste time fixing something you should've gotten right the first time) and squeeze in more clients as a result.

REMEMBER

Whenever you have a question as to what requirements your system will be held to, check in with the local building department for clarification.

## Providing Proper Working Clearance

When you install different pieces of electrical equipment, you need to make sure each component has the proper amount of working clearance around it. Article 110.26 of the *NEC*® defines the general requirements for proper working clearances. The basic idea is that someone should be able to walk up to a piece of electrical equipment and have proper access to service the item. In addition, many inverter manufacturers have a minimum clearance requirement to help keep cool air flowing around the inverter.

REMEMBER

The requirements for working clearance include a minimum width, a minimum height, and a minimum depth that must be kept clear for access. Generally, electrical components need to have at least 30 inches of width, no less than 6½ feet of height (with some exceptions for residential installations), and 3 feet of depth. (*Note:* Some of the requirements vary based on the operating voltages of the equipment, so be sure to check the *NEC*® to figure out exactly what's required for your installation.) You can imagine this space as a cube floating in front of your equipment that must be kept clear at all times.

TIP

If you install the equipment in a location that your client may use as a storage space, mark the area to eliminate the potential for clutter. It can be very tempting to use "unused" space for storage, but when the day comes that someone (usually you) needs to access any of the system components, you don't want to spend half the day clearing the space of the client's junk just to get to the equipment (especially in an emergency situation).

## Supplying the Right Structural Support

REMEMBER

No matter how you mount a PV array — on a roof, on the ground, or on a pole — all the components need correct support. To determine exactly what kind of support is necessary, always refer to both the manufacturer instructions that come with the PV modules and those that come with the racking system. These two documents note the basic requirements for installing an array correctly.

- Module manufacturers help you make sure the module frame is properly supported by the racking system. These requirements are independent of the racking type and ensure the modules' weight is distributed properly.

» Racking manufacturers point out the maximum distance you can have between supports for the racking system; they also tell you what length of rail can be run past the last support (the *cantilever*) and the maximum allowable length between supports along the rail length.

When you mount an array on a roof, you must review the roof truss or rafter system to make sure the roof is capable of handling the new loads imposed on it by the array. (Chapter 16 walks you through the major considerations for this analysis.) As for ground and top-of-pole mounts, you must examine the foundation that will hold the array down for the soil conditions you'll install it in.

## Keeping Water out of Buildings with Flashing

No surprise here, but people are generally unhappy with water on their ceilings. This fact is all the more reason to keep water out of the buildings you work on. Water leaks are a real concern for any installation located on the roof of a building, but they're also a consideration for installations where wiring and conduit are brought from the outside to the interior of the building. Always use appropriate sealants for the building material you're penetrating.

For roof-mounted PV systems, use flashing to create a watertight connection between the racking and the roof. *Flashing* is a term used to describe a mechanical device that's installed in conjunction with a roofing material to keep water out of roof penetrations (I describe flashing in more detail in Chapter 16). If you take a look at your roof, you'll see flashings on all the plumbing vents coming out of your house.

## Ensuring All Conductors Have the Necessary Ratings

PV systems offer the challenge of installing multiple *conductor* (wire) types in multiple environments. To meet this challenge, you must make sure all the conductors have the right ratings. For the PV source circuits, for example, you must use either underground service entrance (USE-2) wire or the relatively new PV cable. These conductors aren't appropriate to run inside buildings, though, so you have to transition to a building-safe conductor, such as moisture- and heat-resistant thermoplastic (THWN), when running wires inside your client's home or business.

TIP

For the building wire, I highly recommend you only buy the cable that's marked THWN-2. The -2 in the name indicates that this conductor is rated at 90 degrees Celsius in both wet and dry locations. When ordering the wire, make sure to specify THWN-2; this way you know that, regardless of the conductor's location, it'll have the proper temperature and moisture ratings (although conduit is still required). See Chapter 10 for details on several different types of conductors.

## Managing the Conductors on Modules

No matter where the modules are installed (on a rooftop or on the ground), conductors must be properly supported all the way to the junction box or combiner box. If they're not, they may become damaged and eventually fail.

When PV module manufacturers added quick-connect plugs on the backs of their modules, they eliminated a lot of the labor required to connect modules in series. With quick-connect plugs, all you have to do is plug one module into the next one, and the connection is made.

What the manufacturers also did was introduce a new challenge for PV system installers to overcome. Before, all the conductors were nicely contained inside conduit; now that conduit isn't necessary (or easily accommodated because there are no longer places to terminate conduit), installers have to secure the conductors between the modules. To do this, you must neatly tuck any excess wire along the module frames and racking structure. Fortunately, specially designed cable clips that attach to the module frame and hold onto one or two PV source circuit conductors are available. You can also use plastic or stainless steel wire ties to hold the conductors to the racking system if you desire.

WARNING

Using plastic ties often seems like an attractive and easy option, but remember that PV modules will be out in the environment for a very long time and that plastic degrades. By using wire cable clips in combination with wire ties, you can keep the conductors in place and out of harm's way.

See Chapter 17 for more information on managing conductors on modules.

## Selecting the Correct Conduit

During the electrical portion of the installation, you must install some of the conductors inside conduit. For residential applications, you usually run the conductors from a rooftop array down through the building and to the electrical equipment within properly rated conduit. Article 690.31(E) in the *NEC*® sets the requirements for such conduit runs. Be sure to verify the requirements based on the *NEC*® version the local jurisdiction uses. (The *NEC*® committee has made changes in this section in each of the last two Code cycles, with more changes to come in 2011.)

TIP

When you install conductors in other areas, such as in trenches for ground-mounted arrays or alongside the exterior of buildings for rooftop arrays, you need to make sure the conduit is rated for that location. PVC conduit can usually be installed in these locations, but be sure to verify the temperature limitations and sunlight-resistance qualities of the conduit for your installations.

Flip to Chapter 10 for a rundown of the different types of conduit and where they're appropriate.

# Locating the Disconnects

Disconnect locations are an important detail to keep in mind when designing and installing a PV system. The *NEC*® requires that the disconnecting means for the conductors be grouped together and readily accessible. To satisfy this requirement, you must install disconnects for the PV circuits, battery circuits, and AC circuits used in the system. These disconnects must be grouped together and within sight of the inverter. Turn to Chapter 17 to find out about the most commonly used approaches for meeting the disconnecting requirements.

# Grounding the Equipment

The subject of grounding is an ongoing source of frustration and friendly debate. But no matter your opinion, the *NEC*® is fairly clear on at least a few grounding requirements, one of which is that all the exposed, non-current-carrying, metal parts of a PV system must be connected to ground. This requirement is telling you that you have to install what's known as the equipment grounding conductor (EGC) for your racking system; PV modules; and all the disconnect boxes, charge controllers, and inverters.

REMEMBER

Be sure to reference the module manufacturer's and racking manufacturer's installation instructions for specific grounding requirements. (Also, head to Chapter 17 for full coverage of the different methods for installing EGC.) ***Note:*** Some jurisdictions may only allow certain methods, so be sure to check in with the local inspector if you have questions.

## Grounding the System

Although the *NEC®* is perfectly clear about equipment grounding requirements (see the preceding section), its system grounding requirements most definitely aren't. The *NEC®* makes it clear that your PV system requires a connection to a grounding electrode (like a ground rod driven into the ground outside the building), but the exact methods and requirements for making this connection are less than straightforward.

Of all the system grounding requirements, the one that has you drive a supplementary grounding electrode for the array is the one that's argued about the most. The more you read this requirement (found in Article 690.47 of the *NEC®*), the less sense it makes. In short, you have to install a new grounding electrode (like a ground rod but separate from the existing grounding electrode) and connect the PV array racking system directly to this new electrode. This can become a difficult task because you don't want to run the conductor from the array to the new grounding electrode through the house with the other conductors. For a ground- or pole-mounted array, adding this grounding electrode and connecting the array to it is no big deal because the array has clear access to the ground. As with a lot of the Code, exceptions to the rule exist, and if you can meet these exceptions, your life will be much easier. (I cover the most common methods for system grounding in Chapter 17.)

Because grounding is such a debated topic, I suggest that you talk through your strategies with the local inspector before you go down a path that he feels is inappropriate.

## Labeling the System Properly

Of all the Code requirements that are violated, system labeling seems to be the most common one. I suggest you spend some time going through Articles 690 and 705 of the *NEC®* to familiarize yourself with the label requirements as part of your design process. All too often, system labels are an afterthought, and they either don't get installed or get installed incorrectly.

In Chapter 18, I outline the major system-labeling requirements for you. Most of the labels are "stock," meaning you can buy a batch of labels that indicate the same information and can be used on any of your installations. On the other hand, some labels are custom for each job. One important example of such a label is the one that must go on the PV disconnecting means; this label requires you to calculate and list system-specific voltage and amperage levels.

IN THIS CHAPTER

» Taking the time for proper evaluation and orientation

» Using the right system components

» Keeping an eye on the array and its associated equipment

# Chapter 20

# Ten Ways to Maximize Energy Production for Your Clients

Regardless of their varying reasons for installing a PV system, there's one thing all people want to see from their investment: the maximum amount of energy production possible. Your job as the system designer and installer is to help deliver this maximum energy yield. Yes, each system will be slightly different, and each client will have varying needs, but at the core, the processes are similar for all PV systems. The tips presented in this chapter focus on a variety of stages, from system design and preplanning to installation and long-term maintenance, so you can give your clients the value they're looking for.

## Select the Right Site

It may seem like a silly (and obvious) thing to say, but selecting the proper site is one of the most important steps you can take to maximize a PV system's energy production. All too often I see situations where someone just picked a random location for an array and never thought to stop and look around a bit for a different location. After all, some locations work better than others due to a variety of shading, structural, mechanical, and electrical issues.

With the components that are available today and the wide variety of system configurations you have to work with, you shouldn't need to compromise on an array's site. Take the time to walk around the site and gauge which installation location will work best. At the very least, keep an open mind when it comes to array locations and flip to Chapter 5 for advice on how to properly select a site for a PV system.

## Orient the Array Correctly

An array's *orientation* is the direction in which it points. Conventional wisdom says to orient the array to true south (when you're in the Northern Hemisphere) and tilt it at an angle equal to your latitude. Although this strategy may yield a high amount of energy, it may not give you the maximum amount of energy the system can produce. A client's PV system should be in place and operating for more than 25 years, so small gains in energy production can have a large effect when measured over the system's lifetime. (To determine the best orientation for an array in your client's location, turn to Chapter 5.)

When orienting a PV array, you need to evaluate both the structural and aesthetic effects. A rooftop PV array that raises the modules off of the roof plane in an attempt to point them in some direction other than that of the existing roof may boost the energy output, but it may also cause serious damage to the client's roof due to the extra weight if the location is subject to high winds and/or heavy snow loads. And even though maximizing energy production is the main goal, you also want to help promote PV installations by installing good-looking arrays.

## Configure the Array Properly

For maximum energy production, the PV array should be properly matched to the charge controller and/or inverter, which means you need to consider both the voltage and the current relationships.

All the components connected to a PV array have a voltage window that they must operate in. As I explain in Chapters 11 and 12, you must account for changes in voltage based on temperature; otherwise, the charge controller(s) and inverter(s) may shut down. On top of that, PV modules degrade over time, so you need to properly account for that eventual energy loss due to aging; if you don't, you run the risk of the PV array falling outside of the voltage window due to natural degradation.

WARNING

The amount of power a PV array can generate in comparison to the amount of power the electronics connected to it can process is another important comparison to make. Each manufacturer of charge controllers and inverters has a recommended amount of power you should put through its machine. Exceeding the recommended values results in excessive heat production and shortened life spans for the components. Chapters 11 and 12 show you how to properly account for these limits.

# Work within the Limits of the Utility Voltage

Grid-direct systems are really at the mercy of the utility grid because their inverters have to follow the lead of the utility and go offline if the utility goes outside of the acceptable limits. Utility-interactive, battery-based systems have a little more flexibility, but the grid is still critical for them too. (See Chapter 2 for an introduction to both of these system types.)

When it comes to the utility grid and voltage, inverters are important for both PV system types thanks to the "follow the leader" approach the inverters have to take. The voltage window on the AC side of the inverters is much smaller than the DC side, and for safety reasons, the AC side isn't adjustable. For all utility-interactive inverters, the acceptable range of voltages is −12 percent to +10 percent of the nominal line voltage. (I note the voltage windows available to inverters in Chapter 11.)

On top of that, the voltages delivered by the utility aren't under your direct control and can fluctuate over time. When the grid's voltage is off of the nominal value and the inverter has a long wire run before connecting to the utility, that narrow voltage window can disappear. In Chapter 13, I show you how to look at these conditions and minimize their effects on your clients by reducing the voltage drop in the *conductors* (wires).

# Choose the Correct Inverter

REMEMBER

The inverter is the brains of the whole PV system. Given this fact, you need to make sure the inverter's power output level properly matches what the array can deliver. The proper inverter for a PV system that you're installing is the one that delivers the right type and amount of information for the end user.

At a minimum, the client should be able to see the power and energy values on the inverter. Some people want to see voltage and current levels as well; if your client is one of these individuals, you can install a more sophisticated monitoring system for her. If the system owner is left in the dark in terms of the system's operation, no one will be able to make sure all is well with the system (after all, the system owner is the person who has the ability to check the system on a daily basis). Chapter 9 introduces the basics of inverters; Chapters 11 and 12 present the inverter selection processes for different system configurations.

## Size Conductors Appropriately

Conductors (which I introduce in Chapter 10) are the arteries that carry the precious cargo of solar-generated amperes down to the brains and bodies of PV systems so your clients can benefit from the power and energy. Yes, that was a little dramatic, but it made you think, right?

REMEMBER

The conductors in PV systems have to effectively deliver power. If they don't, then you've installed a rather expensive power-producing system that can't deliver to its potential. A *National Electrical Code®* (*NEC®*) requirement is to size conductors so that they have the proper *ampacity* (ability to carry current) for the conditions in which they'll be used. As I explain in Chapter 13, this means you have to estimate temperatures and derate the conductors' ability to carry current due to the presence of multiple conductors in a single conduit.

*Voltage drop,* when voltage (and therefore power) is lost between the power source and the load, is the other part of the conductor-sizing equation. The *NEC®* makes no mention of voltage drop as it relates to PV systems, but the PV industry has come up with generally accepted standards of no more than 2 percent voltage drop on the DC side of the system and 1½ percent on the AC side. You can evaluate these percentages for your specific application, but 2 percent for DC and 1½ percent for AC are good starting points.

## Keep the Components Cool

All PV systems use some form of electronics to process the power they generate, and as with all electronics, they're much happier when they're kept cool and able to get rid of the heat they generate. In addition to their general happiness, you want to keep PV systems cool because as they get hotter, they have a reduced capacity to do the work they're designed for. For example: An inverter rated at

3,000 W at 25 degrees Celsius (77 degrees Fahrenheit) may only be able to have a maximum output of 2,500 W at 40 degrees Celsius (104 degrees Fahrenheit).

REMEMBER

Use proper mounting locations and techniques to keep inverters and charge controllers as cool as possible. In other words, don't install a PV system in a location where it'll be exposed to full sun.

TIP

For equipment with the proper ratings, installing components on the north side of a building is a good option. If the south or west sides of the building are the only options, put some sort of shade structure over the inverter to protect it in the afternoons when the temperatures are the hottest and the array is producing energy at maximum power.

# Advise Clients to Monitor Their System

One of the reasons PV arrays are so great is that they produce power all day long without visible signs of operation. Most other forms of electricity generation have some sign of operation (spinning blades, humming motors, and the like). Then again, this plus also makes it difficult to tell when a particular component of a PV system isn't working.

To ensure that any PV system you install is working at maximum power, advise your clients to monitor their systems regularly. A quick check of the daily energy production gives them the best monitoring option. Most inverters show that day's performance as well as cumulative performance over various time periods. If a client says that checking the daily production isn't an option, suggest she try a Web-based monitoring program offered by the inverter manufacturer. The information such programs provide give your clients exactly what they need: energy production values. (Refer to Chapter 9 for more on inverter monitoring and communications.)

# Clean the Array Periodically

The amount of current (and power) produced by a PV array is directly proportional to the intensity of the sunlight striking the array. As dirt and dust (as well as leaves and other debris) begin to accumulate on an array, the intensity of the sunlight striking that array is reduced. This layer of dust covering the modules is described as *soiling;* in extreme cases, soiling can reduce an array's output by nearly 20 percent. The level of soiling typically varies among installation sites and throughout the year. It therefore deserves consideration as part of an array's ongoing maintenance.

Soiling on a residential roof-mounted array in an urban environment will be less than a ground-mounted array on farmland. The exact schedule and level of cleaning necessary is therefore dependent on the array's location — as well as the value of the energy the array generates.

Generally, the system owner is responsible for cleaning the array periodically. Your main task for this portion of the maintenance process is simply to instruct your client that whenever she cleans the array, she should make sure to do so early in the morning. Although the modules use safety glass that probably won't break when hit with a sudden blast of cool water, it's not worth testing them. (Flip to Chapter 18 for more details on basic PV system maintenance.)

# Inspect the Array Annually

PV arrays often get the "out of sight, out of mind" treatment. After an array is up and running, people begin to ignore it — at least in the sense of going out or up to it and inspecting it. What they forget is that the PV array is constantly being exposed to extremes. It sees extreme temperature swings on a daily basis, its metal components are constantly expanding and contracting, its conductors are blown around by the wind, it gets weighted down with snow, and so on. For all of these reasons, I strongly encourage you to annually (at a minimum) inspect the PV arrays you install, along with their associated equipment.

As the PV system designer and installer, you should be responsible for performing the annual inspection; be sure to tell your clients the importance of such an inspection so they're aware of the additional annual cost of the system. During the inspection, you should look for conductors that have come loose and conduit that doesn't seem properly supported; you should also put a wrench on the array mounts to check for proper tightness in the mechanical connections. If you have the proper tools, go ahead and verify the array's power output too (I tell you how to do this in Chapter 18).

I suggest a nice, cool, summer or fall morning for your inspection. This way you can visually look over the array, get down on your hands and knees to check the underside, and bring a bucket of water to clean the modules. (You're already up there, so why not give your client a break from the cleaning routine? That's customer service right there.)

# Index

## Numerics



## A



## B

# C

# D

# E

# F



# G

# H



# I



# J



# K



# L

# M

# N



# O



# P

# Q



# R



# S

# T



# U

## V



## W



## Z

# About the Author

**Ryan Mayfield** discovered his passion for renewable energy shortly after beginning college at Humboldt State University in Arcata, California. While pursuing his degree in environmental resources engineering, Ryan was exposed to the exciting possibilities of solar power by his professors and the university's Campus Center for Appropriate Technology. This desire to learn more about and be involved in the solar industry only increased as Ryan continued his studies.

After getting his hands on an actual installation during a class with Solar Energy International, Ryan knew exactly where he wanted to spend his workdays. Early in his career, he worked as an installer, provided technical support for a renewable energy retailer, and moved up to Engineering Manager at a national wholesale renewable energy distributor. It was during his time at the distributor when Ryan's knowledge of PV systems grew exponentially. During this time, he read every *Home Power* magazine he could get his hands on from cover to cover.

Now, Ryan lives in Corvallis, Oregon, with his wife and two children. He is a Certified Affiliated Master Trainer for the PV Design and Installation courses he teaches at Lane Community College in Eugene, Oregon. He's also a North American Board of Certified Energy Practitioners (NABCEP)-Certified PV Installer. Ryan is one of a handful of people who hold both certifications. In the state of Oregon, Ryan holds a Limited Renewable Energy Technician's license and chairs the apprenticeship committee for that license, helping to bring up the next group of licensed PV installers in the Beaver State.

Ryan is currently the President of Renewable Energy Associates, a consulting firm that provides design, support, and educational services for architectural and engineering firms, contractors, and government agencies. The courses he teaches take him across the United States working with various individuals looking to increase their knowledge of PV systems.

In the spirit of going full circle, Ryan also serves as Photovoltaic Systems Technical Editor for *SolarPro* magazine and regularly authors feature articles in *SolarPro* and *Home Power* magazines. Occasionally, he's able to go back and teach with Solar Energy International, the same organization that got his hands on some equipment and provided his first true installation experience.

# Dedication

I would like to dedicate this book to my children, Aidan and Lauren, in hopes that I can make the world just a little bit better for them while giving them the opportunity to make a positive impact themselves; my dear wife, Amy, who's my biggest fan and supporter; my parents and sister for guiding me to become the person I am; and all of my friends and colleagues — your support to me personally and your contributions to the solar industry are invaluable, and I thank each and every one of you.

# Author's Acknowledgments

I feel very fortunate for the people who have come into my life. On a professional level, I would like to make special recognition of the following people: Bob Maynard, who opened up so many doors for me and allowed me to flourish; Joe Schwartz, whose work was highly influential to me long before we met and began collaborating; David Brearley, a true friend who always has a good answer for any question I have — solar or otherwise; Roger Ebbage, for giving me the opportunity to discover my passion for teaching; Paul Farley, for always keeping me on my toes and reminding me to have fun; and Eric Maciel, for giving me a chance to get started in the industry.

There are so many more people that have influenced and helped guide me over the years. Without their help, lessons, and guidance, I'd still be walking around bright-eyed and clueless: John Berdner, Bill Brooks, Justine Sanchez, Tobin Booth, Kirpal Khalsa, Christopher Dymond, Christopher Freitas, Jon Miller, Jsun Mills, John Wiles, Carol Weis, Wes Kennedy, Jason Sharpe, and Glenn Harris — just to name a few.

I have to give recognition to the solar pioneers that made all of this possible for so many, some of whom I've never met personally but want to thank anyway: Richard Perez, Bob-O Schultze, Windy Dankoff, Allan Sindelar, Michael Welch, and Johnny Weiss.

Thanks to all the wonderful people at Solar Energy International, *Home Power*, and *SolarPro*. The work you do to spread the word to so many never ceases to amaze me.

My editors at Wiley, Mike Baker, Erin Calligan Mooney, and Jen Tebbe, helped turn an idea into reality. Georgette Beatty earns a special thank you for her ability to keep me on task; without her input and perspective, I couldn't have written this book. Thanks to all the people behind the scenes at Wiley, especially the folks in the composition department. Thanks to Alex Jarvis for the insightful technical review and to Precision Graphics for taking ideas and turning them into works of art.

On a personal level, my extended family, friends, and heroes also deserve a special nod: Grandma Eva, your lessons continue to direct me to this day; Grandma Joan, for your never-ending support; Matt Minkoff, for helping me discover who I am; Beth Baugh, thanks for all you do for us; Jim Meyer and Luke Nersesian, I wish I could have just a moment to share this with you two; Jennifer Fisher for her always-unique perspective; John Panzak, world's greatest teacher; Jeff Lebowski, hey, that's your name, Dude; and the Oregon State Beavers and the Milwaukee Brewers, who make rooting so much more fun.

Given the long list of people I've mentioned as directly supporting and influencing me on this project and in life, I can only take partial credit for the pages in this book, yet I'll take all the blame. If there are any errors or omissions, that's due to an oversight on my part, and I regret any errors.

## Publisher's Acknowledgments

**Senior Project Editor:** Georgette Beatty

**Senior Acquisitions Editor:** Mike Baker

**Copy Editor:** Jennifer Tebbe

**Technical Editor:** Alexander D. Jarvis

**Production Editor:** Mohammed Zafar Ali

**Editorial Manager:** Michelle Hacker

**Cover Image:** © Watchara Kokram/iStockphoto

# Index

## Numerics



## A



## B

# C

# D

# F



# G

## H



## I



## J



## K



## L

# M

# N



# O



# P

# Q



# R



# S

# T



# U

# V



# W



# Z

# About the Author

**Ryan Mayfield** discovered his passion for renewable energy shortly after beginning college at Humboldt State University in Arcata, California. While pursuing his degree in environmental resources engineering, Ryan was exposed to the exciting possibilities of solar power by his professors and the university's Campus Center for Appropriate Technology. This desire to learn more about and be involved in the solar industry only increased as Ryan continued his studies.

After getting his hands on an actual installation during a class with Solar Energy International, Ryan knew exactly where he wanted to spend his workdays. Early in his career, he worked as an installer, provided technical support for a renewable energy retailer, and moved up to Engineering Manager at a national wholesale renewable energy distributor. It was during his time at the distributor when Ryan's knowledge of PV systems grew exponentially. During this time, he read every *Home Power* magazine he could get his hands on from cover to cover.

Now, Ryan lives in Corvallis, Oregon, with his wife and two children. He is a Certified Affiliated Master Trainer for the PV Design and Installation courses he teaches at Lane Community College in Eugene, Oregon. He's also a North American Board of Certified Energy Practitioners (NABCEP)-Certified PV Installer. Ryan is one of a handful of people who hold both certifications. In the state of Oregon, Ryan holds a Limited Renewable Energy Technician's license and chairs the apprenticeship committee for that license, helping to bring up the next group of licensed PV installers in the Beaver State.

Ryan is currently the President of Renewable Energy Associates, a consulting firm that provides design, support, and educational services for architectural and engineering firms, contractors, and government agencies. The courses he teaches take him across the United States working with various individuals looking to increase their knowledge of PV systems.

In the spirit of going full circle, Ryan also serves as Photovoltaic Systems Technical Editor for *SolarPro* magazine and regularly authors feature articles in *SolarPro* and *Home Power* magazines. Occasionally, he's able to go back and teach with Solar Energy International, the same organization that got his hands on some equipment and provided his first true installation experience.

# Dedication

I would like to dedicate this book to my children, Aidan and Lauren, in hopes that I can make the world just a little bit better for them while giving them the opportunity to make a positive impact themselves; my dear wife, Amy, who's my biggest fan and supporter; my parents and sister for guiding me to become the person I am; and all of my friends and colleagues — your support to me personally and your contributions to the solar industry are invaluable, and I thank each and every one of you.

# Author's Acknowledgments

I feel very fortunate for the people who have come into my life. On a professional level, I would like to make special recognition of the following people. Bob Maynard, who opened up so many doors for me and allowed me to flourish; Joe Schwartz, whose work was highly influential to me long before we met and began collaborating; David Brearley, a true friend who always has a good answer for any question I have — solar or otherwise; Roger Ebbage, for giving me the opportunity to discover my passion for teaching; Paul Farley, for always keeping me on my toes and reminding me to have fun; and Eric Maciel, for giving me a chance to get started in the industry.

There are so many more people that have influenced and helped guide me over the years. Without their help, lessons, and guidance, I'd still be walking around bright-eyed and clueless: John Berdner, Bill Brooks, Justine Sanchez, Tobin Booth, Kirpal Khalsa, Christopher Dymond, Christopher Freitas, Jon Miller, Jsun Mills, John Wiles, Carol Weis, Wes Kennedy, Jason Sharpe, and Glenn Harris — just to name a few.

I have to give recognition to the solar pioneers that made all of this possible for so many, some of whom I've never met personally but want to thank anyway: Richard Perez, Bob-O Schultze, Windy Dankoff, Allan Sindelar, Michael Welch, and Johnny Weiss.

Thanks to all the wonderful people at Solar Energy International, *Home Power,* and *SolarPro.* The work you do to spread the word to so many never ceases to amaze me.

My editors at Wiley, Mike Baker, Erin Calligan Mooney, and Jen Tebbe, helped turn an idea into reality. Georgette Beatty earns a special thank you for her ability to keep me on task; without her input and perspective, I couldn't have written this book. Thanks to all the people behind the scenes at Wiley, especially the folks in the composition department. Thanks to Alex Jarvis for the insightful technical review and to Precision Graphics for taking ideas and turning them into works of art.

On a personal level, my extended family, friends, and heroes also deserve a special nod: Grandma Eva, your lessons continue to direct me to this day; Grandma Joan, for your never-ending support; Matt Minkoff, for helping me discover who I am; Beth Baugh, thanks for all you do for us; Jim Meyer and Luke Nersesian, I wish I could have just a moment to share this with you two; Jennifer Fisher for her always-unique perspective; John Panzak, world's greatest teacher; Jeff Lebowski, hey, that's your name, Dude; and the Oregon State Beavers and the Milwaukee Brewers, who make rooting so much more fun.

Given the long list of people I've mentioned as directly supporting and influencing me on this project and in life, I can only take partial credit for the pages in this book, yet I'll take all the blame. If there are any errors or omissions, that's due to an oversight on my part, and I regret any errors.

## Publisher's Acknowledgments

**Senior Project Editor:** Georgette Beatty
**Senior Acquisitions Editor:** Mike Baker
**Copy Editor:** Jennifer Tebbe
**Technical Editor:** Alexander D. Jarvis

**Production Editor:** Mohammed Zafar Ali
**Editorial Manager:** Michelle Hacker
**Cover Image:** © Watchara Kokram/iStockphoto